Advance Praise for *Cyber Persistence Theory*

"Michael Fischerkeller, Emily Goldman, and Richard Harknett have once again made an incredibly valuable contribution to the development of American cyber policy and strategy through the writing of *Cyber Persistence Theory*. The authors push its readership to think beyond classical deterrence theory to new concepts for engaging and defeating undeterred adversaries in cyberspace. In short, this book argues the need for change and to take more risk to close an increasingly larger risk in our defense and national security as well as our public safety posture as American citizens. To do so, the authors argue will require not only persistent engagement, but a 'whole-of-nation plus' effort. A must-read for both national and cyber security professionals!"

—**Robert J. Butler**, former Deputy Assistant Secretary of Defense for Cyber and Space Policy

"Time will tell whether cyberspace operations can have coercive effect, but it is unambiguously true that to date, nations have used cyberspace mostly to gain advantage in competing with other nations. Understanding how they do so is a new challenge that scholars of international relations would do well to take on, and this book is a superb point of departure for them."

—**Herb Lin**, Hank J. Holland Fellow in Cyber Policy and Security, Hoover Institution, Stanford University

"This book helps to fill a crucial gap in strategic thinking about the fundamentals of cyberspace and sets out a clear course of action for the US government. It is a must-read for students, analysts, and policymakers."

—**Max Smeets**, Senior Researcher ETH Zurich, Center for Security Studies, and author of *No Shortcuts: Why States Struggle Develop a Military Cyber-Force*

BRIDGING THE GAP

Series Editors
James Goldgeier
Bruce Jentleson
Steven Weber

The Logic of American Nuclear Strategy: Why Strategic Superiority Matters
Matthew Kroenig

Planning to Fail: The US Wars in Vietnam, Iraq, and Afghanistan
James H. Lebovic

War and Chance: Assessing Uncertainty in International Politics
Jeffrey A. Friedman

Delaying Doomsday: The Politics of Nuclear Reversal
Rupal N. Mehta

Delta Democracy: Pathways to Incremental Civic Revolution in Egypt and Beyond
Catherine E. Herrold

Adaptation under Fire: How Militaries Change in Wartime
David Barno and Nora Bensahel

The Long Game: China's Grand Strategy to Displace American Order
Rush Doshi

A Small State's Guide to Influence in World Politics
Tom Long

Cyber Persistence Theory: Redefining National Security in Cyberspace
Michael P. Fischerkeller, Emily O. Goldman, and Richard J. Harknett

Cyber Persistence Theory

Redefining National Security in Cyberspace

MICHAEL P. FISCHERKELLER, EMILY O. GOLDMAN,
AND RICHARD J. HARKNETT

OXFORD
UNIVERSITY PRESS

Oxford University Press is a department of the University of Oxford. It furthers the University's objective of excellence in research, scholarship, and education by publishing worldwide. Oxford is a registered trade mark of Oxford University Press in the UK and certain other countries.

Published in the United States of America by Oxford University Press
198 Madison Avenue, New York, NY 10016, United States of America.

Library of Congress Control Number: 2022934428

ISBN 978–0–19–763826–2 (pbk.)
ISBN 978–0–19–763825–5 (hbk.)

DOI: 10.1093/oso/9780197638255.001.0001

Paperback printed by Marquis, Canada
Hardback printed by Bridgeport National Bindery, Inc., United States of America

CONTENTS

FOREWORD

As the Department of Defense (DoD) established United States Cyber Command in 2010, the United States was only beginning to understand the complexity and challenges of how the military would operate in cyberspace. Over the next six years, much was accomplished in fielding forces, training, and planning; however, there was still a need to develop an operational approach suited to the domain. By 2016, Michael Fischerkeller, Emily Goldman, and Richard Harknett were laying the foundation for the Command's approach of Persistent Engagement.

US Cyber Command drew on this thinking in its 2018 Vision, just as forces, policy, and training had matured to enable effective operations. The results: a growing operational tempo in defense of US elections and in support of US military operations, as well as improved cybersecurity for the DoD. These results in turn promoted collaboration between government agencies and private-sector companies to address common threats in cyberspace.

In their book, the authors demonstrate the value of a deep grounding in the scholarship of international relations theory and thorough analysis of the attributes and emerging dynamics of the cyber domain. As we gain experience and learn through action, this volume offers a framework for understanding that can improve operational effectiveness moving forward. Our understanding of cyberspace and how it is exploited has grown since Cyber Command's founding, as it undoubtedly will in the years ahead. The authors have made an important contribution to this understanding, and I look forward to seeing the role it plays in the continued development of strategy, national security, and cybersecurity scholarship.

—General Paul M. Nakasone, US Army

The views expressed are the writer's own, and do not necessarily represent the views of the US Department of Defense or the US Army.

ACKNOWLEDGMENTS

Fischerkeller—I would like to thank the senior leadership, recent past and present, of the Institute for Defense Analyses for their unwavering support of my research behind this work, with a special thank you to Dr. Margaret Myers, General (retired) Larry Welch, Dr. David Chu, General (retired) Norty Schwartz, and Mr. Phil Major. To the many colleagues who've offered expert commentary and strong encouragement over the past four years, I'm thankful for that support and look forward to many more engagements. To my colleagues, Emily and Richard, it is my hope that all research professionals, no matter their field, experience such an extraordinary collaborative effort. And to Naomi, I have boundless gratitude for creating an environment in which my contribution to this volume was inspired, nurtured, and matured.

Goldman—I want to thank the women and men of US Cyber Command and the National Security Agency, who increase my knowledge of cyberspace each and every day. My deepest gratitude to Michael Warner and Steve Peterson, whose insights, support, optimism, and faith inspire me to persevere. Jake Bebber, Ryan Symonds, Gary Corn, and TJ White are treasured colleagues who continually broaden and deepen my understanding. LTG Steve Fogarty's vision and support were crucial to these ideas taking root inside the Command. Without the support of ADM Mike Rogers, I would not have had the opportunity or latitude to challenge conventional thinking and without GEN Paul Nakasone I would not have been able to continue those efforts which lie behind this book. A special thanks to Lou Nolan who shepherded the manuscript through security review. To my coauthors, Michael and Richard, this has been a remarkable intellectual journey, a destination no one of us would have reached without the others. And finally, my deepest thanks and love to Catherine, Alex, and JR, the lights of my life, my greatest sources of strength and resilience.

Harknett—This book does not develop as it has without the initial bridge-builder, my coauthor Emily, who recognized the opportunity to broaden perspective in the policy community and who, with the support of ADM Mike Rogers, launched the initiative that allowed me to freely cross back and forth between the scholarly and policy worlds, seeking new ways to map, exploit, and eventually advance thinking about cyberspace. Along with those mentioned above, I would add policy and academic colleagues in the United Kingdom, who added great perspective as well as opportunity to explore how new conceptualization would impact security, as I developed a new relationship with Oxford University through the ever-important Fulbright program. I would like to thank among many who have pushed and critiqued, Jelena Vicic, Monica Kaminska, Florian Egloff, Lucas Kello, Mustafa Sagir, Graham Fairclough, Jim Miller, and my colleagues at the Center for Cyber Strategy and Policy at Cincinnati, particularly Stephanie Ellis. Along the way, Michael and Emily have been the ideal co-thinkers, persistently giving and taking. Finally, my enduring faith that things can always get better flows from the positive unwavering support I am blessed to receive every day from Kathryn and Margot—the two reasons for everything.

1

The Misapplied Nexus of Theory and Policy

> Theory should cast a steady light on all phenomena so that we can more easily recognize and eliminate the weeds that always spring from ignorance: it should show how one thing is related to another, and keep the important and unimportant separate. . . . The insights gained and garnered by the mind in its wanderings among basic concepts are benefits that theory can provide. Theory cannot equip the mind with formulas for solving problems, nor can it mark the narrow path on which the sole solution is supposed to lie by planting a hedge of principles on either side. But it can give the mind insight into the great mass of phenomena and of their relationships.[1]

Developing a new theory to shine light on complex emerging phenomena is no easy task. Finding acceptance for a new theory that questions a dominant paradigm is more challenging. Translating that new theory into a noteworthy change in strategy and policy pushes the envelope of the improbable. This book strives for all three, despite the herculean nature of the task, because these objectives have been intertwined for the three authors of this book over the past several years.

We offer a structural theory of cyber security that explains the core logic driving cyberspace competition and conflict and that reveals the existence of a distinct strategic environment to which all States are subject. It is a theory that is applicable to all State, and potentially non-State, behavior. We posit that alignment to the structural features and strategic opportunities of the strategic environment that emerged from the creation of global networked computing will, in large measure, determine how well States and non-State actors leverage cyberspace to advance their interests and values.

To place the bottom line up front, cyber persistence theory posits that cyberspace must be understood primarily as an environment of exploitation rather than coercion. Achieving strategic gains in the cyber strategic environment does

Cyber Persistence Theory. Michael P. Fischerkeller, Emily O. Goldman, and Richard J. Harknett, Oxford University Press.
 DOI: 10.1093/oso/9780197638255.003.0001

not require concession of the opponent. We recognize this is no minor assertion.[2] Nevertheless, we demonstrate how States can reset the cyber playing field to their advantage without shaping the decision calculus of the opposing side. All actors in the cyber strategic environment have this opportunity and as a result, States must continuously anticipate the persistent resetting of the security conditions in cyberspace by others as they seek to do so in turn.

The logic captured by cyber persistence theory presented in this book does not amount to just a competing explanation of cyber security dynamics in the early twenty-first century. Rather, it reaches the level of dissonance with dominant thinking in national security studies that satisfies the often referenced, but not often met, criterion Thomas Kuhn established for paradigms and their changes. As Kuhn notes in his classic work, *The Structure of Scientific Revolutions*, a paradigm provides a community with its basic assumptions, key concepts, and methodology.[3] Once established, a paradigm becomes very difficult to dislodge, even in the face of empirical evidence that the assumptions, concepts, and methods do not align with observed behavior.

However, paradigms do not fall simply due to the friction between the expected and the observed. For a shift or "change in worldview" to occur, there must first be a realization of the misalignment between theory and reality.[4] This must be coupled with an alternative way of thinking, one resting on different and more compelling assumptions, concepts, and methods. Paradigms are stubborn, and there is no guarantee that new concepts will win the day; in fact, Kuhn's analysis suggests there will be much resistance.

For our purposes, this is relevant on both sides of the bridge—growing academic acceptance of the principles of cyber persistence does not guarantee effective policymaker adoption, nor does prescriptive adaptation shut down academic disagreement.

The purpose of this book is to establish cyber persistence theory and position it for greater development by both the academic community of security studies scholars and the policy community managing national security strategy in the digital age. In the following chapters, we examine the limitations of traditional security paradigms and offer policy prescriptions derived from cyber persistence theory. In order to bridge the gap between theory and policy, we offer in our closing chapter an analysis of the United States as an example of actual policy adjustment that requires ongoing shifts in fundamental conceptual thought that constitute paradigm change.

Global digital connectivity is now a feature of modern human interactions, and we hope a contribution of this book is a broad rethinking of international relations theory and practice. Mis-framed theory or misapplied policy are equally troublesome, and this book rests on the assertion that both are present in this early intersection of global networked computing and international relations theory and policy. Theorists have defaulted to framing State-driven cyber dynamics in terms of traditional notions of coercion and war. Policymakers in many countries have defaulted to a strategy of deterrence to solve the inter-State challenges posed by global networked computing. Our theoretical explanation challenges the primacy of both coercion theory and deterrence strategy for understanding and mitigating the strategic impact of cyberspace on international security.

War, Coercion, and Deterrence

From the outset of global networked computing, policymakers and strategic thinkers were legitimately concerned about the consequences of cyber capabilities for war and viewed cyberspace through the lens of war. In "A Brief History of Cyber Conflict," Michael Warner explains that cyberspace became a military matter when "governments and institutions began storing and moving wealth and secrets in the form of digital data in and among networked computers having international connections" and when "those same enterprises also began maneuvering to protect their secrets and wealth against opponents who wanted to steal or impair them."[5] The highest levels of the US government acknowledged the national security risks from converging telecommunications and automated information systems in 1984. A presidential directive foresaw that US and foreign national security data could be not only exploited by foreign adversaries but also corrupted or destroyed, with strategic implications, to include the security of both superpowers' nuclear command and control systems.[6] To wit, in the 1980s the Democratic People's Republic of Korea (DPRK) was laying the foundations of its strategic cyber program, by, among other things, establishing the Pyongyang Informatics Center (PIC) and reportedly hiring twenty-five Soviet instructors to train military students in "Cyber warfare."[7] Interestingly, these early 1980s examples reveal one State that saw the national security consequences of network computing as a threat, while another saw it as an opportunity.

Views of cyberspace also became intertwined with concepts of information war, particularly after the 1991 Persian Gulf War. Information at the tactical and strategic levels proved to be critical to the US-led international coalition's swift victory, and pundits quickly dubbed it "the first 'information war.'" Not long after, prominent defense intellectuals John Arquilla and David Ronfeldt published their seminal article, "Cyberwar Is Coming!"[8] In it, they describe how information had become as important to victory on the battlefield as capital, labor, and technology, altering the character of conflict and extending the battlefield beyond geographic terrain to the electromagnetic spectrum.[9]

In 2010, US Deputy Secretary of Defense William Lynn III declared cyberspace a new domain of warfare, "as critical to military operations as land, sea, air, and space. As such, the military must be able to defend and operate within it."[10] Other countries would follow suit.[11]

Seeking to understand if and how States could engage in cyber war, the academic and policy communities have paid particular attention to the disruptive and destructive nature of cyber operations.[12] For more than two decades, cybersecurity literature debated if, when, and how cyber war would occur.[13] The compelling work of Arquilla and Ronfeldt in the context of the First Gulf War spurred scholarly debate on how military forces' information and communications systems could be disrupted or even destroyed in a future militarized crisis or conflict.[14] In the late 1990s, the debate turned to the potential of cyber operations to cripple a society's critical infrastructure.[15] In the decade that followed, perceptions of the impact of the Distributed Denial of Service (DDoS) attacks against Estonia (2007) and Georgia (2008) reinforced this shift.[16]

The academic and policy communities' focus on cyber war did not reflect a consensus about its practicality or potentiality. Some argued that concerns about cyber war were well founded and that it may already be upon us. In 2009, Mike McConnell, former Director of National Intelligence and the National Security Agency (NSA), wrote, "[W]e have entered a new age of threat, defense, deterrence and attack equivalent in some ways, to the atomic age. Cyberattacks have the potential to damage our way of life as devastatingly as a nuclear weapon."[17] In 2010, Richard Clarke and Robert Knake argued that "cyberwar is real" and "has begun."[18] In 2013, Gary McGraw opened his article with the following:

> Information systems control many important aspects of modern society, from power grids through transportation systems to essential financial services. These systems are riddled with technical vulnerabilities. Consequently, our reliance on these systems is a major factor making cyber war inevitable, even if we take into account (properly) narrow definitions of cyber war.[19]

Other scholars and analysts, for various reasons, remained skeptical of the notion of cyber war and the argument that cyber operations or campaigns that are not "war" could be strategically consequential. For example, the absence of physical violence in reported cyberattacks encouraged Thomas Rid to argue that cyber war has not and will not occur.[20] According to Rid, no cyberattack meets all three of Clausewitz's criteria of war as "violent," "instrumental," and "political."[21] Instead, Rid concluded that "all past and present political cyberattacks are merely sophisticated versions of three activities that are as old as warfare itself: subversion, espionage, and sabotage."[22] Erik Gartzke shared Rid's perspective in arguing that cyberattacks have not transformed States' pursuit of strategic advantage and dubbed cyberwar a "myth."[23]

Gartzke also argued that cyber operations could only be relevant in "grand strategic terms" or "pivotal in world affairs" if they independently "accomplish tasks typically associated with terrestrial military violence."[24] These include deterring or compelling, maintaining or altering the distribution of power, and resisting or imposing disputed outcomes. Gartzke referenced Clausewitz in his arguments, concluding that "[t]he internet is generally an inferior substitute to terrestrial force in performing the functions of coercion or conquest. Cyber 'war' is not likely to serve as the final arbiter of competition in an anarchical world and so should not be considered in isolation from more traditional forms of political violence."[25] Adam Liff, exploring the implications of the proliferation of cyberwarfare capabilities for the character and frequency of inter-State war, arrived at the same conclusion: "[c]yberwarfare appears to be a tool for states to pursue political (strategic) and/or military (tactical) objectives at relatively low cost only under very limited circumstances. Although Stuxnet manifests cyberwarfare's potential to become a useful brute force measure, no examples of irrefutably effective coercive CNA [Computer Network Attack] exist."[26] Martin Libicki similarly doubted that "strategic cyberwar"—cyberattacks that determine the outcome of war or "state policy"—would occur in the future.[27]

Despite differing perspectives regarding the salience and potential of State cyber behaviors for war, these groups share a common paradigm. Scholars and analysts who are focused on cyber and war construct their arguments through the lens of coercion theory. The study of war, particularly in the nuclear era, has been anchored firmly on coercion. Robert Art and Kelly Greenhill succinctly describe how coercion focuses on change in the behavior of an opponent (inducing change in how they calculate benefits and costs through compellence or deterrence) and "always involves some cost or pain to the target or explicit threat thereof, with the implied threat to increase the cost or pain if the target does not concede."[28]

Many who argue that the notion of cyber war is valid and should be a central concern of cyber strategy adopt the coercion paradigm when they call for a strategy of deterrence to be the central feature of States' cyber strategies. This outcome is not surprising, since it seemed logical to apply the strategic approach of coercing others not to attack through the threat of response—deterrence—that had been successful in the physical domains to the information environment, in general, and cyberspace, in particular.

The "deterrence default" was reinforced by a national security enterprise dominated for nearly two generations by deterrence thinking. The vast majority of contemporary national security practitioners and senior academics were schooled during or immediately after the golden years of Cold War scholarship, which produced a trove of classics focused on coercion theory.[29]

Over many decades, deterrence proved to be conceptually well aligned with the Cold War strategic environment. Nuclear deterrence was associated with the strategic stability and absence of major war between the United States and the Soviet Union during the unprecedented historical period from the end of World War II through the Cold War—what John Lewis Gaddis called the "Long Peace."[30] It attracted an enduring group of scholars and practitioners away from examining how to fight and win war and toward how to deter war. As a result, in the first decades of the twenty-first century, the United States and other Western democracies assumed cyberspace was a deterrence strategic environment and that prospective response and operational restraint would produce positive norms and stability.

Without question, the seventy-year history of deterrence serving as the central security strategy for the United States and its allies influenced policymakers immediate gravitation toward the same strategy for cyberspace. However, their inclination was further exacerbated by the fact that scholars had failed to provide policymakers with an alternative paradigm (to coercion theory) for understanding State cyber behaviors and developing strategy aligned to the cyber strategic environment. In the end, we are not surprised that a fixation on coercion, militarized crisis, and war in cyberspace led to a "high-and-right" bias in the cyber literature. For over two decades, practitioners and academics have been debating if, when, and how cyberwar will occur, why coercion appears ineffective and can be made more effective, and why cyber deterrence fails and must be fixed.[31]

It's Strategic Exploitation

The reality of State behavior and interaction in cyberspace over the past two decades has been quite different from the model of war, catastrophic attack, and coercion upon which the cyber strategy and policy of many countries is based. Most adversary State-sponsored cyber activity occurs outside armed conflict and has not been primarily coercive in application.[32] Instead, we have seen the persistent use of cyber operations and continuous nonviolent campaigns for a variety of purposes—to circumvent sanctions; to increase economic competitiveness through cyber-enabled illicit acquisition of intellectual property and research and development (R&D) at scale; to erode an opponent's military capabilities through supply-chain manipulation; and to weaken domestic political cohesion, undermine confidence in government institutions, and erode international alliances through disinformation and information manipulation.

Moreover, increasingly assertive activity in cyberspace has not escalated into armed conflict. This book offers an explanation as to why this is the dominant reality. In simple terms, we argue that States are persistently active in sustained campaigns within cyberspace deliberately calibrated to remain below a threshold that would likely elicit an armed response, seeking instead to produce cumulative gains over time. Each intrusion, hack, or technical action—although not strategically consequential on its own—often cumulatively results in effects that, in past generations, required armed conflict or a threat thereof. The theory of cyber persistence argues this is due to a combination of structural features, strategic incentives, and, importantly, the emergence of a non-coercion-based primary mechanism for achieving strategically relevant outcomes. In the chapters ahead, we develop the argument that exploitation of cyberspace vulnerabilities and opportunities and not coercion is the primary route toward gain. This is a significant theoretical proposition, one that opens the aperture of security studies from its previous foci and suggests that States may pursue the same ends of war through other ways and means than what we have known for millennia.

The academic and policy default to coercion has assumed that its absence in cyberspace results in a sideshow. If cyber operations cannot be used to coerce another for strategic gain, cyber operations inherently are strategically inconsequential.[33] An important distinction between coercion theory and cyber persistence theory lies in the recognition that while all strategic bargaining is a form of competition, not all strategic competition requires bargaining. What if the absence of coercion as a dominating behavior in cyberspace is due to the inherent features of cyberspace itself and to States having figured out how to exploit those features with an expectation (and achievement) of strategic advantage?[34]

Purpose of This Book

This is a book about aligning theory with reality to create a basis for properly applied policy. To achieve this outcome we take seriously Clausewitz's call for "wanderings among basic concepts." From a theory construction standpoint, we introduce the core notion of strategic environments, which we argue are defined by structural features and security dynamics that are distinctive in character. Much of previous academic literature and the policy reaction focused on cyberspace took their cue from the nuclear and conventional strategic environments. Theories and strategies that work in those realms do not necessarily fit into a third strategic environment—cyberspace—whose features set it apart (just as the nuclear environment stood distinct from the conventional strategic environment).

Throughout this book, we try and strike a balance in introducing new terminology only when analytically and prescriptively necessary to bring clarity to the phenomena we examine. Where appropriate, we use the lexicon of conventional and nuclear studies when clean breaks are not required.

In the end, States must manage all three of these strategic environments simultaneously, so using common language when appropriate is our default. However, the first several decades of dealing with cyberspace and security followed the expectations of Thomas Kuhn, with new phenomena squeezed into existing concepts, thinking, organizing, and acting. This has produced analytical and prescriptive misalignment, and obscured blind spots that require change.

Cyber persistence theory provides a framework for the expansion of academic research and adoption of more effective policy and strategy to address strategic competition in and through cyberspace. We hope that the theory drives policy outcomes in order to achieve a more cyber secure and stable international environment. The first step to such an outcome is getting the fundamentals right, which is the task to which we now turn.

2

The Structure of Strategic Environments

> The transition from a paradigm in crisis to a new one from which a new tradition of normal science can emerge is far from a cumulative process, one achieved by an articulation or extension of the old paradigm. Rather it is a reconstruction of the field from new fundamentals, a reconstruction that changes some of the field's most elementary theoretical generalizations as well as many of its paradigm methods and applications.[1]

This chapter focuses on the structure of strategic environments to explain the dynamics of security-seeking in contemporary inter-State relations. It establishes that security in and through cyberspace rests on a distinct definition of security, one that differs from the dominant security paradigms of the twentieth and early twenty-first centuries associated with nuclear and conventional weapon environments—deterrence and warfighting. The theory of cyber persistence requires a reconstruction of how we think about security in the digital space. This is because the structures of the strategic environments in which warfighting and deterrence are most logically salient and rationally practiced differ from the structure of the cyber strategic environment created by ubiquitous networked computing. This reconstruction leads to novel prescriptions for strategy, doctrine, operating concepts, organization, resourcing, and legal authorities. In other words, it alters how States should practice security-seeking.

The main strength of structural theorizing is its parsimony. It is a form of theory construction and practice that derives explanatory power from a focus on fundamentals rather than comprehensive details.[2] In this spirit, the straightforward proposition to be developed in this chapter is that competition over national security now takes place in three distinct strategic environments. What makes these environments distinguishable is how one conceives of the

Cyber Persistence Theory. Michael P. Fischerkeller, Emily O. Goldman, and Richard J. Harknett, Oxford University Press.
 DOI: 10.1093/oso/9780197638255.003.0002

definition of security itself. That definition aligns with the challenge to security that exists at its most core (structural) level and which is shaped fundamentally by the dominant technology that can be used to challenge security.

Some may view this argument as technology deterministic; we do not conceive it that way. Although technology anchors each of the strategic environments, what makes them distinct is not the technology itself, but how the technology shapes conceptions of security. While the Truman administration understood that the atom bomb was a distinctive weapon, they employed it twice in a manner that was not significantly different from all the other strategic bombing raids employing conventional munitions that had preceded Hiroshima and Nagasaki. Countries could have continued the practice of using atom bombs as weapons of war, but the unique elements of the technology enabled a different concept of security to emerge. Once that different concept was realized, a reconstruction of theory was introduced that translated into new practice, which was then continually reinforced over time through the emergence of a community of strategists and practitioners trained in the theory and prescriptions of nuclear deterrence. At its core, this change in paradigm was related to a strategic environment characterized by the mutual possession of nuclear weapons (and the extension of protection by means of a security commitment made by States possessing nuclear weapons to those that did not). The ends, ways, and means associated with the conventional war paradigm continued to be salient for security relations conducted by States outside of or parallel to the nuclear strategic environment (non-nuclear vs. non-nuclear, nuclear vs. non-nuclear, or nuclear vs. nuclear via proxy).[3]

The thesis presented here is that the technical, tactical, and operational features of conventional, nuclear, and cyber strategic environments are distinct enough that these environments require their own paradigms—that is, exemplars, theories, lexicon, rules of investigation, and practice—with distinct prescriptions for how States should organize themselves effectively to advance their security in and through these environments. By "strategic" we mean that actions taken within and through these environments can directly impact the sources of national power upon which the distribution of power regionally and globally rests.

To support our claim that the cyber strategic environment requires its own paradigm consisting of distinct theory and practice, the first part of this chapter discusses the more familiar environments organized around conventional and nuclear weapons. It then turns to an explanation of the cyber strategic environment and its impact on core security dynamics, which, we argue, requires reorienting the ways and means used in the pursuit of national interests and redefining what it means to be secure in and through cyberspace.

Thinking Structurally

Debates over different approaches to theorizing are rich and vigorous, but we will not justify one approach over another in this chapter.[4] Rather, our purpose is to show how organizing our thinking around structure can help illuminate fundamental aspects of security as pursued primarily by States. Thus, our starting point adopts the organizing concept of the strategic environment and uses that to explore how fundamental elements are organized in the pursuit of security.

We define a strategic environment as a concept that describes core features of a technology or composite of technologies capable of independently maintaining or altering the international distribution of power that generate distinct systemic conditions, and thus distinct security logics, influencing the full spectrum of interstate strategic competition from competition short of armed conflict through militarized crisis and war. The adjective "strategic" is meant to distinguish this set of conditions as driven by an intentional focus on the contest over relative power. Although international relations are generally shaped through an overall structure characterized by an absence of centralized power (anarchy) and self-reliance that applies across all strategic environments, that recognition only tells us that a contest over power is possible. It does not tell us much about how States organize themselves to pursue security in the specific conditions to which they are subject at any given time in history. Those specific conditions, based on the interplay of technical, tactical, and operational features, ultimately drive security thinking, organization, and behavior. That is the consequence of the strategic environment's inherent structure.

Although the complexity of these conditions is significant, we adopt the most parsimonious starting point to explain how States think about security, organize for it, and act to achieve it in and through cyberspace. Our contention is that the core features of networked computing and the digital interfaces that have developed around it combine to produce a set of conditions distinguishable from the two other strategic environments that States must navigate to achieve security. Before we discuss the cyber strategic environment, it is important to outline the core logics associated with the two other coexisting strategic environments that shape the pursuit of security.

The Structure of the Conventional and Nuclear Strategic Environments

Throughout most of human history, challenging the core sources of power of a unit—be it a tribe, city-state, empire, or nation-state—required direct access to those sources of power. As those sources existed primarily in the physical control of land, adjacent sea, and eventually the air, territoriality has been a dominant conceptualization of how one should organize to secure one's most important resources. Specifically, territoriality consists of the set of State policies and organizations constructed to deny the extension of direct political control over, and to ameliorate indirect political influence of one's territory from, hostile external forces.

Understood as such, territoriality supports and is reinforced by the modern manifestation of sovereignty, in which political authority structures are territorially bounded.[5] Thus, one's control over territorially based or designated sources of power ultimately has been the principal metric in determining relative security. The more one is in control, sustains that control, and, in certain circumstances enhances that control over territory, the more secure the unit of political organization could consider itself to be. Although measures of power in absolute terms could tell us something about a State (we will use this modern form of unit henceforth in this chapter as our default), understanding how that power measured up against the power of others is key to measuring security.[6] It does not necessarily matter if one has significant sources of power to control territory and hold on to power if someone else has more capacity to take those sources of power away from you.

The potential to lose or gain sources of power, fundamentally, follows from the fact that sources of power (arable land, accessible water, energy sources, populations, resources for the dominant tools of the day) are unevenly distributed throughout our planet. If everyone had everything that they needed, being concerned about what others had would be unnecessary. It is the variance in the distribution of power across the globe that sets the stage for international politics and enables a contest over controlling sources of power. This uneven distribution of power has driven States to seek specific capabilities whose main purpose is to contest control over sources of power.[7] Primarily, States have sought capacity to defend control over sources of power, but in seeking that capacity, they have created a condition that undermines the confidence and ability of others to control their sources of power. A dilemma regarding the pursuit of security emerges because one needs the capacity to protect sources of power, but because power is distributed unevenly, that very capacity to protect one's own power has the potential to undermine someone else's relative capacity to feel or be secure.[8] The dilemma rests on the notion that if one does not pursue that capacity, they themselves become exposed relative to others, but the pursuit necessitates others to act to further their now uncertain capacity. Some have described this fundamental interaction as a "tragedy."[9]

Flowing out of this uneven distribution of sources of power, the history of military conflict and military studies rests on the basic conditionality of what emerged as a conventional strategic environment. The pursuit of security in such an environment depends on the alignment of strategy to the relative advantage or disadvantage of engaging in defensive (protecting) and offensive (extending) control over sources of power anchored around territory. Across military theorists from Clausewitz to Mahan to Douhet, the richness of military studies revolves fundamentally around the analytical concepts of defense and offense. At any given time, the combination of military technology, tactics, and operations may favor the offense or defense to the point of a structural advantage. Under such defined conditions, if military strategy, preparation, and execution are not aligned with the advantage, a State can be punished severely. Alternatively, strong alignment with the structural advantage brings reward. To be clear, adopting the frame of a distinct conventional strategic environment does not mean assessing the ability of one State to defend or attack relative to another State. Rather, it means thinking about how the overall environment that springs from the combination of conventional technical, tactical, and operational features impacts how a State should organize for and employ technology, tactics, operations, and strategy.

Across human history, the relative combinations of these features have created conventional strategic environments that essentially range between combinations that advantage the defense to combinations that advantage the offense. Although this range has remained fluid historically, it has been remarkably stable as a range regardless of the technical base (be it horse, chariot, or armored tank, for example). If the combination at the time of contest advantages the offense and a State has organized itself for a defense-advantaged environment, the likelihood of defeat rises. The classic example of this is the French military in the 1920s and 1930s, which created the most impressive defensive works of its time in the Maginot Line,[10] drawing on the lessons of the First World War in which the defense had been advantaged. The technical, tactical, and operational innovation of the blitzkrieg was the unsettling manifestation that proved to the French that the conventional strategic environment had shifted to a relative advantage for the offense, which persisted through the remainder of the Second World War.[11]

The First World War is the exemplar of how misalignment with the prevailing structural conditions of the conventional strategic environment can be catastrophic. By 1914, the General Staffs of each of the great powers about to commence in hostilities built their thinking, organizing, and actions in pursuit of security on the assumption that the combination of the technology, tactics, and operational features so advantaged the offense that to even conceive of defense was to concede defeat. So committed were they to this assumption that as political leaders broached the possibility of a more defensive approach, whether in the case of Czar Nicholas of Russia raising the prospect of a partial mobilization or Kaiser Wilhelm speculating about a blocking action against the French, the conclusion, as articulated succinctly by German Chief of the General Staff Helmut Von Moltke, was that it could not be done.[12] Importantly, the view that the offense was advantaged was so pervasive among leaders that the conduct of the war persisted with a commitment to offensive assaults despite the mounting evidence that the conventional strategic environment of the time was defense-advantaged. Being misaligned to the structural conditions of the strategic environment at the time meant a catastrophic loss of life and very little shift of territory on the battlefield.

Our intent is not to analyze the history of world wars, but rather to make the conceptual point that security in a conventional military environment rests on how well one aligns to the relative advantages of the moment between defense and offense. At its most basic, the security of the State depends on being able to fight and win military engagements in a relative struggle between offense and defense.

August 6 and 9, 1945, changed this equation. The technical achievement of the atomic bomb, creating the practical outcome of one bomb, one plane, one city, so overshot the destructive potential of even industrial warfare that in short order, the State possessing the new weapon and those that followed fundamentally reoriented their thinking, organizing, and acting in the pursuit of security. In 1946, Bernard Brodie captured the distinction between the conventional and nuclear strategic environments almost immediately when he concluded, "Thus far the chief purpose of our military establishment has been to win wars. From now on its chief purpose must be to avert them. It can have almost no other useful purpose."[13]

The emergence of this second strategic environment, one in which the pursuit of security no longer rested on the range of offense to defense advantage, was remarkable in its comprehensive introduction of not only new technological developments but also new bureaucracies, authorities, and even lexicon. In military strategy terms, the recognition of the nuclear strategic environment fits Thomas Kuhn's notion of a paradigm change.[14] What is noteworthy is that the United States pivoted so rapidly to recognizing this new strategic environment despite just completing a war in which the relative advantage it had in bringing to bear offensive operations on two fronts had produced significant victory that transformed its position in global politics. We have become so accustomed to the logic and approach of nuclear deterrence that we have perhaps forgotten what a radical departure it was to organize principally around deterrence from the several millennia of human history that preceded, particularly in the context of just winning a war.

Imagine for a moment that, upon the unconditional surrender of Germany on May 8, 1945, Congress held a hearing on how the United States should organize itself to pursue its security (making the assumption that victory over Japan would also occur in due time). Imagine that an academic strategist proposed that Congress should prepare to spend trillions of dollars over decades to build a military capability whose main purpose would be not to be used. Note this advice would have been provided in the context of a war just won in which military capability was produced, moved, and used on the battlefield as fast as possible. The scale, scope, and speed of that production and use was in large measure why the United States won. How quickly would that strategist have been thrown out of the hearing room? The mushroom clouds over Hiroshima and Nagasaki so totally refocused thinking that, what would have been dismissed as illogical months before, became a necessity from that moment forward. In contrast to the extreme mismatch between thinking and strategic environment that existed in 1914, the post-1945 pursuit of security in the nuclear strategic environment stands out for its significant and logical alignment.

The conventional strategic environment is structurally conditioned around a shifting range of offense to defense advantage. The structural condition of the nuclear strategic environment rests on the dominance of the offense. Here we insist on a greater precision in the use of the term "dominance" than has heretofore been used in security studies literature. There is an extensive literature that examines the balance between what has been called offense- or defense-dominant conditions. The logic and debate associated with that balance is best understood as a balance (what we call a shifting range) been offense- and defense-*advantaged* environments. The term "dominance," we argue, as an analytical term for theoretical development should be reserved for a condition in which the outcome between offense and defense is not contestable but assured. To suggest that the offense is dominant is to assert that the consequences that flow from the offense will always overwhelm the mitigation that can flow from defense. Anything short of that should be understood as an offense-advantaged environment in which mitigation is still possible at some level (denoting the advantage is to imply that ultimate success will follow if you are aligned with the capacity that is advantaged at the time of contest, but leaves room for the contest to still play out in which relative skill and other factors can impact the outcome). The conventional strategic environment, in fact throughout most of history, has been fluid between the offense and the defense. If there are periods of conventional structural advantage in either the offense or defense direction, they tend to be relatively short and open to shift between conflicts as well as shift even during a conflict. Thus, offense dominance is, in fact, rare and found currently only as a structural condition of the nuclear strategic environment.

In the conventional strategic environment, one's planning revolves around having the right alignment with offense or defense advantage. The security question boils down to "Can I attack or defend in order to win?" To say that the nuclear strategic environment is offense-dominant is to rest our thinking on the assumption that the destructive potential associated with nuclear weapons is incontestable. The core security question, drawing from Brodie, in the nuclear environment is essentially, "How can I secure when I cannot effectively defend at all?" The presumption of offense dominance necessitated a shift to the logic of deterrence. This radical shift in thought meant that security did not principally rest in my own hands (imposing force through the contest of offense and defense), but in the mind of my opponent. In a nuclear strategic environment, I must convince my opponent not to attack because I cannot rely on physically preventing the consequences of an attack if it were to occur. In this context, deterrence should be understood as a structural imperative—it becomes the necessary security strategy due to the condition that defense can no longer effectively secure the nation.

The literature on deterrence is quite extensive and it is not our purpose to examine deterrence comprehensively as a strategy in this chapter. However, because it has so anchored the approach to managing the nuclear environment and subsequently has become so pervasive in national security thinking, a summary here is appropriate to distinguish organizing around deterrence as a response to offense dominance and the construct of defense and offense advantage that still anchors the conventional strategic environment. We will return to this notion of structural imperative in the subsequent section.

Deterrence

Although "deterrence" as a term is commonplace in the academic field of strategic studies and in the policy community, specific variations of the term abound.[15] Our purpose here is to discuss it in its most commonplace understanding in the policy space and leave the academic nuance to the volumes already published.[16]

We start from the perspective that one only needs to consider the national security form of deterrence (as opposed to criminal forms of deterrence) or, for that matter, offense and defense in a setting in which opposing decision makers are considering how to pursue actions that will directly harm the national sources of power of a country through aggressive action. In this context, deterrence involves delineating the range of actions an opponent may contemplate so that the cost associated with action that directly undermines national security outweighs the benefit the opponent may wish to achieve.[17]

Deterrence is successful when the challenger is convinced that attacking is not a cost-effective option. Success rests on several variables, but most critical are the following elements:[18] the country attempting deterrence must *commit* itself to protect a certain source of power, *communicate* this commitment to its adversary, *possess the credible capability* to threaten costs that may exceed the adversaries' expected gains, and display the *credible resolve* to follow through on the threat. What is key here is the perceptions of the challenger, specifically their assessment of the will of the deterrer to respond and the expected effectiveness of the deterrer's response.[19]

Deterrence strategy is prospective threat. It represents a committed contingency to engage in action that will impose costs that outweigh benefits. Imposing costs directly moves us out of prospective threat and into action—it moves us into a condition beyond deterrence, one involving fighting, contesting, blunting, countering, and other forms of operations.

Thomas C. Schelling, in his 1963 classic *The Strategy of Conflict*, presents a similar definition of deterrence, while raising two additional points that (although obvious) should be noted. "The deterrence concept requires that there be both conflict and common interest between the parties involved."[20] Although the presence of conflict is obvious, what is implied by the situation of common interest? It is best exemplified by the nuclear strategic environment of mutual vulnerability in which nuclear States find themselves. Schelling's concept of common interest suggests that what is being deterred is an action that *both* sides wish to avoid in the long run. The element of common interest adds stability to the deterrence regime. If, however, the avoidance of certain actions is not considered by one State to be in its own interest, then deterrence is unstable and in due time will fail.

Defense and Deterrence by Denial

Patrick Morgan concludes that defense and deterrence are "analytically distinct."[21] Although the distinction was eloquently presented in the early 1960s by Glenn H. Snyder, it is a definitional problem that many contemporary theorists and policymakers have either forgotten or gloss over. In short, military hardware can be described as possessing an offensive/defensive value or deterrent value.[22] Defensive value refers to the capability of mitigating the damage resulting from adversarial aggression. The deterrent value of weapon systems refers to their ability to reduce the likelihood of aggression by an adversary.[23]

This distinction also holds for national policies. Deterrence deals with intentions and the ability to influence them. Defense deals with actions and the capability to thwart, mitigate, or contain them. Simply put, "there is a difference between . . . fending off an assault and making one afraid to assault you, between holding what people are trying to take and making them afraid to take it . . . [it] is the difference between defense and deterrence."[24]

The distinction is significant because deterrence by denial is analytically a deterrence approach, not a defense strategy. As such, it remains an attempt to raise concern (fear) that the aggressor's actions will backfire and cost them more than the action is worth. Deterrence by denial is not simply relying on actual denial of benefits, as many now conceive of it.[25] While during the Cold War, both superpowers tried to enhance the credibility of their deterrence commitments by marshaling conventional forces that would defend their respective interests and thus presumably raise costs to an unacceptable level, this conventional form of deterrence (or its limited nuclear warfighting variant), which is the empirical foundation of the concept of deterrence by denial, remained an uneasy and unstable form of deterrence due to the contestability of conventional weapons.[26] In other words, differing military technology—conventional and nuclear—impacts the effectiveness of deterrent threats and the outcome of deterrence strategies. In fact, the nature of the core technology has an anchoring effect on the structure of each of the three strategic environments. Thus, understanding the features of a strategic environment requires an assessment of the technical foundation of the environment and the tactics, operations, and strategies that then flow from it.

Technology but Not Technological Determinism

British Major-General J. F. C. Fuller stated, "[W]eapons, if only the right ones can be found, constitute ninety-nine percent of victory." This sentiment, as historian Martin van Creveld points out, is based on the presumption that "where once war was waged by men employing machines, more and more war [is] seen as a contest between machines that are served, maintained, and operated by men."[27] Although the debate may be struck over the correct proportion, few would argue with the implication: that basic weapons technology significantly impacts the manner in which wars are fought and their ultimate outcomes.

A distinction needs to be drawn between a weapon's technical capacity (what it can do) and its military capability (its impact and effectiveness in war). *Military capability* should be viewed as the combination of technology with techniques and tactics for employment and utilization.[28] What a weapon can accomplish in time of war is not merely a function of what is technically feasible. Again, to acknowledge Fuller's comment, a weapon's capability is dependent to some extent on being correctly "found," that is, being chosen and utilized in the proper circumstances in the appropriate manner. The French mitrailleuse is an appropriate example. Introduced in the Franco-Prussian War of 1870, this prototype machine gun had the capacity to fire at an unprecedented rate; however, it was employed not at the front as a tool of the infantry, but as a supplement to heavy artillery, thus significantly lowering its military capability.[29]

Aside from the necessity of proper employment and use, military capability is also affected by war itself. Carl von Clausewitz, in his classic military treatise *On War*, begins to broach this broader appraisal of military capability during his discussion of the inherent elements of war that work against the application of force:

> Action in war is like movement in a resistant element. Just as the simplest and most natural of movements, walking, cannot easily be performed in water, so in war it is difficult for normal efforts to achieve even moderate results.[30]

Deterrence is concerned with the calculation and application of costs, typically evaluated based on the amount of expected damage to be incurred versus gains to be accrued. Expected damage cannot be viewed simply as the technical destructive capacity of a deterrent; it must be assessed on a much broader scale. A bullet fired from a rifle has a certain understandable destructive capacity—it can pierce through human skin. A soldier positioned as a sniper has the potential to threaten to shoot anyone who walks within line of sight. Assessing this threat is intricately more complex than simply understanding that a bullet can pierce skin. An opponent contemplating moving through the sniper's line of sight (assuming they have no doubt that the sniper will shoot) will have to consider how good a shot the sniper is (how steady are their hands, how sharp are their eyes), whether their movement is in the effective range of the rifle, and whether there are any evasive movements (such as ducking) or protection (such as a bulletproof vest) that can be employed. Bringing civilians into the contested space might even undermine the will of the sniper to shoot. Ultimately, the effective raising of costs to deter enemy movement is much more involved than simply possession of a rifle; the bullet must hit its mark and the opponent is going to try to ensure that such an outcome does not occur.

In sum, recognizing that military capability (to include nuclear weapons) is a function not simply of technology but of human and environmental interaction is important for understanding and assessing the effectiveness of deterrent threats, managing warfighting threats, and, as we will discuss related to the cyber strategic environment, competition threats. What capability is necessary to effectively shape cost-benefit calculation is different from that which is needed for effective offensive and defensive warfighting, and as we will add to the theoretical mix that which is needed to compete over the exploitation of networked computing vulnerabilities and opportunities.

Structure and Imperatives

The shifting range of offense-defense advantage that flows from the nature of conventional weapons technology structures the conventional strategic environment and necessitates a general imperative—States must be able to engage in offense and defense effectively if they are to be secure. This requires States to organize to fight and protect. The actual strategies employed to secure will vary based on many factors (including conventional deterrence) and can leverage nonmilitary capabilities, including diplomatic, information, and economic instruments of national power.

While in early human history physical proximity was required to attack national sources of power, military technology developed over time to make geographic distance less of a hurdle for more actors. This increased the importance of timely information about potential and imminent threats, the capacity for rapid mobilization if necessary, and the ability to be resilient while under attack. Placing a premium on understanding the other side has been a principle of State action for two millennia. While the core saliency of Sun Tzu's maxim of understanding your opponent better than you know yourself has not varied in the conventional strategic environment, the complexity in now achieving that knowledge would likely amaze him.[31]

Despite increased demands on the necessity of preparation to fight, and at times, because of it, conventional armed aggression has remained a constant of human history. In the past twenty years we have seen great powers attacked, armed aggression between States, including in Europe, where the loss of territory through armed attack is no longer a distant memory of the Second World War.[32] The conventional strategic environment supports a strategic dynamic in which action involves preparing for potential conflict and, unfortunately too often, engaging in it. In sum, in the conventional strategic environment the final arbiter of national security—the protection of national sources of power—is the ability to fight and win wars.

The offense dominance that flows from the nature of nuclear weapons technology, which structures the nuclear strategic environment, necessitates its own imperative. If national security is to be achieved, States must be able to advance their interests, *while avoiding war*. Herein lies the core difference between these two strategic environments: in the conventional environment, I can look to advance interests *despite* war and sometimes through its prosecution. War and advancement of national interest are not incompatible. In a nuclear environment, advancement of interest and the prosecution of nuclear war are incompatible.

The recognition of that latter relationship has led to a wholly different approach to security. Aside from the base strategy of deterrence, crisis management, escalation control, arms control, coercive diplomacy, and sanctions all took on prominence. The new logic and lexicon associated with the nuclear strategic environment became so pervasive that its terms and concepts have been exported to the conventional environment. We must, however, not lose sight of the unique conditions they were originally developed to address.

This does not mean that tactics and operational approaches are not applicable across strategic environments; far from it. What we have learned from managing the nuclear challenge has indeed shaped and impacted how conventional forces are conceived and used. However, the distinctiveness of the thought, organization, and action that emerged after fission and then fusion were weaponized is profound. Nuclear States fundamentally think, organize, and behave differently when confronting each other than when States are only conventionally armed (or when nuclear States confront non-nuclear States or non-nuclear allied States). The fact remains that the only two times the weapon has been used were against a State that did not have them and when they were in short supply. The exponential growth in numbers and lethality as well as the increase in the number of States in possession of them has not led to a third use.

This is a fundamental departure from the behavior we see of States (and some of the same States) in the conventional strategic environment. This absence of use reflects States operating in a distinctly recognized and structured strategic environment with its own organizing principle, logic, and dynamics. Nuclear weapons altered the manner and practice of coercion and the use of force. The focus of coercion and force became the threat to use them, rather than their actual application. The coercive (deterrent) power of nuclear weapons comes from their possession, not their use. Ultimately, conventional security rests in the presence of war; nuclear security in the absence of war.

Existing Strategic Environments and Cyber Activity

For over seven decades, the conventional and nuclear strategic environments have coexisted. A small set of States have had to manage both simultaneously and have done so with different integrative strategies: extending nuclear deterrent commitments to allies while pursuing strategic deterrence of the homeland but use of force globally (the United States and Soviet Union); bolstering conventional forces to deter any level of direct war (the United States, Soviet Union, Russia, North Korea); enhancing deterrence of strategic war while recognizing territorial flashpoints require some flexibility on conventional use of force (India, Pakistan, and Israel's opaque possession); and, finally, strategic deterrence of the homeland and the use of conventional forces globally (France and the United Kingdom). Although there is a relationship between the two environments, the thinking, organization, and behavior within the nuclear strategic environment remain recognizably distinct from the conventional strategic environment. The environments are distinct in logic, but States must also understand how those logics intersect if they are to secure themselves. This intersection-distinction dichotomy is important to keep in mind as we examine the emergence of a third strategic environment.

It is our contention that the fundamental logics of the nuclear and conventional strategic environments do not capture what we are seeing behaviorally in and through cyberspace. At first blush, one might look at the empirical record of cyber operations to date and contend that this is explained through the logic of the conventional strategic environment. National interests are being advanced through a contest of offensive and defensive capabilities designed and executed with the recognition that use of force and the cyber equivalence to armed attack are viable options. Networked computing technology is the same as the militarization of the airplane—a new means to conduct war. This view has certainly dominated the lexicon of the past twenty years, where the term "cyber war" has been used to describe all manners of cyber operations.

And yet, there is an empirical problem with this view—despite *millions* of executed cyber operations, few States have treated and reacted to these operations as a use of force or armed attack. There are three plausible explanations for this remarkable absence of war in the presence of so much activity: (1) the technology only enhances "subversion, espionage, and sabotage";[33] (2) deterrence of war, armed attack, and use of force is stable, so States are choosing to just rely on cyber means to subvert, spy, and occasionally (and in limited degrees) destroy, but if deterrence could be designed around, cyber war would occur;[34] or (3) there is a fundamentally different strategic logic driving the behavior not captured by the logics of coercion and warfighting, and it is of a strategic, rather than tactical, subversive or only intelligence-gathering nature.[35]

Many in the academic and policy communities have applied the logic of the nuclear strategic environment to explain cyber activity. Almost as prevalent as cyber war, the term "cyber deterrence" has been a default for many arguments and policy documents during the 1990s through 2020s.[36] The empirical reality of ubiquitous cyber activity, however, challenges the notion that we should think about, organize, and use cyber means based on the logic that security rests on the avoidance of action. Using nuclear logic to explain cyber behavior relies on a related set of plausible explanations parallel to those stated above: (1) cyber activity is traditional statecraft below war and thus deterrence does not apply, as the activity is not strategic in nature;[37] (2) cyber activity is potentially war-enabling, but such use for coercive purposes is avoided due to fears of escalation to war; or (3) States are deterred from war generally, but the activity we are seeing is a strategic response to opportunity, rather than an acceptance of a limitation.

The reality of continuous cyber operations and campaigns on a massive scale strains the logic associated with existing security studies theory and policy (intelligence, coercion, escalation, deterrence, and war). We must be open to the possibility that, from a scholarly explanatory perspective and from a policy development and execution standpoint, something fundamentally different is occurring and it requires a new explanation and policy prescriptions. Getting this wrong could have dire consequences for international security, as we witnessed in 1914 and 1939.

In the chapters that follow we provide direct challenges to the notion that the cyber strategic environment is old wine in new bottles—that it is just espionage or coercion by other means and/or ways. Here, we introduce the theoretical basis for the standalone argument that the extraordinary amount of cyber activity we are seeing follows a distinct logic. To develop the parameters of a theory of cyber persistence, we must first explain the structure of a third strategic environment.

The Structure of the Third Strategic Environment

States are heavily engaged in the use of cyberspace directly and indirectly to advance their national interests. Networked computing integrates all aspects of governance and the conduct of State relations internationally in and through cyberspace. It is the backbone for commerce and communication globally. This means that the full range of competitive and conflictual interactions that define international security relations must be understood in the context of the digital world.

Cyberspace is becoming so ubiquitous and so integrated into human activity that, at some point not in the distant future, appending the word "cyber' as an adjective will appear redundant. Both the academic and policy communities require the analytical tools to understand the most basic of State activities as they relate to security and to capture what it means to operate and pursue national interests in the digital age. There is much to be gained in both academic research and policy development, if we build a set of assumptions and concepts that map to the reality we face and will face. Relying, instead, on existing paradigms that do not align with fundamentals will result in non-policy-relevant research and, potentially, policy failure.

This reorientation must begin by recognizing the distinctiveness of a third strategic environment, one in which protecting and enhancing national sources of power rests on a set of conditions predicated on the unique features of networked computing and its digital interfaces. These features of networked computing create, as did conventional and nuclear capabilities, distinguishable conditions that require new concepts to align logic, thought, organization, and action for national security.

The nature of cyberspace requires a redefinition of security itself.

Whereas the structural conditions of the conventional strategic environment rest on the interplay of offense and defense advantage and those of the nuclear environment on offense dominance, it is the distinct notion of *initiative persistence* that flows from the features of information communication technology (ICT, broadly understood as networked computing and its digital interfaces) that comprise the cyber strategic environment. This third environment necessitates its own imperative. If national security is to be achieved, States must persistently set and maintain the conditions of security in and through cyberspace in their favor. Those conditions are measured as the relative balance between being cyber vulnerable to exploitation and being able to exploit the cyber vulnerabilities of others.[38] Given the features of this space, a State can only *set conditions* if they are able to anticipate where those conditions will lie in an ever-changing "virtualscape." Thus, we argue, there is an imperative that necessitates persistence in striving for initiative.

The term "initiative persistent" is meant to be as descriptively accurate and analytically useful as possible. The idea that cyberspace is offense-dominant or offense-advantaged or even defense-advantaged has been raised before in the literature—by some of us a decade ago.[39] However, the analysis presented here leads to a different argument—the terms "offense" and "defense" are analytically too limiting and are not explanatorily helpful.

Offense and defense are terms that can still be ascribed to tactics in cyberspace. However, the ever-changing features of the technology and the rapid adaptation of its use at the tactical, operational, and strategic levels are so significant that thinking in terms of building offensive and defensive capabilities, of conducting offensive and defensive operations, and of defining campaigns as being offensive and defensive at the strategic level misses the most crucial point about persistence. That is, we are dealing with fluidity on a scale and scope such that what is meaningful to outcomes is whether or not one has the initiative—whether one is anticipating the exploitation that *will come next* by either you as a defender or another State as an attacker. Tactically, a State may exploit to protect or exploit to advance. However, the cyber strategic environment is defined not by such tactics but instead by the fact that persistent exploitation in setting the conditions of cyberspace is the means to more or less security.

It is not useful to think about cyber operations as basketball on fast forward. It is not just speed of play that is at issue. It is the addition of scale (in number) and scope (in variety) of players as well as playing surfaces (software-hardware-processes that connect them and humans that use them) that combines with speed of play to make this a fundamentally different game. The complexity of this environment and its engagement dynamic are not captured in a simple offense versus defense conceptual frame. Neither offense nor defense is dominant or inherently advantaged. If I track an active breach of my network and simultaneously protect aspects of that network, but allow access to other sectors of the network to understand the techniques, tactics, and procedures of the opponent and then use information gained to enhance a prepositioned set of code and execute my own exploitation of the opponent's systems all in a simultaneous set of maneuvers that take effect in a matter of minutes, if not seconds, at what point am I playing defense and at what point offense?

We assert that the better way to conceptualize this environment is to recognize that what is occurring is grappling over initiative, something initially lost at the breach moment, regained at the detection moment, reversed at the tracking moment, and sustained at the moment the opponent's systems are exploited. These are not episodic linear actions of attack and protect, as many have conceptualized. Rather, it is a fluid set of engagements driven by who has the initiative at any given moment.

Understanding this third strategic environment as an initiative-persistent space captures the essence of the primary cyber behavior of the past twenty years: cyber *faits accomplis* (a concept we develop in Chapter 3). Much of what is happening in cyberspace consists of parallel attempts to gain enough initiative to be able to set the conditions of security and insecurity within and across devices, systems, and networks. Although direct cyber engagement (another concept developed in Chapter 3) can occur between an attacker and a defender, most of what is occurring consists of continuously flowing parallel operations that do not start with any expectation of shaping the other side's calculus, but rather focus on exploiting inherent vulnerability. We contend, therefore, that the cyber strategic environment is a space of exploitation, not principally one of coercion.

What is shared by both the conventional and nuclear strategic environments is a logic associated with war (or war avoidance) and coercion to achieve political ends in which there is direct exchange or expected exchange between protagonists. In these environments, attack or the prospect of attack that shapes behavior and the advancement of strategic ends is transactional. It flows from the exchange (or expected exchange) between mutually engaged and identifiable protagonists.

In the cyber strategic environment, States can advance strategic-level cumulative effects without direct exchange, without coercive shaping of behavior, and without war because they can directly change the virtualscape in which they seek to advance their security through exploitation of vulnerabilities that allow them to access, maneuver, fire, and—at the highest point—control without directly interacting with other States. Even when the actions are not covert or become discovered, the exploitation of vulnerability is not wholly dependent on the target's actions. A security patch can certainly shut down one vector of exploitation, but it does not fundamentally alter a capacity to exploit in general. In some circumstances, patching can send an attacker back to the drawing board for some time, but it does not negate the capacity of an attacker persisting to regain initiative through some alternative exploitative option.

To update the comparative summation provided earlier, conventional security rests in the presence of war, nuclear security rests in the absence of war, and cyber security rests in the alternative to war.

An Alternative Theory

The remainder of this book turns to the construction of a comprehensive theory of cyber persistence to explain a strategic environment based on initiative persistence (rather than offense dominance) that requires States to understand a logic of exploitation (rather than coercion).

Here we discuss the core tenets of this theory and then turn in Chapter 3 to a deeper analysis of new concepts and their prescriptive implications.

The cyber strategic environment's initiative persistence rests on the features of the technology itself and the construct that organizes those features. Specifically, we are referring to recursive simplicity and interconnectedness.

The most fundamental components of computing technology, both the hardware circuit board and the software code, rest on an overall default of iteratively building from a simple starting point, essentially building on simpler versions of the version we have. Although not all hardware and software is specifically built in such a fashion, the inherent nature of computing technology has sought and leveraged recursive simplicity.[40]

Recursion can be defined as self-similarity in structure where symmetry runs across scale—in essence, "pattern inside pattern."[41] Put another way, a recursive structure is one in which the whole is structurally identical to its parts.[42] Although recursive structures exist in nature (e.g., snowflakes and ferns), they were originally discovered as mathematically based constructions. By identifying the principle of self-similarity across scale, mathematicians Helge von Koch and Benoit Mandelbrot showed that fractal patterns could lead to infinite length in a finite space.[43] These observations laid the foundation upon which early software developers constructed large programs out of existing smaller ones. The recursiveness of software tremendously simplifies its development. Instead of regarding each independent part of a design separately, base commands can be reused in similar but broader commands. Thus, one need only know a fraction of the levels (essentially its foundation) to understand and construct multilevel designs.

Recursiveness as a structuring principle allows for exponential growth factors. The microchip (integrated circuit) has been constructed with a similar recursive logic. In a broad sense, the faster chip is simply more transistors pressed into one central component. The ability to construct a chip simply by compressing more transistors into the same space through greater miniaturization means that the basic design of the chip is not radically altered from its slower antecedent. Gordon Moore's "Law" first articulated in 1965 that one could double the number of transistors on a chip every year; it was modified in 1975 to doubling every two years and has remained remarkably prescient for nearly fifty years. While that law may now need a third modification, increasing computing power remains viable.[44] This is due, in part, to the fact that the integrated circuit became a "general purpose technology—one so fundamental that it spawns all sorts of other innovations and advances in multiple industries" and as such created the economic incentive to follow Moore's Law, which in turn has created economic patterns that assume adherence to the law.[45] There are also significant variables that affect the pace of the changing rate of computing power, including, for example, coding efficiency.[46]

Understanding recursiveness broadly as inherent to the nature of computing technology creates some specific consequences that we will discuss later. As important as this base nature is, the organizing construct that undergirds the cyber strategic environment, itself, is of equally profound importance. What makes the cyber strategic environment distinct is how the combined computational and communicative power of the micro-processing silicon-chip-powered computer has been networked. While individual computers in isolation represent powerful tools, it is the connecting of these devices that has proven so fundamental to the creation of the cyber strategic environment.

Computer networks support everything from local, regional, and national banking systems to telephone switching systems and transportation structures. ICT has enabled the stitching together of people, platforms, and performance in a dense interconnectedness that is aptly conceived of as a "web." The Internet—the network of networked computers—while not comprehensive of cyberspace is both the conceptual and physical driver of this organizing principle. Initially conceived of as part of an American defense plan to improve communications during a nuclear attack, the Internet transformed computer usage.[47]

While the creation, accumulation, and manipulation of information has always been a central part of human activity (warfare in particular), the computational and communicative power of the networked computer is creating distinctive consequences—a qualitative shift anchored on quantitative factors that moved incrementally during previous periods of human history.[48] In isolation, none of these variables themselves require discontinuity in strategic thought—a paradigm change—but in combination, they open such a possibility. In this context, four variables seem most important: accessibility, availability, speed, and affordability.

Accessibility. The networking of personal computers has led to the networking of individual networks. The universe of these networks, cyberspace, carries data and information in all its forms and enables action and interactions that previously were not possible at the scale and productivity rates now being achieved. Accessibility constrained by geographical proximity is changing, not in that geography is simply becoming less a limitation or constraint, but rather that the relationship between geography, information, and individuals is fundamentally shifting. Whereas throughout human history I had to travel to access information, people, and places, I now bring information, people, and places to me. Interconnectedness means that physical location now has little or no impact on the ability to access information.

Availability. Traditional terrestrial-based systems of information retrieval depended on geographic proximity for access and thus required an enormous amount of duplication to guarantee efficient availability of information. In this system, availability depends on how many copies can be made and stored in relation to how many information retrievers are at work. The difference between accessibility and availability is important. Living near a library might mean one has access to a book, but if someone has already checked out the only copy, the book will not be available for some specified time. Cyberspace offers a significant advance in availability by creating the opportunity for simultaneous retrieval of information. The limitation on availability is dependent not on how many copies of a particular instrument exists, but on the server's capacity to manage users simultaneously accessing the digitized database. While servers can crash, we are fast approaching the stage where availability is not a vexing or cost-prohibitive problem due to interconnectivity. Increasing a computer network's ability to handle more users is proving to be more cost-effective than having to duplicate the actual source of information by the same number of users.

Availability can also be discussed in quantitative terms with regard to the impact on managing available resources. There is, at a base level, too much information available via cyberspace. The systems meant to manage availability of the past were built on an assumption of scarcity—for example, a waiting list to be the next person at the library to get the one copy of the book when it was returned by the previous user a month later. One of the challenges at the start of the 2020s, with serious national security implications, is that societies have yet to figure out systems to manage *availability abundance.* Debates around free speech on social media platforms, for example, miss this more fundamental qualitative shift that has occurred. Managing availability abundance requires a fundamentally different system of laws, organizations, and social expectations/norms than environments of availability scarcity. In the realm of national security, States can wreak havoc on each other in the absence of such a management system through a variety of information manipulation and distortion opportunities that undermine trust in data, information, institutional authority, and leaders.

Speed. A third variable profoundly impacted by interconnectedness of network computing is computational and communicative speed. Enormous increases in both directly boost the ability to rapidly cycle through the basic sequence of observation, orientation, decision, and action (OODA loop).[49] When vast sums of accessible available information can be disseminated and processed in seconds, time relative to managing the OODA loop may shrink to the point of collapsing the loop. This has serious security implications.

Counterintuitively, traditional models of efficiency defaulted to the compression of time to achieve greater productivity or effectiveness. Interconnectedness enables such an extreme level of time compression that the challenge becomes finding mechanisms for time expansion within the OODA loop that do not cede initiative to the other side while retaining ones' capacity to observe, orient, and effectively decide before acting. The entire debate about automation and the emergence of algorithmic decision-making rests on recognizing this profound shift in time management and efficiency. For many human interactions, being "in the loop" is not efficient—the computer can process the OODA loop more effectively. This creates a deepening of reliance on code to take action, whether it is directing people via GPS, driving one's car, or flying a plane. Such reliance has knock-on effects, such as a current generation of youth who are less proficient at understanding the concepts of north, south, east, and west because they no longer need (or believe they need) to understand such directional constructs. Their software tells them the best route and they simply follow the arrow (or voice).[50]

The pursuit of efficiency through time compression raises ethical, social, and national security concerns, including reliance on algorithmic medical assistants rather than doctors to diagnose, the use of predictive modeling for conditional crime proclivity, and reliance on autonomous weapon systems and robotic soldiers that will resist being "out of the loop." Managing speed will require models of decision-making that allow humans to be not in or out, but critically, "on the loop."[51] States' ability to manage and manipulate time compression and expansion relative to action is at a premium in the cyber strategic environment.

Affordability. A fourth distinctive feature of interconnecting networks of computers is that the resource base required to exploit the advantages of this technology is relatively low and has continually declined over time. In terms of computing power, the median-income family in the United States has more computing power in its home in 2020 than many countries could afford or deploy only a few decades ago. The nature of the technology itself also lowers cost in terms of skill development. There has been an inverse relationship between increases in the complexity of the technology and what it can do, and the ease at which the technology can be used. The time, energy, and cost necessary to effectively use increasing complex platforms has continually declined. ICT is affordable in the broad sense of that term. Training, possession, and use require less time, less money, and fewer resources.

Due to its recursive simplicity, accessibility, availability, speed, and affordability, networked computing creates a profoundly different organizing principle on which the relationship between people, platforms, and place rests. The density at the global scale is truly remarkable. It took fourteen years to reach one billion users of the Internet, but only three years to add the last billion (2020 estimates suggest 4.5 billion users, or about 60 percent of the world's population).[52] It took approximately twenty years to create 250 million websites (distinct hostnames). It took fewer than two years to double that and less time still to double it again. In 2019, the estimate was about 1.75 billion publicly accessible websites.[53] As with Koch's snowflake, the growth in these features is driving the creation of a vast and ever-expanding virtualscape even though the principal device we use—the computer—represents a finite space inhabiting a finite terrestrial globe.

Thus, interconnectedness, as we present it here, is not simply the technical feature that follows from the Internet's backbone; it is not synonymous with networking. Rather, understanding interconnectedness conceptually as a structural feature of the cyber strategic environment requires us to see it as the sum of the four variables discussed above that, in combination, creates a virtualscape of continuous flux and sustained linkage.

In cyberspace, one can "be" anywhere at any time, in which "be" means "sufficient presence to take meaningful action." In terms of State relations, to be interconnected means to be in constant contact with one another at a level in which the potential to influence or affect sources of national power exists. This condition within the cyber strategic environment is different from that found in the conventional and nuclear environments, in which contact is episodic, potential, or imminent, but not constant. This is because terrestrial space is organizationally segmented (vice interconnected)—defined geographically and reinforced through international law's principle of sovereignty.

From the perspective of structural theory, constant contact must be understood as a condition, not a choice. It is a circumstance that follows logically from being in an environment organized by interconnectedness. If the system is not interconnected—that is, it is segmented—then the base condition is not a connection that is constant. The two are inextricably linked, and that linkage is profoundly significant.

One would be hard-pressed to find a State strategy document, policy statement, or corporate report that does not refer to cyberspace as "global" and "interconnected."[54] However, what mostly follows in those documents does not treat interconnectedness as a distinctive structural feature that drives an inherently different logic from segmentation. Particularly when it comes to State relations over security, the working assumptions of most State strategy relative to cyberspace have been grounded for decades in a logic that assumes security is to be found in barriers and separation (firewall thinking, for example). To be interconnected is to be in constant contact, which cannot be solved by denying this fundamental feature. Yes, I can disconnect from the network, but that does not achieve security within a cyber strategic environment. Removing oneself from the situation is not a sustainable solution because it precludes one from leveraging the beneficial outcomes of networked computing. National cyber security must solve the challenge of interconnectedness and constant contact working within, not outside or in spite of, the unique features of cyberspace. Digital life is a reality; securing it requires we accept that reality.

The implications of this combination alone raise some distinct security concerns. One must, as a planning principle, allow for the prospect that it is possible for an adversary to persist on the networks of critical infrastructure (electricity, water treatment, transportation, healthcare), the networks of leading industries, the networks of government agencies, and the core communication networks within society. This constant contact may position the adversary for exfiltration or manipulation of data resident on or traversing those systems in such a manner as to create adverse effects cumulatively over time. This can occur without some overt crossing of a terrestrial boundary that, in the past, had been the demarcation between peace and war.

Interconnectedness and constant contact, however, only create the potential. Unfortunately, the nature of the technology itself makes that potential a continuous reality. At its most fundamental starting point, the notion of the network was meant to replace the vulnerability of a single point failure that existed in the communication bureaucratic hierarchy. Specifically, as it related to solving the threat of a surprise nuclear attack that might decapitate the top of a hierarchically based decision-making system of communication, the ARPANET's innovation was to create centralized control without a center. The magnificence of the Internet is that its recursive structure created systemic wide redundancy at a level of efficiency out of reach of industrial age technology. It did this essentially through a portal system, where there is always a route around closed doors. It was not built to deny access, an essential ethos behind security, but rather to expedite it.

This base default to access is exacerbated by the way cyberspace has expanded. Although there have been many benefits from releasing software version updates—constantly evolving code builds on previous configurations—the sheer size of software packages alone is staggering. It is estimated that the Windows 10 operating system stands on 50 million lines of code, while the Google platform leverages nearly 40 times that number, at 2 billion lines. Regardless of system, a basic reality is that this accessible terrain is of such vastness that any mistake, unanticipated use, or truly novel application can enable an unauthorized user (or insider threat) to create terrain, which they can configure, and thus control, to advance their interests.[55]

Within the conventional and nuclear strategic environments, the technology of war is effectively distinct from the terrain in which war is conducted. The plane is different from the air in which it flies, the ship from the water it sails, and the tank from the land it traverses. In the cyber strategic environment, computer code is simultaneously the means to maneuver and the space through which one maneuvers. Although cyber physical systems—the integrated circuit, for example—reside physically in some device and are thus distinct from code, the processing that it accomplishes is all driven by the code that not only activates programmed action through the integrated circuit, but can add functions previously not present for that integrated circuit to process. Those new functions effectively become an addition to the virtualscape—new terrain that is being traversed by the new code that has created it. In this sense, every new software update, new hardware version, and new process that links them together reconfigures the space that had existed previously. The scale, scope, and form of maneuver is ever shifting in cyberspace.

At the tactical level, you indeed can defend in cyberspace, but you only defend in the moment—in the configuration of software, hardware, and processing that existed at the time you deployed a configuration you thought was secure. As noted earlier in the discussion of deterrence by denial, the effects of that defensive mitigation are not sufficient to attrite capacity to the point of denying an adversary another way around. This highlights the mismatch between the structural conditions of the cyber strategic environment and what is necessary for deterrence by denial to succeed. It is one reason deterrence by denial should not be relied on as the primary strategy to achieve security in cyberspace. Deterrence by denial relies on a calculus on the part of the prospective attacker that it will expend more force in attacking than can be sustained to achieve or hold a gain. The prospect of attrition is how significant defense and resilience capabilities might dissuade an attack. That other vectors of intrusion are likely available and that exploitative code can be produced (or otherwise acquired), manipulated, and repurposed with relative ease undermine that prospect. Thus the loss of an avenue of intrusion or of an exploit's effectiveness, where and when it occurs, is surmountable. Attrition cannot anchor security in such an environment.

The scale and scope of the technical backbone and speed at which it can reconfigure mirror the scale and scope of the actors that can engage in consequential action in cyberspace. This is a province not exclusive to great powers, who are interconnected with States that typically could not engage in similar activities, but can do so in cyberspace.[56] To be clear, funding sophisticated ICT development costs money, it requires trained skilled operators, and great powers tend to have the resource base that provides them some advantages. However, the nature of ICT does not have the same barriers to achieve levels of ICT possession and skill in operation required to produce nuclear weapons or sustain conventional force. At the level of episodic consequential action, sophisticated individuals and small groups with open source cyber technology can have strategically consequential impact.[57]

From a national security planning standpoint, interconnectedness means that States' sources of national of power are constantly connected not only to more States than in the past, but to non-State actors, private industry, organized crime, and others, all of whom may act from a different motivation base. Their pursuit of interests might not be directly adversarial to mine, but in the cross-cutting density that is cyberspace, their actions may impact me quite negatively. In January 2018, a student in Australia started tweeting comments about heat maps released by the fitness company Strava that showed the exercise patterns of what clearly became recognized as forward-deployed military bases of several Western countries. Although many of them were known locations, others were not. More troubling is that the data could enable military intelligence agencies to track individual soldiers' movements and potentially discern deployment patterns. It is likely that when the marketing team at Strava met to look at their heat maps, they probably said, "This is really cool—we are even in the most remote places of the world. Let's put that up on the web." It is reasonable given their motivation (profit-seeking through increased visibility and user support) that they never once thought they were undermining US Special Forces' operational security. The detail of global military personnel movements that this fitness company published would have been inconceivable for even the largest and best intelligence agencies in the Second World War. In a classic case of interconnectedness and constant contact, US Central Command now had to address the "threat" posed by a San Francisco–based fitness company.[58]

The virtualscape of cyberspace is, therefore, vast in the scale and scope of its technical base, user base, and, most important for assessing behavior, motivation base. Taken as a whole, the reconfiguration of the space to advance one's interests is not simply potential, it is a continuous reality. Somewhere, someone at any given moment is both capable and motivated to shift cyberspace to align with their interests. In business activities, that is a normal and legal practice in the competitive pursuit of profit. In the world of statecraft, it can be normal and legal practice through regulation, such as the EU's 2018 General Data Protection Regulation, which required significant shifts in how data are held in and traverse across cyberspace.[59] Such shifts might also be direct attempts to create conditions to enhance security in one's favor that might have a neutral or negative effect on adversaries' security or may be direct attempts to create conditions of insecurity for an opponent to unbalance them or set back advantages they may have been seeking to gain.

The structural feature of interconnectedness and the condition of constant contact combined with the inherent capacity to reconfigure introduce a significant incentive to be in some control of setting the conditions within and across networked computing and the digital interfaces that knit together cyber activity in one's favor, rather than ceding that capacity and action to others. We are left as a result with structurally induced persistence, defined through a continuous willingness and capacity to seek initiative. As such, in security terms, persistence is a structural imperative. In a strategic environment in which the conditions of security and insecurity can be configured directly, States must persist in ensuring as best they can that their constant contact with this vast interconnected set of actors is configured in a way that their core sources of national power can leverage the interconnectedness for growth (be it economic wealth, social cohesion, improved national health, education, informed public policy, or military might) while remaining secure.

This requires a premium to be placed on anticipating exploitation of software, hardware, and the processes that link them together and on translating that anticipation into a favorable advancement of their interests. Exploitation can take the form of simply using the technology lawfully in unexpected ways, but from a planning perspective one must also anticipate exploitation of vulnerabilities through illegal and unauthorized ways. Cyberspace is littered with vulnerabilities that leave all actors, including the most powerful States, exposed to exploitation. In fact, the States most dependent on cyberspace are at once both very powerful and very vulnerable cyber actors.

Systemically, we are left with the realization that cyberspace is macro-resilient (and thus stable) and micro-vulnerable (and thus inherently exploitable). The inherent vulnerability to exploitation is what raises the potential for cyber activity to have strategic effect, because the opportunity exists for cumulative effect achieved through setting and resetting the configurations of the virtualscape to yield strategic gains in relative power over time. Blood loss from a thousand nicks can be as devastating to capacity as the loss of blood from a single massive wound.

We now have the outlines of a theoretical framework. Cyber persistence theory, as a structural theory, posits that the combination of a core structural feature (interconnectedness), core condition (constant contact), and reinforcing structural features (e.g., macro-resilience and micro-vulnerability) forms the basis of a distinct logic (exploitation) that carves out a form of strategic behavior (initiative persistence) among States in the pursuit of security that is distinguishable from the conventional and nuclear strategic environments.[60]

The cyber strategic environment is, therefore, the product of interconnectedness, constant contact, and an inherently reconfigurable terrain and capacity to act across and through that terrain. The structural features of the cyber strategic environment reward those States that succeed in initiative persistence, where success is measured as being able to effectively anticipate and persistently set the conditions of security in their favor in and through cyberspace. Those that do not persist and cede the continuous reconfiguration of cyberspace to others, minimally, will suffer from a lack of alignment with the ever-changing virtualscape. In adversarial relations, States that cede the initiative can assume that opponents will directly set conditions that will increase their insecurity and ultimately risk degradation of their sources of national power. Over time, the cumulative effect of such action, in relative terms, will begin to shift the relative distribution of power among States. In the context of cyberspace relations, this opens the prospect that State relations, strategically, will be defined not through the prosecution of war, or the avoidance of war, but by an alternative to war.

Conclusion

Bernard Brodie, in two highly enlightening explanatory essays on the great work of Carl von Clausewitz, *On War*, concluded that what makes that work enduringly relevant to modern security studies was the approach that Clausewitz took to his daunting task of explaining war. Clausewitz's main achievement, Brodie argued, "was to get to the fundamentals of each issue he examined beginning with the fundamental nature of war itself."[61]

Clausewitz's own analysis was that war is "a true political instrument, a continuation of political intercourse, carried on with other means. What remains peculiar to war is simply the peculiar nature of its means."[62] Clausewitz understood war to be "different from anything else," and thus worthy of intense study, but also nothing more than a subset of a larger category—the core politics between international actors.[63] This was his single most insistent point—that war had to be studied and practiced as the subjugated instrument of high politics. It was not war for war's sake that motivated Clausewitz's inquiry, but rather the desire to understand how war is used to acquire and advance power relative to your opponents of the day.[64]

The logic of cyber persistence theory suggests that it is best to reorient our thinking about national security as it relates to cyberspace to the point of considering that a third strategic environment exists in which States pursue relative power to secure—an environment that does not follow the logic of coercion and war that Clausewitz and so many others that followed have illuminated. As we address in the remainder of this book, cyber persistence both in theory and in practice resonates through a different logic captured by a different lexicon. Explanatory power will be gained in academic research if we recognize that the cyber strategic environment rests on *interconnectedness,* not segmentation; that constant contact is a condition, not a choice; that cyber activity of consequence should be understood primarily as *campaigns,* not incidents, intrusions, or hacks; that the inter-State dynamic is primarily one of competitive interaction, not escalation; and that *initiative* rather than restraint is necessitated.

Policy prescription will strengthen if we reorient thinking toward *rules of engagement,* not contingency planning options; *seizing targets of opportunity* rather than holding targets at risk; *being active and anticipatory* not as aggressive or offensive inherently but as primarily defensive in orientation; that we must execute *continuous operations,* not episodic ones; that costs and benefits are *cumulative,* not event/episode-based; that cost imposition can be considered an *effect of changing the cyberspace environment,* not as a strategy to influence adversary decision cost-benefit analysis or decision-making; that effective cyber operations and campaigns are primarily *exploitative,* not coercive; and that *competition below the level of armed conflict is just as consequential strategically* as war and territorial aggression.

Cyber persistence theory assumes that while the conventional and nuclear strategic environments remain essential to State politics, the "peculiar nature" of the cyber strategic environment may support a different logic and, thus, require a different approach to competing over relative power. Whereas security requires one to win war in the conventional environment and avoid war in the nuclear environment, States in the cyber strategic environment may have a true alternative through which to achieve strategically relevant outcomes. In the next chapter we turn toward developing this theoretical reframing further.

3

Cyber Behavior and Dynamics

In Chapter 2, we argued that States face a structural imperative to persist in seizing and maintaining the initiative to set the conditions of security in and through cyberspace by exploiting adversary vulnerabilities and reducing the potential for exploitation of their own. We now turn our attention to the strategic choices States face and their behavior in an environment of initiative persistence.

Due to the features discussed in Chapter 2, States are presented with a fluid opportunity-laden environment in and through which they can experiment, test, plan, and achieve gains that cumulatively shift the overall distribution of power without seizing territory or directly destroying an opponent's sources of power. Within the cyber strategic environment, States, therefore, can choose to advance interests through alternative mechanisms to war. Furthermore, they are strategically incentivized to operate in and through cyberspace at scale to accumulate strategic gains without causing armed-attack equivalent effects.

We posit that the structural imperative and strategic incentive in tandem reward certain State cyber behaviors, create a specific inter-State phenomenon, and produce a particular dynamic. The observed interactions, engagements, and exchanges do not follow the patterns associated with escalation dynamics, which are central to managing security in the nuclear strategic environment and are a recurring concern of policymakers. Finally, we consider the potential impact of advances in artificial intelligence (AI) on behaviors and dynamics.

State Behaviors

How a State responds to the cyber strategic environment's structural imperative reflects its approach to setting the conditions of security in and through cyberspace by either seeking to avoid or engaging in exploitation.[1] All theories of international politics presume that States engage in some degree of internal-facing security activity, such as arming or "internal balancing."[2] What is novel about

Cyber Persistence Theory. Michael P. Fischerkeller, Emily O. Goldman, and Richard J. Harknett, Oxford University Press.
 DOI: 10.1093/oso/9780197638255.003.0003

cyber persistence theory is its argument that a State's external-facing measures will leverage cyber capabilities for unilateral exploitation, *not* brute force or coercion.[3]

Exploitation, thus, is the dominant State behavior in cyberspace. An exploit is generally understood as computer "code that takes advantage of a software vulnerability or security flaw."[4] Herb Lin defined cyber exploitation in espionage terms as a "cyber offensive action conducted for the purpose of obtaining information."[5] Exploitation is described similarly in the US Department of Defense's cyber doctrine as "actions [that] include military intelligence activities, maneuver, information collection, and other enabling actions required to prepare for future military operations."[6]

Both of these definitions are overly narrow, treating State exploitative behaviors in and through cyberspace merely as an intelligence contest. Instead, cyber persistence theory argues that cyber exploitation represents a strategic competition and therefore should be understood as one State gaining advantage by making use of another's cyberspace vulnerabilities.

We expect status quo and revisionist States alike to respond similarly to cyberspace's structural imperative, albeit with different objectives. Both should seek to reduce the potential for exploitation of their own vulnerabilities through internal-facing measures like patching, firewalls, and intrusion detection systems highlighting anomalous behaviors. Both may also pursue external-facing measures, chiefly, exploitation of adversary vulnerabilities to discern intentions and capabilities, gain a foothold to preclude or constrain adversaries' opportunities to operate in and through cyberspace, or expand their own opportunities to support their objectives.[7]

This chapter focuses on States' use of cyber capabilities at the operational and tactical levels, thereby generating effects that are cumulatively strategic in their impact on the international distribution of power.[8] It is at the operational and tactical levels that cyber capabilities provide unique strategic value.[9] At the grand strategic level, initiative persistent behavior should be coordinated with operational and tactical exploitation efforts. For example, China's digital Silk Road Initiative aims to control the global digital backbone to (as some argue) enable operational and tactical exploitation on a massive scale. Several States decided to exclude Huawei equipment from their 5G networks to preclude that opportunity from China's.[10] And, in 2020, US actions that might be classified as a cyber industrial strategy included a "Clean Network" policy to safeguard sensitive public and private sector information from aggressive intrusions and their exploitation by malign actors.[11]

Cyber operational and tactical exploitation targets vulnerabilities in one or more of cyberspace's three layers: the physical, logical, and cyber persona.[12]

Exploitation of the physical and logical layers generally occurs through direct hacking by seeking out open ports or other external network or system access points to exploit known vulnerabilities in software and hardware. Exploiting the cyber persona layer relies instead on the unwitting complicity of users. Examples include (1) social engineering and credential or spear phishing to gain access to passwords and login information and/or to facilitate malware downloads granting unauthorized access to a network; and (2) "water holing" or drive-by download techniques in which attackers estimate (or note) websites visited by organizations or individuals and implant malware that is downloaded when the sites are visited.[13]

Rather than create from whole cloth new concepts embodying the primary manifestation of exploitative inter-State cyber behavior in cyberspace, we leverage those developed for conventional and nuclear strategic environments—the *fait accompli* and agreed battle—but adapt them to the distinctive features of the cyberspace environment. Both concepts in their academic heyday were largely empirically derived, perhaps viewed as describing anomalous, less-consequential, or lesser-included cases. As a result, they received far less theoretical attention. It seems that the worm has turned, as the cyberspace environment has elevated the importance of those concepts for describing State behavior and, consequently, has diminished the centrality of theories like coercion and its related concepts, which were developed for the nuclear and conventional environments.

The Terrestrial *Fait Accompli*

In his research on the *fait accompli* in terrestrial disputes, Dan Altman noted how James D. Fearon, when reviewing the literature on strategic interaction during crises, drew a basic distinction between crises as competitions in risk-taking and crises as competitions in tactical cleverness (i.e., as attempts to outmaneuver the adversary).[14] Fearon argued for the importance of both but focused on the former.[15] International relations theorists leveraged Fearon's insights on competitions in risk-taking to develop a strategic bargaining paradigm that places central emphasis on the concepts of coercion, signaling resolve, brinkmanship, and escalation.[16]

With the advent of cyberspace, perhaps it was natural to adopt these concepts to describe and explain State cyber behaviors.[17] However, as we show later in this chapter, those concepts fail to explain most State cyber strategic behavior short of militarized crises and armed conflict. Fearon's less-explored alternative better describes this behavior; its premise is captured in the strategic bargaining concept of the *fait accompli*.

The *fait accompli* is described with little variance in international relations strategic bargaining literature. Altman says the "*fait accompli* imposes a limited unilateral gain at an adversary's expense in an attempt to get away with that gain when the adversary chooses to relent rather than escalate in retaliation."[18] Alexander George describes it as altering the status quo in one's favor through a "quick decisive transformation" of the situation that avoids unwanted retaliatory escalation.[19] A recent illustrative example is Russia wresting the Crimean Peninsula from Ukraine in February 2014. Altman uncomfortably categorizes *faits accomplis* under coercive bargaining, but only because they are considered to represent the failures of deterrence in the conventional strategic environment (i.e., the terrestrial frame).[20]

The strategic logic behind the *fait accompli* in terrestrial disputes hinges on finding vulnerabilities in "red lines." Altman defines red lines as the part of a coercive demand that distinguishes compliance from violation.[21] When red lines are viewed as arbitrary, imprecise, incomplete, or unverifiable, States are incentivized to act unilaterally to achieve their limited desired gain.[22] In terrestrial disputes, red lines are usually anchored on a disputed border. India and Pakistan, for example, have clashed over Kashmir's status and border several times, with both making claims to the whole of Kashmir; today, they control only parts of it—territories recognized internationally as "Indian-administered Kashmir" and "Pakistan-administered Kashmir." Altman concludes that, when States do act, "*faits accomplis* are more likely to succeed at making a gain without provoking war when they take that gain without crossing use-of-force red lines."[23]

Finally, although the *fait accompli* may fail to achieve the desired outcome for several reasons—for example, the defender chooses not to relent and marshals superior forces to take back the gain made—it fails in execution for only one reason: the defender anticipates the unilateral action and takes steps in advance to set the conditions of security in its favor. This contrasts with the multiple ways that coercive strategies can fail: lack of commitment, ambiguity of demands, or non-credible capability.

The Cyber *Fait Accompli*

We define the *fait accompli* in the cyber strategic environment as a limited unilateral gain at a target's expense where that gain is retained when the target is unaware of the loss or is unable or unwilling to respond.[24] The immediate "gain" in or through cyberspace is the setting of security conditions in one's favor, essentially a reconfiguration of cyberspace technically, tactically, operationally, or strategically that has an immediate impact in advancing an interest and/or positioning a State for further advancement of other interests.

The United States' military Cyber Command's (USCYBERCOM) reported effort to secure the 2018 US midterm elections can be understood in this way. USCYBERCOM reportedly took the initiative to exploit vulnerabilities in the cyber infrastructure of Russia's Internet Research Agency (IRA) to constrain their ability to act against the US 2018 elections through that infrastructure.[25] This campaign resulted in an initial benefit gained for the United States. It also created sufficient organizational friction in Russian system since time, talent, and treasure had to be redirected toward figuring out whether the obvious American exploitation was the only exploitation underway, thereby changing the conditions of security in its favor within that space in which the initiative Russia might have had in electoral interference was now lost.[26]

What emerged in early 2021 as the "SolarWinds" campaign can be understood similarly as a cyber *fait accompli* in which the initial exploitation of technical update functionality managed through clever tactics and sophisticated operational planning provided Russia with an initial benefit gained of wide-ranging real-time access to government and private sector networks at a scale that positioned Russia to realize strategic effects. The initiative seized through this initial exploitation, however, and importantly, was not lost upon discovery of the intrusion. In fact, its discovery put the US government on its heels as it could no longer trust the confidentiality and integrity of data and communication flowing across its unclassified networks in a number of agencies. Therefore, while seeking to mitigate the intrusion, the United States also had to readjust its thinking, planning, and policies toward what Russia could do with the information it had required, for example, internal communications on sanctions and other policies.[27]

Cyber *faits accomplis* often occur in series. An initial gain is often followed by the pursuit of subsequent gains, further expanding the scope and/or scale of the condition of security or acquiring a "good" (such as intellectual property) that is perceived to be of value by the attacker for maintaining or altering the international distribution of power.[28] For example, Russia's security service reportedly leveraged the technical gain from its SolarWinds campaign to exfiltrate sensitive tools that FireEye, Inc. uses to find vulnerabilities in clients' computer networks.[29] Similarly, the United States leveraged its initial technical gain against the IRA in 2018 to further enhance the condition of security in its favor by subsequently revealing its presence on IRA networks. This reportedly resulted in IRA organizational friction and Russia shifting its focus and efforts toward defense, both of which served a US strategic objective of taking Russia's focus away from cyber-enabled information operations directed at US elections.[30]

In 2014, Chinese cyber operators obtained login credentials for the US Office of Personnel Management (OPM) networks by first breaching the networks of KeyPoint Government Solutions, an OPM contractor. This initial gain set the conditions of security in their favor vis-à-vis OPM. These credentials were subsequently used to log into OPM networks and install malware that exfiltrated sensitive data on approximately 22 million key US personnel, data China could conceivably leverage in a strategic effort to maintain or alter the international distribution of power.[31] Additionally, beginning around 2014 Chinese cyber operators initiated a campaign ("Cloud Hopper") premised on compromising managed service providers (MSPs)—companies that remotely manage their clients' information technology infrastructure—as a way to establish footholds for follow-on exploitations of the MSPs' clients.[32] After compromising a MSP, the operators mapped out the network topology to find the credentials of the system administrator who controlled the company "jump servers," which act as a bridge to client networks. With that information, they then "jumped" to client networks, mapped those topologies, and exfiltrated information that served their strategic interests.[33]

With these examples in mind, we eschew George's "quick" adverb in our cyber *fait accompli* definition because, although initial gains can be realized quickly through cyber exploitation, subsequent efforts to realize follow-on gains can extend to minutes, hours, days, months, or even years. To wit, analyses indicate that the time from an attacker's first action in an event chain to the initial compromise of an asset is typically measured in minutes.[34] Crowdstrike reported that after an initial beachhead has been established, the average "breakout time" for the most competent State cyber actors ranges from 19 minutes to around 10 hours.[35] Mandiant noted that Chinese cyber operators engaged in significant exfiltrations of intellectual property by maintaining a persistent presence on targeted networks for an average of 356 days.[36]

Within the definition of the cyber *fait accompli*, "unilateral" means that the exploitative action is pursued independently of any decision made by the targeted entity. Thus, the *fait accompli* is distinct in principle from coercion, which depends on demands, commitments, and signaling expressly for the purpose of influencing the target's decisions.[37] Moreover, making gains at the expense of an adversary is not the same as threatening to impose costs or actually doing so in an effort to change an adversary's decision calculus. *It can, however, be equally strategically consequential for international politics*—an important point we expand on later in this chapter and in Chapter 4.[38] Once a benefit or gain is realized, it may subsequently serve as a foothold for future coercive strategic bargaining, depending on the target's political value; however, the cyber *fait accompli* is first and foremost about seeking unilateral operational and/or tactical gains through exploitation.

As in the terrestrial frame, States pursuing cyber *faits accomplis* have a strategic incentive to pursue their desired gains in and through cyberspace in ways that do not invite escalatory retaliation. That said, cyber persistence theory argues that the opportunity for reward, not fear of retaliation, is the primary incentive driving States to engage in cyber operations short of armed-attack equivalence. States recognize that strategic opportunities flow from cyberspace's abundant vulnerabilities and resilience to persistent, exploitative cyber operations/campaigns with effects short of armed-attack equivalence. Given opportunities, States are taking them.

The strategic logic behind the *fait accompli* in cyberspace also hinges on finding vulnerabilities. However, unlike the terrestrial frame, those vulnerabilities do not lie in the ambiguity of a coercive demand, but rather *in the very fabric of cyberspace itself.*[39] Cyberspace has been described as a vulnerable yet resilient technological system,[40] organically offering an "abundance of opportunities to exploit user trust and design oversights."[41] When considered along with a condition of constant contact, the prevalence of vulnerabilities provides a strategic incentive for States to pursue unilateral gains in and through cyberspace, persistently.[42] This incentive is further enhanced because of cyberspace's resilience. The *fait accompli* in physical space returns a marginal, episodic gain—often a small piece of territory. Cyberspace, by contrast, encourages the accumulation of gains to levels of strategic significance through series and/or campaigns of *faits accomplis* at scale because its resilience mitigates concerns that such campaigns might put at risk the digital environment's systemic functionality.[43]

There is a second notable way in which the cyber *fait accompli* diverges from its counterpart in the conventional strategic environment. George describes the *fait accompli* as a strategic bargaining concept that States use to change the status quo in the international system. Although cyber *fait accompli* campaigns could conceivably change the status quo in the international system, they are unilateral, independent actions and thus are not indicative of strategic bargaining.[44] For example, North Korea has unilaterally pursued an aggressive campaign of cyber-enabled theft from financial institutions and cryptocurrency exchanges and has used the proceeds to expand its nuclear weapons program and develop intercontinental ballistic missiles.[45] The $2 billion Pyongyang reportedly accumulated from 2016 to 2019 to support those programs while evading sanctions is more than three times the amount of currency it was able to generate through counterfeit activity over the four decades prior.[46] Thus, pursuit of cyber *faits accomplis* is more accurately described as a unilateral, independent strategic choice in the cyberspace environment rather than as a mechanism for strategic bargaining.[47] As noted earlier, an important distinction between coercion theory and cyber persistence theory lies in the recognition that *while all strategic bargaining is competition, not all strategic competition is bargaining.*

The cyber *fait accompli* is a useful concept for describing and explaining States' primary operational and tactical cyber behaviors—how they persist in seizing and maintaining the initiative to set the conditions of security in and through cyberspace and, through their success, further seek to maintain or alter the international distribution of power. It accounts for both unilateral operations seeking gains from often significantly disparate targets, as well as efforts that routinely avoid operations that could justify armed retaliation.

Former US Secretary of Defense Leon Panetta's concerns regarding a cyber Pearl Harbor (a *fait accompli*) were met with both skepticism and support.[48] By referencing the *fait accompli*, Panetta highlighted an important concept upon which policymakers should focus in cyberspace. However, Panetta fixed only on the *fait accompli* as a coercive strategic bargaining mechanism in cyberspace, which is currently infrequent in the cyber environment. He failed to recognize or anticipate the cyber *fait accompli* as the primary and highly consequential strategic choice for both attackers and defenders pursuing an alternative strategic approach.

Direct Cyber Engagement

It may seem counterintuitive that, in a strategic environment whose central feature is interconnectedness, the primary State cyber exploitative behavior is unilateral, independent action (the cyber *fait accompli*) rather than mutually dependent behavior. This is a consequence of the vastness of potential uncontested opportunities. Most State cyber operations are running parallel to each other, occurring in and through an interconnected space that rewards persistence. States must assume there are a multitude of parallel operations and campaigns being run at any given time by a variety of actors that can adversely affect them. Arguing that the cyber *fait accompli* is the primary behavior, however, does not preclude mutually dependent behavior, which we call *direct cyber engagement.*

Direct cyber engagement is a cyber exploitative action short of armed-attack equivalence in a mutually dependent competition for control of key cyberspace terrain that peer cyber States mutually perceive as strategically, operationally, or tactically significant for setting the conditions of security in their favor (e.g., the command and control infrastructure of a State's widely dispersed malware enterprise or its nuclear command and control critical infrastructure).[49] The cyber strategic environment rewards States pursuing direct cyber engagement (in addition to cyber *faits accomplis*) as distinct from operations of armed-attack equivalence.

To reset the conditions of security, States may communicate that certain behaviors are unacceptable by assertively responding through direct cyber engagement to the detection of such behaviors.[50] A direct cyber engagement could also take the form of States deliberately aligning their actions around a shared objective of seeking mutually dependent understandings of acceptable behaviors. For example, States may seek tacitly and/or explicitly to communicate that compromising the confidentiality but not integrity or availability of certain critical infrastructure systems is acceptable behavior.

Direct cyber engagement is currently scarce, a frequency consistent with its status as the secondary State behavior in the cyber strategic environment. Scarcity is driven by the extraordinary abundance of opportunities for unilateral exploitation (cyber *fait accompli*) given the vastness and ever-changing substance of cyberspace, which make achieving gains possible without direct cyber engagement.[51] Additionally, the improbability that competitors will have equal situational awareness of opportunities for unilateral exploitation allows States with initiative to avoid direct cyber engagement when seeking gains.

Current scarcity could give way to increasing frequency of direct cyber engagement. States are in the early stages of strategic cyber competition. Managing this competition through more predictable engagements that hold a pattern around mutual dependence and interest can and will likely occur. Nevertheless, the sheer opportunity expanse along with other factors currently reinforce the cyber *fait accompli* as the primary default behavior.

For example, competitors' domestic laws and national cyber policies and capacities may constrain the degree to which they are able to secure all manifestations of their national interests. Or, States may limit the operational scope of their primary cyber forces to their own national defense department/agency systems when adversaries see opportunities in a wider scope of government systems and private sector or other non-governmental infrastructure representing national interests.[52] These variances will likely complicate intentional efforts to converge on behaviors.

Additionally, national interests and strategic cultures will skew how States might fear or value exploitation of the same vulnerability—a vulnerability appearing as a low threat to one State may be considered a valuable opportunity for another.[53] Further, even when hostile intrusions are detected (or attempted) or cooperative engagements are offered, States may struggle to respond in a timely enough manner to overcome the high "noise" level in cyberspace that undermines the success of tacit bargaining in any environment, let alone cyberspace.[54] Lack of timeliness could result from policy, human, and/or technological constraints.[55] Finally, as cyberspace is a global phenomenon, potential operational constraints may also derive from interpretations of international law as applied to cyberspace.

The distinction between cyber *faits accomplis* and cyber direct engagement is also a very important observation for most global cyber norms approaches. Historically, these have assumed the prominence of the latter and are not geared to managing a strategic environment incentivized toward the former (an issue we address in Chapter 5).

In sum, the primary and secondary behaviors of States in and through the cyber strategic environment—the cyber *fait accompli* and direct cyber engagement—are consequences of a structural imperative to persist and of a structurally derived strategic incentive to pursue gains through cyber exploitation short of armed-attack equivalence. That these two behaviors comprise most State cyber behaviors has significant implications for the inter-State phenomenon and dynamic that best characterize the environment.

Inter-State Phenomena and Dynamics

In efforts to understand dynamics in cyberspace, the security studies community has primarily defaulted to an escalation dynamics framework, a consequence of presuming that coercion theory explains State cyber behavior.[56] It is not contentious to say that much of that scholarship is heavily influenced by modern thinking regarding escalation dynamics introduced in the work of Herman Kahn, who defined escalation as "an increase in the level of conflict in international crisis situations."[57] Kahn's focus on managing escalation was a direct result of the nuclear strategic environment's imperative to avoid nuclear Armageddon. Beginning with the assumption of some sort of limited conflict or agreed battle, Kahn proposed a framework populated by three mechanisms in which a would-be escalator could increase, or threaten to increase, his efforts: "increasing intensity," "widening the area," and "compounding."[58] Thus, escalation in Kahn's framework describes both the mechanisms and the resulting dynamic of their employment. The State that could employ these mechanisms to achieve escalation dominance, he argues, could gain strategic advantage while avoiding all-out nuclear war. Kahn specifically noted that his escalation theory was developed with a focus on giving primary attention to the threat or reality of force or coercion as a factor in negotiation. This may explain its centrality to coercion theorists seeking to explain cyberspace dynamics.

Kahn argued that there are two basic classes of strategies that each side can use from his origination point of limited conflict or agreed battle. One class is coercion-based, using the risk or threat of escalation and resulting in an escalation dynamic. This class, he noted, refers to deterrence strategies and is clearly the focus of his escalation framework. The second class is not coercion-based and does not include escalation mechanisms. Rather, it is a class in which States *unilaterally make use of the factors relating to particular levels of escalation in order to gain or maintain an advantage* (i.e., factors organic to agreed battle itself). This class receives far less theoretical attention, yet is better aligned with the tenets of cyber persistence theory.[59] Exploring it in greater depth provides insights into the operational and tactical dynamics among States populating the cyber strategic environment, the mechanisms of gaining or maintaining advantage in the same, and the inter-State phenomenon that results. As a first step, we explain the concept of agreed battle by reflecting back on the Cold War strategic context in which it was birthed.

Agreed Battle

Agreed battle is a concept rooted in factors relating to particular levels of escalation. It emphasizes that, in a potential escalation situation in which both sides are accepting limitations, there is, in effect, an "agreement" on acceptable and unacceptable behaviors based on a shared strategic interest, whether or not an overall "agreement" is explicit or mutually understood. "Thus the term does not have any connotation of a completely shared understanding, an intention of containing indefinitely with the limitation, or even a conscious quid pro quo arrangement."[60]

Kahn introduced the concept to describe a phenomenon of great power strategic competition during the Cold War that was short of all-out war, but included limited war and proxy war. Agreed battle, he explained, was and would continue to be a principal approach through which great powers had "agreed" to compete to avoid escalation into a "central" war. This aligns with our contention that both the conventional and nuclear strategic environments rest on coercion and the prospect of war (prosecuting it or avoiding it). Consider, for example, that despite their frequent and significant involvement in proxy wars around the world, both superpowers unilaterally and independently avoided any overt direct engagement of military forces.[61] Indeed, such proxy wars are the central concern of Kahn's analysis and are captured in the escalation mechanism of "compounding," which he defined as "consisting of an attack on an ally or client of the principal opponent."[62]

Agreed battle describes a phenomenon that has both tacit structural and substantive features. For example, agreed battle characterized by proxy war comprises a strategic space with a lower structural bound of militarized crisis and an upper bound of a limited war. Within this space, Soviets and Americans held many common but independent understandings of unilateral behaviors they did and did not consider acceptable. These included common views on who does and does not use force, where it is and is not used, and/or how it is and is not used. Thomas Schelling might refer to these as factors informing focal points around which tacit coordination could emerge.[63] Schelling observed that limits in limited war are arrived at "not by verbal bargaining, but by maneuver, by actions, and by statements and declarations that are not direct communication to the enemy. Each side tends to act in some kind of recognizable pattern, so that any limits that it is actually observing can be appreciated by the enemy; and each tries to perceive what restraints the other is observing."[64] Over time, continued observation and assessment of unilateral, independent patterns result in mutual understandings of acceptable and unacceptable behaviors.

Great power competition during the Cold War did not occur only in a strategic space bounded on the low end by militarized crises and on the upper end by limited wars. Strategic competition was also intense in a competitive space bounded on the low end by restraint and on the upper end by militarized crises. Schelling noted that, "in peacetime," the United States and Soviet Union both unilaterally and independently abstained from harassing actions on each other's strategic forces, did not jam each other's military communications, put at risk each other's populations with fallout from weapons tests, or wage surreptitious undersea wars of attrition.[65] Robert Osgood noted that both countries also refrained from aiming their missiles even in the general direction of the other side while testing them.[66] Thus, in the early Cold War, underpinned by common but independent understandings of acceptable and unacceptable behaviors, the Americans and Soviets formed mutual understandings through observation and assessment of each other's actions—the substance of agreed battle.

An important characteristic of these mutual understandings was the element of restraint motivated by a shared strategic interest in avoiding the kind of false alarm, panic, misunderstanding, or loss of control that may lead to unintended or non-deliberate escalation.[67] Thus, mutual understandings comprised understandings of inaction as well as limited action. This accords with Kahn's description of agreed battle as a bounded space within which States seek gains without inviting escalation. Kahn also never suggested, nor do we endorse, a normative connotation with the term "agreed battle." States need not assent or accord to the normative substance of a phenomenon for it to be present.[68]

In sum, agreed battle describes a phenomenon comprising a tacitly bounded space within which adversaries seek to maintain or gain advantage based on mutual understandings of acceptable and unacceptable behaviors in a coercive environment in which fear of costs reinforce such understandings. A tacitly bounded space also exists in the cyber strategic environment, but it results from a different dynamic and thus is analytically distinct from agreed battle.

Cyber Agreed Competition

The conventional, nuclear, and cyberspace strategic environments share (in varying degrees) a strategic incentive to self-limit behavior. In the nuclear environment, according to Kahn and others, the incentive for the great powers derived from a desire to avoid escalation from limited war to "central" war and its most extreme manifestation of nuclear war.[69] There was a premium on finding mechanisms of self-limitation, while still making gains. This is what escalation dominance was all about.

In the conventional environment, the incentive for some self-limitation derives from a desire to use force efficiently, since it always entails cost (both self-created and imposed by the opponent). Thus, States have historically sought to apply just the right amount of force needed to achieve one's ends and not escalate from militarized crisis into limited exchanges to full-scale conventional war unless necessary. Prosecuting war in all its forms is part of the dynamics of seeking security in and through the conventional strategic environment, but it does not mean that States plan to fight war without restraint.

For the cyberspace strategic environment, self-limiting behavior takes the form of setting aside escalation strategies and coercion, essentially because they are not necessary to configuring and reconfiguring cyberspace in a State's favor. Cyber persistence theory argues that opportunities for reward, not fears of retaliation or costs from exchanges, are the primary incentive driving States to engage in cyber operations short of armed-attack equivalence. This strategic incentive to self-limit behaviors through an alternative to coercion results in a phenomenon we have named "cyber agreed competition," which we differentiate from the agreed battle phenomenon of the conventional and nuclear strategic environments. The cyber agreed competition phenomenon's tacit lower and upper bounds are inclusive of operational restraint and exclusive of operations causing armed-attack equivalent effects.[70]

Cyber persistence theory argues that unilateral, independent action—the cyber *fait accompli*—is the primary inter-State behavior for States wanting to persist in seizing and maintaining the initiative to set the conditions of security in and through cyberspace and, by doing so, maintain or alter the international distribution of power. Therefore, cyber *faits accomplis* are the primary behaviors driving the cyber agreed competition phenomenon. This unilateral, independent, and parallel-conducted behavior is the response to the challenge of finding security in an initiative persistent environment filled with opportunity.

Within the bounds of the cyber agreed competition phenomenon there exists among the most significant actors a core set of tacit, mutual understandings of acceptable and unacceptable behaviors that currently manifest as a "voluntary and non-binding" consensus.[71] Given the ingenuity of States and the ever-changing technology of the cyber strategic environment, there is the potential for a larger set of common but independent understandings that may, over time and through observation and assessment, supplement the existing core set of tacit mutual understandings.[72] Cyber persistence theory argues that a core tacit mutual understanding will minimally converge on the permissibility of persistence in seizing the initiative.

There also exists a set of uncommon, independent (and divergent) understandings that, through mutually dependent interactions in cyber agreed competition, States can also come to recognize what the "legitimate and illegitimate moves are," which moves are "within the rules," and which moves would breach the upper boundary.[73] In other words, when States observe positive-novel or discordant-frictional behaviors inconsistent with patterns of observed behaviors that formed the basis of existing mutual understandings, they can choose a course of mutually dependent interaction (rather than continued unilateral, independent action) to communicate their observation and acceptance or unacceptance thereof.[74] This behavior describes perfectly the secondary manifestation of State exploitative behavior in cyberspace discussed previously: direct cyber engagement.

Thus, given that the cyber agreed competition phenomenon encompasses both the primary and secondary behaviors hypothesized by cyber persistence theory, we argue that it is the central inter-State phenomenon in the cyber strategic environment, with important implications for the central dynamic of "competitive interaction."[75]

Competitive Interaction Dynamic

To summarize, we have argued that the dominant behavior in cyberspace is exploitation; that its primary and secondary manifestations as State cyber behaviors are cyber *faits accomplis* and direct cyber engagements, respectively; and that these behaviors result in the inter-State phenomenon of cyber agreed competition, a tacitly bounded strategic competitive space inclusive of operational restraint and exclusive of operations causing armed-attack equivalent effects. However, cyber *faits accomplis* and direct cyber engagements do not merely describe State behaviors—they also serve as the mechanisms through which States seek to achieve strategic gains in and through cyber agreed competition. We describe the dynamic resulting from States leveraging these mechanisms as *competitive interaction*. Although States may pursue cyber *faits accomplis* and direct cyber engagements that are analogous to Kahn's mechanisms for escalation—"increasing intensity" could be construed as increasing the scale of cyber operations, "widening the area" could be increasing the scope, and "compounding" could be targeting a State's extraterritorial infrastructure—such behaviors would only be mechanisms of escalation if their effects breached the tacit ceiling of cyber agreed competition. Given that States are not strategically incentivized to breach that ceiling, the dominant dynamic in the cyber strategic environment is competitive interaction, not escalation.

A Consideration of Escalation Actions and Dynamics

Cyber persistence theory posits that core features of the cyber strategic environment do not incentivize coercive action and escalation. Rather, they incentivize cyber *faits accomplis* and direct cyber engagements, which are mechanisms for achieving strategic gains in competitive interaction. This diverges from much of the existing cyber security literature and from the core assumptions that underlay policy planning. The nuclear paradigm has been so profoundly embraced that security thinking has assumed escalation management as a universal security strategy concept without considering its conditional nature. It is indeed *essential* in an environment in which security is defined by the avoidance of a definable set of actions (nuclear war), and it is *important* in an environment in which the prosecution of a definable set of actions (conventional war) is costly. But it is *not a useful construct* analytically or prescriptively in an environment in which security is defined by an alternative (exploitation rather than coercion) to what escalation seeks to control.

The dynamic of competitive interaction reinforces the distinct definition of security we have offered for the cyber strategic environment. Security cannot be measured as the absence of proscribed action as so defined in the nuclear environment. The imperative and incentive of the cyber strategic environment all reinforce continuous action and, thus, without a redefinition we would consistently have to assess our national security strategies as failing. Similarly, security measured by the success in prosecuting war (effective offense and defense) is not applicable to an environment in which the primary behavior, the cyber *fait accompli*, is an alternative to war.

Our analysis of the cyber strategic environment focuses on behavior in and through that environment. It is, of course, necessary to consider cyber capabilities as an integral means to warfighting. In fact, there is no future war that will be fought that does not leverage some aspect of cyberspace as an enabler. But this implies breaching the armed-attack equivalent threshold, which then opens the door for States to legitimately bring to bear cross-domain, conventional, kinetic weapons in self-defense. Under these conditions, the traditional escalation dynamic associated with militarized crises and war supplants the competitive interaction dynamic.[76] Once breached, the logics of the other two strategic environments—nuclear and conventional—will prevail as States supplement their militarized crisis, warfighting, and deterrent behaviors with cyber means. Actors engaged in multi-domain war to influence relative State power will engage in cyber operations, but they will be of a fundamentally different sort from those in and through the cyber strategic environment that seek to influence relative power without having to resort to violence.[77]

A spiraling escalation dynamic in cyberspace is a concern of policymakers.[78] To address this, we adopt Kahn's definition of escalation dynamic: "an increase in the level of conflict in international crisis situations."[79] We describe a cyber escalation action as a cyber behavior causing armed-attack equivalent effects, thereby breaching the upper bound of cyber agreed competition and supplanting the competitive interaction dynamic with an escalation dynamic. With this understanding, we offer hypothetical scenarios in which the previously described incentives for not escalating could be overwhelmed and lead to deliberate cyber escalation action. The scenarios derive from the substantive immaturity of cyber agreed competition, accidental or inadvertent cyber escalation actions, and a perception that competitive interaction is resulting in an unfavorable shift in relative power.

Deliberate cyber escalation actions are understood to have specific purposes in mind. Broadly speaking, a State may seek to gain advantage in armed conflict, preempt armed conflict, penalize an adversary for some previous militarized action, signal an adversary about its own intentions and motivations regarding competition, or avoid defeat in competition. The first three describe contexts of the other two strategic environments in which an escalation dynamic is already present (militarized crisis or armed conflict) and are well covered by the existing literature. The latter two are novel and deserve additional scrutiny.

In these early days of State cyber activity, misinterpretation of cyber *faits accomplis* or direct cyber engagements is possible due to incomplete information or lack of shared reference frames. The substance of cyber agreed competition is currently immature—mutual understandings of acceptable and unacceptable behaviors and of cyber key terrain are maturing slowly and narrowly. Thus, a conceivable scenario today is that a cyber *fait accompli* or direct cyber engagement target is unexpectedly viewed by a defender as highly sensitive or threatening, respectively, thereby inadvertently prompting a deliberate cyber escalation action to signal an intention and motivation to aggressively secure that asset.

Accidental or inadvertent operations or effects that may encourage a cyber escalation action may occur if, for example, rules of engagement are ambiguous, cyber forces are undisciplined, or a high-level command decision is not received properly by all relevant cyber units.[80]

It is also conceivable that perception of an emerging or enduring significant imbalance of cyber *faits accomplis* or direct cyber engagement outcomes portending a relative shift in power between adversaries or a relative decline of a State across the global distribution of power could prompt a cyber escalation action to redress the competitive imbalance and ultimately avoid defeat.

In all of these hypothetical scenarios, it is conceivable that an escalation action could launch a cyber escalation spiral, but such a dynamic should not be presumed. An escalation action need not *ipso facto* portend the emergence of an escalation dynamic.[81] It is also conceivable that a deliberate cyber escalation action could lead to a cross-domain escalation dynamic fueled by non-cyber military capabilities.

Recognizing the plausible conditions under which States may deliberately breach the upper bound of the cyber agreed competition and potentially fuel an escalation dynamic is important. This moves the discussion to militarized action and warfighting with cyber means, and into the two other strategic environments.[82] Moreover, the empirical record reviewed in Chapter 4 reveals little or no support for these scenarios in which cyber *faits accomplis* and direct cyber engagement are misconstrued and inadvertently lead to cyber exchanges that escalate out of cyber agreed competition and into crisis and conflict. Cyber persistence theory offers an explanation for why spirals might be unlikely—the exploitative rather than coercive intent behind cyber operations tempers the action-reaction that might follow a mistake.

Potential Impact of AI on Behavior and Dynamics

When considered alongside the other strategic environments, the technology in the cyberspace environment changes far more rapidly.[83] And this feature of the environment may impact primary behavior, the central dynamic, and their trends. Several books, expositions, and journal articles have been written on this topic, but none from the perspective of cyber persistence theory.[84] We scratch the surface in this section yet we are not convinced that AI will change the current primary behavior and central dynamic in the cyber strategic environment.

AI might expand the breadth of the bounded space within which cyber *faits accomplis* could be executed. For example, by crawling Internet-facing applications with "fuzzers" and/or "symbolic execution" tools, attackers may discover significantly more vulnerabilities to exploit at scale.[85] "Smart" malware could decrease "breakout" time after intrusion by increasing the efficiency of lateral movement and the probability of post-compromise additional infections. In a 2016 cyber operation against a UK bank, the malware infected multiple additional machines within minutes of an established foothold with a clear pattern of target selection. Not only was the malware able to scale its attack at some speed, it also selected targets based on a "smart" analysis of prospective success in further infection. Although the malware did not truly utilize AI techniques that many experts herald as coming in the near future, its selective programming likely foretells that future.[86] Alternatively, deep learning may improve the effectiveness of defensive efforts by constraining the ways vulnerabilities can be exploited and, consequently, restricting opportunities for cyber *faits accomplis*.[87]

In either case, AI does not alter the imperative of initiative persistence. AI does not clearly portend moving to dominance of the offense or the defense.

AI could potentially increase the prevalence of direct cyber engagements by increasing opportunities for States to engage in tacit bargaining, potentially leading to more rapid and granular mutually dependent understandings of acceptable and unacceptable behaviors. For example, from a securer's perspective, reducing mean time to detection of an initial compromise would facilitate direct cyber engagements. Absent rapid detection, no timely effort to tacitly bargain is possible, whether the dyadic interests are divergent or common. Machine learning can substantially improve time to detection through the application of real-time correlation engines across various intrusion detection system product's logs, thereby supporting timely alerts to system administrators.[88] Such systems may even go one step further to thwart the attacker by changing the configuration of other security controls to disrupt an attack or, alternatively, communicate tacit acceptance of the behavior.

Common examples of thwarting include reconfiguring a network device (e.g., firewall, router, switch) to block access from the attacker or to the suspected target and altering a host-based firewall on a suspected target to block incoming attacks. Some systems could even cause patches to be applied to a host if they detect that the host has vulnerabilities.[89] Alternatively, as machine learning detects patterns in behavior, a State seeking to communicate with others its preferences of acceptable behavior around targets of interest could leverage that capability by consistently and repeatedly (perhaps millions or billions of times) acting to consistently communicate its preference in regard to a system or network. Thus, a future in which States possess and implement AI to secure their national interests in and through cyberspace conceivably portends a cyber strategic environment in which the primary behavior is direct cyber engagement rather than the cyber *fait accompli*.

Finally, some have expressed a concern that great power competition has created an "arms race" in AI, possibly resulting in a "race to the bottom" on AI safety, in which firms or States compromise on safety standards while trying to innovate faster than the competition.[90] This speaks directly to the potential for accidental or inadvertent escalation actions and, potentially, escalation dynamics.

At this time, all that can be offered regarding potential AI cyberspace futures is conjecture. It has been argued that AI may not be significantly or widely adopted by attackers because "attackers are only as clever as they need to be."[91] As long as attackers are able to achieve gains without using AI, they will continue to do so.[92] Bruce Schneier sums it up nicely by noting that "No one doubts that artificial intelligence (AI) and machine learning (ML) will transform cybersecurity. We just don't know how, or when."[93] That said, it is prudent to consider the potential adoption by States; in Chapter 5 we return to this issue to examine AI's potential impact on cyber stability.

Tacit Coordination, Tacit Cooperation, and Tacit Bargaining

While introducing the behavioral concepts of cyber *faits accomplis* and direct cyber engagements and the phenomenon of cyber agreed competition, we have made several allusions to the importance of tacit understandings and tacit bargaining in the cyber strategic environment. In Chapter 5, we leverage the scholarship on tacit coordination and tacit bargaining to consider under what conditions States pursuing these behaviors impact the prospects for norms construction and cyber stability. To set the stage for that discussion, this section aims to make more explicit the connections between our behavioral concepts and tacit coordination and tacit bargaining.

When States pursue cyber *faits accomplis,* the concept of tacit coordination captures well how States may come to mutual understandings of behaviors that populate cyber agreed competition. Tacit coordination describes common but independent behaviors around a shared strategic interest. As cyber *faits accomplis*—characterized by unilateral, independent behaviors—are the primary State behavior in and through cyberspace, States' observations of others' cyber *fait accomplis* and the common strategic interests they may indicate can serve as a basis for establishing mutual understandings of the substance of cyber agreed competition.

Importantly, the phenomenon of cyber agreed competition should not imply that tacit *cooperation* is being pursued by adversaries over a shared strategic interest. In tacit cooperation, one party in effect takes a chance in the expectation that another will take an equivalent chance, leaving both better off. Tacit cooperation describes a process of mutually dependent behaviors intending to arrive at mutual understandings, in contrast to the unilateral, independent behavior characterizing tacit coordination (and cyber *faits accomplis*).

Agreed battle during the early Cold War provides a fine example of the distinction between tacit coordination and tacit cooperation. During this period, tacit coordination was the primary basis of mutual understandings, as neither party took chances with expectations when maximizing unilaterally and independently around a shared strategic interest; each merely pursued its best strategic choice, given the other's independent choices. A deeper understanding of other parties' behaviors occurred through observation, but there were no implicit agreements to adjust actions on a mutual or contingent basis. During the

Korean War, Stalin's decisions to restrict the number of fighter planes provided, where the planes would be stationed, and the roles and airspace in which they could fly were informed by his perspective that any evidence of a shared fear of escalation in the West should not be overvalued.[94] Soviet limited actions were not made under an expectation of American reciprocity; rather, Stalin was acting on Soviet interests unilaterally and independently of expectations of American decisions.

When States observe positive-novel or discordant-frictional behaviors inconsistent with patterns of observed behaviors sustaining tacit coordination, they may engage in a tacit bargaining process through direct cyber engagement to communicate their observation and acceptance or unacceptance. Tacit bargaining describes a process of mutually dependent behavior (direct cyber engagement), in contrast to unilateral, independent actions (cyber *faits accomplis*) that characterize tacit coordination. Tacit bargaining may occur over common or divergent interests (with the former characterizing tacit cooperation as described above). A State bargains tacitly with another State when it attempts to communicate its own policy preferences or manipulate the latter's policy choices absent a reliance on explicit, formal, or informal diplomatic exchanges. The process is tacit because actions, rather than rhetoric, constitute the critical medium of communication;[95] it is bargaining, but not coordination or coercive bargaining, because the actions aim to influence an outcome that can only be achieved through some measure of joint, voluntary behavior.[96] Coercive bargaining, instead, speaks to the core concern of coercion theory, changing decision calculus, which was discussed in Chapter 1.

Conclusion

In this chapter, we have argued that the structural imperative in cyberspace—persistence in seizing and maintaining the initiative to set the conditions of security in and through cyberspace in one's favor—encourages exploitation as the dominant State behavior in the cyber strategic environment. States recognize that strategic opportunities flow from cyberspace's abundant vulnerabilities and resilience to persistent exploitative cyber operations/campaigns with effects short of armed-attack equivalence. We introduced two concepts representing the primary and secondary manifestations of such exploitative operations/campaigns—the cyber *fait accompli* (unilateral, independent behavior) and direct cyber engagement (mutually dependent behavior)—and linked these concepts to tacit coordination and tacit bargaining, respectively. Further, we described the dynamic resulting from States pursuing these behaviors as competitive interaction and how it produces the phenomenon of cyber agreed competition.

Direct cyber engagement and, therefore, tacit bargaining, are currently scarce in the cyber strategic environment due to several factors, including an abundance of opportunities for unilateral exploitation and asymmetries in cyber situational awareness. However, AI platforms might mitigate the constraining effect of some factors. In the next chapter we put cyber persistence theory to the test, literally, by assessing its arguments against the empirical record of State cyber behaviors, inter-State dynamics, and observed outcomes.

4

Theory and the Empirical Record

In Chapter 3, we introduced the core concepts of cyber persistence theory that describe and explain State cyber behavior and dynamics. We also raised the analytical distinction between the cyber strategic environment that revolves around exploitation from the two strategic environments—the nuclear, which rests on coercive threat, and the conventional, which rests on warfighting (offense and defense). In this chapter, we investigate cyber persistence theory's explanatory power by deriving hypotheses from the theory and evaluating them based on the best open source evidence available.

Based on cyber persistence theory, we expect that:

- States act persistently in and through cyberspace, rather than engage in episodic hacking or breaching of devices, systems, and networks.
- The dominant State behavior in cyberspace is exploitation, rather than coercion.
- State action changes the conditions of security and insecurity directly through cyber *faits accomplis,* rather than shapes the calculus of opponents through compellence or deterrence.
- Campaigns of cyber *faits accomplis* can independently generate strategic outcomes, which exceed the more limited effects associated with espionage, subversion, and sabotage.
- Competitive interaction is the dominant dynamic.
- Escalation within cyberspace to armed-attack equivalent effects is rare.
- Cross-domain escalation from cyberspace is rare.[1]
- Direct cyber engagement is present but scarce in cyberspace.

Some may argue that conclusions from studies based on empirical data of State cyber behaviors and dynamics are tenuous at best because the data examined are skewed. It may not be representative because the breadth of data in open source materials is not comprehensive. It may be incomplete because it

Cyber Persistence Theory. Michael P. Fischerkeller, Emily O. Goldman, and Richard J. Harknett, Oxford University Press.
 DOI: 10.1093/oso/9780197638255.003.0004

excludes business proprietary or secret (or higher) classified data gathered by private sector cyber security vendors and government military or intelligence agencies or departments. To hedge against a potential lack of representativeness in open source materials, we apply a multi-method approach to open source assessment. We discuss findings from a medium-*n* quantitative analysis of State cyber behaviors, in-depth case studies, unsealed criminal indictments, open source government reporting, and cyber security industry surveys, case studies, experiments, and trend reports.[2] To address concerns that open source materials may not be representative of business proprietary and classified data, we consider a counterfactual based on this premise and draw conclusions regarding how it would, were it true, impact the value and risk of basing cyber policy and strategy on cyber persistence theory.

Operational Persistence

Cyber persistence theory argues that States persist in seizing and maintaining the initiative to set the conditions of security in and through cyberspace in their favor, thereby enabling the generation of strategic effects that can maintain or alter the international distribution of power in and through cyberspace. This central argument rests on a basic assumption that States have the capacity and willingness to operationally persist.

To assess this expectation, we explore two evidentiary tracks: the degree to which States are building infrastructure in support of persistent cyber operations and evidence of operational tempo itself. The first is a necessary condition for the second. Therefore if we have evidence of the latter, we can presume the former is also true. The evidence does indeed support cyber persistence theory's base assumption that this is a space of persistent activity and challenges the traditional default mindset, particularly in policy circles, that describes cyber activity in terms of episodic hacks, breaches, attacks, and war (a term used regularly to describe a single incident).

Coercive environments are characterized by definable states of peace, militarized crisis, and war. Periods of nonaggression are typically the norm and are punctuated with aggression on an episodic basis—we declare war and live in peace. Cyber persistence theory assumes a fundamentally different strategic environment in which action that aims to exploit is continuously in play. This assumption neatly aligns with the terminology commonly used among network security, computer science, and engineering specialists. In 2010, the descriptor "advanced persistent threat" (APT) was used to describe cyber adversaries as follows:

- *Advanced* means the adversary can operate across the full spectrum of computer intrusion. They can use the most pedestrian publicly available exploit against a well-known vulnerability, or they can research new vulnerabilities and develop custom exploits, depending on the target's posture.
- *Persistent* means the adversary is formally tasked to accomplish a mission. They are not opportunistic intruders; they receive directives and work toward those ends. *Persistent* does not necessarily mean they constantly execute malicious code on victim computers. Rather, they maintain the level of engagement needed to execute their objectives.
- *Threat* means the adversary is not simply mindless code. The adversary is a threat because it is organized, funded, and motivated toward malicious ends.[3]

In 2013, Mandiant published a report detailing the cyber activities of China's PLA Unit 61398 and designated the group Advanced Persistent Group 1 (APT1). APT1, the report argues, was responsible for a sustained campaign targeting US intellectual property from 2010 to 2013.[4] FireEye, parent of Mandiant, has designated 20 APT groups as receiving direction and support from an established State, including Iran, North Korea, Russia, China, and Vietnam.[5] Additional groups have been designated by others. In March 2018, the US Department of Homeland Security (DHS) published a Technical Alert to provide "information on Russian government actions targeting U.S. Government entities as well as organizations in the energy, nuclear, commercial facilities, water, aviation, and critical manufacturing sectors."[6] The group behind these actions is designated "Allanite" by cyber security vendor Dragos.[7] DHS characterizes Allanite's activity as a multi-stage intrusion campaign by Russian government cyber actors.

Our intention is not to provide an exhaustive list of State-directed and supported APT groups.[8] Rather, the research referenced above illustrates that government agencies and cybersecurity vendors, drawing upon their respective data sources, have concluded that a number of noteworthy States have dedicated significant resources to developing and sustaining postures of cyber operational persistence and are employing their capabilities persistently.

Government reports offer additional evidence of operational persistence. In February 2021, the French government reported a multi-year exploitation campaign by a group within Russia's military intelligence agency targeting numerous French organizations.[9] A US government annual report tallies government-wide cyber incident reports to convey the threats agencies face every day and the persistence of threat incidents.[10] In 2019, 28,581 incidents were reported by US federal agencies.[11] The average number of annual reported incidents between 2016 and 2019 was approximately 32,000.[12] The German government's Federal Office of Information Security reported that German government computer networks were attacked almost 5,500 times in 2015, with hundreds of those being traced back to foreign agencies.[13] Another US government report covering the 2009–2013 period noted that "East Asia and Pacific cyber actors (read: China) constitute a serious danger because of their 'relentless' targeting and continuous innovation [of and against US cleared defense contractors]."[14]

The empirical record of the past twenty years supports the hypothesis that States have built the infrastructure and employed it to act persistently in and through cyberspace. The presence of this capacity (capability and will) positions States to engage in the behaviors consistent with the expectations of cyber persistence theory.

Exploitation and the Cyber *Fait Accompli*

The cyber *fait accompli* is a limited unilateral gain at a target's expense where that gain is retained when the target is unaware of the loss or is unable or unwilling to respond. We hypothesize that exploitation is the dominant behavior in cyberspace and that cyber *faits accomplis* are the primary manifestation of that behavior.

There is substantial case evidence that, at the time of intrusion, many targets are unaware that they have been compromised (exploited). The OPM breach was discovered in 2015, over a year after the initial compromise.[15] The aforementioned French government report of a Russian campaign noted that exploitative behavior began three to four years before its discovery.[16] In early 2020, cyber security investigators uncovered evidence of "longstanding compromises" at unnamed German companies, according to a May 2020 memo that German intelligence and security agencies sent to critical infrastructure operators.[17] In December 2020, initial forensic analyses of the reported APT29 group's intrusions into the networks of several US departments via the SolarWinds platform, including Treasury, Homeland Security, and Commerce, concluded that the intrusions occurred seven months prior.[18] Further, in April 2019, the Chinese security firm Qihoo reported its discovery of more than 200 VPN servers that had been compromised by a vulnerability in the Sangfor SSL VPN servers used to provide remote access to enterprise and government networks. They reported that the attacks began a month prior and included 174 servers located on the networks of government agencies in Beijing and Shanghai, and in Chinese diplomatic missions operating abroad.[19]

The vulnerability in Sangfor servers was a zero-day vulnerability, defined as a product vulnerability discovered by a threat actor before the developer has discovered it. A zero-day exploit is created when a threat actor writes and implements exploit code while the vulnerability is still open and available (i.e., before the developer discovers the vulnerability and issues a patch). By definition, zero-day exploitation gives attackers complete advantage at initial compromise.

The Chinese group APT3 used a zero-day exploit in targeted attacks in 2016.[20] The North Korean group APT37 conducted a 2017 campaign by exploiting a zero-day Adobe Flash vulnerability. From December 2017 to January 2018, multiple Chinese APT groups used a zero-day exploit in a campaign targeting multiple industries throughout Europe, Russia, Southeast Asia, and Taiwan. Russian groups APT28 and Turla exploited multiple zero-day vulnerabilities discovered in Microsoft Office products. In December 2020, the US National Security Agency reported that "Russia state-sponsored actors" exploited a zero-day in VMware products.[21] In March 2021, Microsoft reported that a Chinese APT was exploiting four zero-day vulnerabilities in Microsoft Exchange.[22] As of September 7, 2021, 62 zero-day exploitations had been documented.[23]

Major cyber powers routinely seek out zero-day vulnerabilities to leverage for exploitation.[24] Indeed, the Chinese government passed a law stating that "Starting September 1, 2021, the Chinese government will require that any Chinese citizen who finds a zero-day vulnerability must pass the details to the Chinese government and must not sell or give the knowledge to any third-party outside of China (apart from the vulnerable product's manufacturer)", thus allowing the government to accumulate a stockpile of zero-day vulnerabilities.[25]

Zero-day attacks are rarely discovered promptly. In fact, it often takes not just days but months and sometimes years before a developer learns of the vulnerability that led to an attack.[26] Once the developer discovers the vulnerability, it takes additional time to develop a patch.[27] And once a patch is developed, it takes additional time for customers to implement it.

The term "patch fatigue" has been shared among information technology managers, who find patch management extremely time-consuming, requiring hundreds of hours every month. Moreover, patches requiring system restarts (more a concern for server patching) result in significant downtime and potential lost business.[28] Most organizations take an average of 100–120 days to patch vulnerabilities.[29] The probability of a vulnerability being exploited hits 90 percent between 40 to 60 days after discovery. This means that the remediation gap, or time that a vulnerability is most likely to be exploited before it is closed, is nearly 60 days.

In many cases, however, APT groups pounce far more quickly than that. A trend has emerged with States redirecting resources from discovering zero-day vulnerabilities to the development of "one-day" exploits. APT groups can often reverse engineer software updates to quickly develop attacks before the fixes are widespread. For States, one-day exploit development is a less expensive and time-consuming process than searching out vulnerabilities from scratch.[30]

The 2017 Equifax breach is a case in point. The first 2017 Apache Struts vulnerability, which led to the breach, was discovered in February 2017. Apache released a patch in March 2017, but Equifax's systems remained unpatched for months.[31] Equifax has stated that, on March 10, 2017, company systems were scanned by unidentified actors to determine if they were susceptible to the Apache Struts vulnerability, a vulnerability that the US Computer Emergency Readiness Team had publicly identified just two days earlier. A server running the vulnerable software was discovered and promptly exploited, thereby allowing unauthorized access to the Equifax portal.[32] On July 29, Equifax discovered the breach of more than 145 million records of personally identifiable information of its customers. The intruders spent seventy-six days exfiltrating data from Equifax's network before they were detected.[33]

These cases are consistent with previously referenced cyber security industry data noting that the mean time to discovery of an intrusion is measured in months. It also aligns with an experiment conducted by Mandiant, in which they assessed security controls' effectiveness across the multiple stages of attack life cycles within eleven global industries. Their research design included executing thousands of tests comprising real attacks, specific malicious behaviors, and actor-attributed techniques and tactics. Security controls failed to detect or prevent 53 percent of intrusions.[34] Moreover, only 33 percent were either detected and/or prevented.[35]

All of this serves to support the hypothesis that cyber *faits accomplis* characterized by a lack of awareness on the securer's part are extensive. There is also evidence that a significant number of targeted entities, having been made aware of intrusions, nonetheless are either unable or unwilling to respond. We consider the frequency of *cyber* responses when assessing the direct cyber engagement hypothesis. It is also important to consider the prevalence of *non-cyber* responses. Non-State or private sector companies or organizations that are frequent targets—for example, companies comprising the US Defense Department's Defense Industrial Base (DIB) or global infrastructure consortia—have little, if any, cyber capability and no vested government authority or legal protections to "hack back." Their primary recourse is reporting incidents to law enforcement.[36] Evidence suggests that this occurs infrequently due to a lack of will more than lack of awareness.

Consider, for example, comments from David Hickton, former US Attorney for the Western District of Pennsylvania (2010–2016) who, after receiving an "avalanche" of complaints from several companies who populate the DIB, launched investigations leading his team to APT1.[37] Hickton's team discovered hundreds of victims, but when he asked companies to become plaintiffs, few wanted any part of it, given the amount of money they had on the line in China.[38] This inclination to relent is consistent with estimates of "non-reporting" activity. A 2015 survey of British companies noted that only 28 percent of their cyberattacks were reported to government authorities.[39] Absent a legal requirement to report an intrusion or breach, companies are reticent to do so.[40] Their priority tends to be in remedying the situation for themselves, shoring up any internal deficiencies revealed, and fulfilling legal obligations by notifying affected parties and regulators.[41]

This cost-benefit calculus is not limited to companies. In response to an intrusion into Norwegian lawmakers' email accounts by APT28—a group linked with Russia's GRU military agency—Norway's national police agency spokesman Martin Bernsen said that "PST [the Norwegian national police agency] see no purpose in investigating further, because we will never reach the goal of bringing someone to court in Norway," and that "[i]t would also have been enormously resource-intensive."[42]

In sum, there is ample evidence supporting the hypothesis that States are primarily acting unilaterally and independently in the cyber strategic environment and that their targets are mostly unaware they are being compromised or they are aware but are unable or unwilling to respond. Exploitation is the dominant behavior in cyberspace, and the cyber *fait accompli* is the primary manifestation of that behavior.

The significant number of cyber intrusions that occur unbeknownst to defenders poses a challenge to arguments and policies that assume the dominant State behavior in cyberspace is coercive—aimed at shaping the decision calculus of the opponent through threat of force (deterrence) or application of contingent violence (compellence). Both deterrence and compellence strategies require the clear communication of a threat and conditions associated therewith, communication that is obviously absent in these intrusions.

One category of cyber activity that is coercive in application is criminal extortion through ransomware malware installed for the purpose of encrypting a victim's files to subsequently coerce the victim into paying a bounty to regain control of their system or the data thereon.[43] Ransomware attacks, however, can only occur if the intruder has changed the conditions of security in their favor. In other words, the cyber *fait accompli* must be the initial act in any malware-infection based ransomware scheme. Only after first exploiting a system and reconfiguring its network functionalities and organization is the intruder in a position to encrypt files and subsequently act to coerce the system's owners. Thus, while ransomware attacks are prevalent, they are and always will be a subset of cyber *faits accomplis*. More important, the empirical record of ransomware activity to date has largely been an extortion technique of criminal organizations seeking financial profit from other non-State actors, not of States seeking to coerce other States to change their decision calculus.[44] The eleven most significant ransomware attacks in the first half of 2020 were against municipal governments, universities, and private businesses, all involving financial gain motivations.[45]

Strategic Outcomes

To assess whether campaigns of cyber *faits accomplis* can independently generate effects or outcomes that are strategically significant, we adopt Erik Gartzke's criteria introduced in Chapter 3. He argues that cyber operations can only be relevant in "grand strategic terms" or "pivotal in world affairs" if they "accomplish tasks typically associated with terrestrial military violence."[46] These include deterring or compelling, maintaining, or altering the distribution of power, and resisting or imposing disputed outcomes. Since we do not expect to see deterrence or compellence as they are strategies of coercion, what remains is a determination of whether cyber *faits accomplis* independently support resisting disputed outcomes or maintaining/altering the distribution of power. Toward that end, we present two "illustrative" case studies: Democratic People's Republic of Korea (DPRK) cyber exploitations of international currency platforms, and China's illicit cyber-enabled acquisition of US intellectual property (IP).[47] The findings support cyber persistence theory's hypothesis regarding the potential independent strategic significance of cyber *faits accomplis*.

Resisting a Disputed Outcome—DPRK and International Sanctions

The DPRK's response to the "toughest and most comprehensive sanctions regime ever imposed" by the UN Security Council via Resolution 2321 (2016) is an example of a State employing a campaign of cyber *faits accomplis* to resist a disputed outcome and achieve a strategic gain.[48] Between January and November 2016, the DPRK conducted two nuclear tests and at least twenty-five launches using ballistic missile technology; both sets of activities represent violations of previous UN resolutions. In announcing Resolution 2321, the UN Secretary-General declared that DPRK's nuclear and ballistic-missile activities were one of the "most pressing peace and security challenges of the present time" and emphasized that it "must reverse its course and move onto the path of denuclearization."[49] Kim Jong-un, unsurprisingly, rejected the validity of these sanctions and continues to successfully resist a disputed international outcome threatening his ability to maintain his country's status as a credible nuclear-capable power.

DPRK leadership responded to this challenge not through terrestrial military violence, but rather through persistent cyber-enabled exploitation of the international banking system and other financial digital manipulation. In August 2019, the UN Panel of Experts charged with assessing the sanctions regime imposed on DPRK concluded that the North Koreans generated an estimated $2 billion for its weapons of mass destruction programs, including efforts to continue enhancing its nuclear and missile programs, through "sophisticated use by the Democratic People's Republic of Korea of cyber means to illegally force the transfer of funds from financial institutions and cryptocurrency exchanges, launder stolen proceeds and generate income in evasion of financial sanctions."[50] The report further notes that UN sanctions enforcement experts are investigating "at least 35 reported instances of DPRK actors attacking financial institutions, cryptocurrency exchanges and mining activity designed to earn foreign currency" in some seventeen countries.[51]

Analyses suggest North Korea began its sustained exploitation of cyber vulnerabilities in the global banking system in October 2015 by generating fraudulent Society for Worldwide Interbank Financial Telecommunication (SWIFT) transactions.[52] SWIFT is a consortium operating a trusted and closed computer network for communication between member banks around the world. The consortium is overseen by the National Bank of Belgium and a committee composed of representatives from the US Federal Reserve, the Bank of England, the European Central Bank, the Bank of Japan, and other major banks. The SWIFT platform has some 11,000 users and processes about 25 million communications a day, most of them money transfer transactions.[53]

On February 4, 2016, members of the "Lazarus Group," a group affiliated with DPRK, used SWIFT credentials of Bangladesh Bank employees to send more than three dozen fraudulent money transfer requests to the Federal Reserve Bank of New York asking the bank to transfer millions of the Bangladesh Bank's funds to bank accounts in the Philippines, Sri Lanka, and other parts of Asia.[54] The hackers managed to get $81 million sent to Rizal Commercial Banking Corporation in the Philippines via four different transfer requests and an additional $20 million sent to Pan Asia Banking in a single request. The $81 million was deposited into four accounts at a Rizal branch in Manila on February 4 and was withdrawn shortly thereafter.[55] Over the following four years, DPRK amassed approximately $2 billion following similar cyber exploitation playbooks.

This is but one example of a sustained effort whose scope and scale was supported by many cyber actors operating under the Reconnaissance General Bureau, a top North Korean military intelligence agency.[56] The UN Panel of Experts concluded that additional "large-scale attacks against cryptocurrency exchanges allow the Democratic People's Republic of Korea to generate income in ways that are harder to trace and subject to less government oversight and regulation than the traditional banking sector."[57] John Hultquist, the director of intelligence analysis at the cybersecurity firm FireEye, notes that the "sheer scale [of these operations] suggests that they are a financial lifeline for a regime that has long depended on illicit activities to fund itself."[58]

DPRK's cyber operations to resist and circumvent international sanctions rests on a capability decades in the making.[59] The foundations of its current cyber force, estimated to number 6,800 operators and consuming 10–20 percent of the North Korean annual budget, were laid as early as the 1980s with consistent support from regime leaders.[60] In a detailed analysis of this history and activity, Stephanie Kleine-Ahlbrandt notes that "North Korea appears to have recognized early on the opportunities that cyber offered, and to have focused with strategic intent."[61] As applied recently, this strategic intent is primarily, but not exclusively, focused on sanctions evasion.[62] Kleine-Ahlbrandt's 2020 analysis reveals how deceptive tactics can effectively support evasion:

> The key way in which cyber attacks on financial institutions allow the DPRK to evade financial sanctions is by rendering ineffective one of the most powerful tools in the financial sanctions toolkit: the assets freeze. Such attacks also render futile the requirement in UN resolutions for Member States to prevent the transfer of assets which could contribute to the DPRK's WMD. When a designated entity—such as the Reconnaissance General Bureau, which plays an important role in many of the DPRK's cyber attacks—is able to launch a successful cyber attack on a bank (often through the SWIFT network) to steal funds, it doesn't have to rely on the traditional steps of using front companies, false documents or complicit foreign nationals to use the traditional tools to request a transfer, which might trigger the bank's compliance mechanisms. Instead, it directly hacks into bank computers and infrastructure to send fraudulent SWIFT transfer messages—and then destroys the evidence. The destruction of the evidence is primarily anti-forensic, i.e., to prevent detection long enough to ensure that the funds are transferred.[63]

In sum, DPRK's unilateral and persistent exploitation in and through cyberspace through cyber *faits accomplis*, rather than coercion and/or terrestrial military violence, allows it to successfully and continuously resist a disputed outcome. The targets of highest political value for the regime are those that facilitate the acquisition of currencies today, not those that might provide future coercive value. These efforts are producing strategic effects and outcomes that are arguably pivotal to world affairs—undermining strategic objectives of both the

United Nations and the United States while bolstering a nuclear capability—the highest end of contemporary day "high" politics.

Maintaining/Altering the Balance of Power—China and Cyber-Enabled Theft of IP

Many pronouncements have been made regarding the strategic significance of Chinese cyber-enabled theft of IP.[64] Some argue that theft of sensitive military information (weapons systems designs and military plans) is strategic because it supports a Chinese objective of degrading US military overmatch, thereby placing US regional allies and interests at greater risk.[65] Others argue the economic impact on the United States is strategic—with estimated losses ranging from $250–600 billion annually—posing a significant, immediate "bottom line" (GDP) threat to the US economy and a potential longer-term threat of disincentivizing investments in innovation.[66] Neither perspective argues that China's cyber campaigns have had a "strategic" effect on the international system in terms of maintaining or altering the international balance of power.[67] On the contrary, it has been argued China's theft of military technology IP has not been strategically significant.[68]

This case study addresses the strategic significance of China's cyber-enabled theft of US IP, specifically commercial IP from 2010 to 2013, from the perspective of Chinese Communist Party (CCP) leadership. By 2010, leadership recognized that China was approaching the limits of its time, based on historical patterns, to transition from a middle- to a high-income economy and that the historical consequences of being "trapped"—enduring economic slowdowns or stagnation (in the middle income range) and sociopolitical upheavals (often referred to as the "middle income trap" [MIT])—would threaten the legitimacy of the CCP and China's status as a great power.[69] Consequently, around 2010, the CCP adopted a centrally directed and substantial campaign of cyber-enabled IP theft in a gambit to avoid falling victim to this trap and its consequences.

Arguing that the CCP has employed China's cyber forces in extraordinary ways is not contestable, or novel.[70] But the focus on China's strategic intentions is new. Securing China's hard-won great power status, a consequence of China's significant economic rise over the previous two decades, speaks directly to intent to maintain the international balance of power. Therefore, it directly addresses whether China's cyber-enabled IP theft campaign is "strategic" according to Gartzke's aforementioned criteria.[71]

In October 2010, Liu He, a vice minister at the central government's Office of the Central Leading Group on Financial and Economic Affairs of the Central Committee and key author of China's 12th Five-Year plan, stated, "We're concerned about how to avoid the so-called "middle-income trap," further noting "only a few countries such as Japan, Korea and Singapore have crossed this bridge easily since World War II. Most countries have stagnated."[72] In highlighting social upheaval associated with the MIT, Liu expressed urgency—"It can be said that we have reached a critical point for transforming the economic development model"—and seriousness—"President Hu Jintao, Premier Wen Jiabao, Vice President Xi Jinping and Vice Premier Li Keqiang all emphasized the need to transform the economic development model in early 2010."[73] It has been reported that Xi Jingping often uses the term "middle income trap." When newly appointed President Xi met with President Barack Obama in 2013, he "stressed his hope of avoiding the middle-income trap."[74] CCP economic research reveals the same concerns. In 2011, the Development Research Center of the State Council published a compendium of MIT research and, in 2013, coauthored with the World Bank *China 2030: Building a Modern, Harmonious and Creative Society*, with a section on structural reforms that repeatedly raises concerns about falling victim to the MIT.[75]

Under the shadow of these concerns, in 2012, Xi Jinping nonetheless invited President Obama to collaborate with him in developing a "new type of great power relationship." Xi's initiative was an announcement that China—in its own eyes—had achieved great power status and expected to be treated as such.[76] In a September 2015 speech before business leaders in Seattle, Xi was steadfast in repeating what had since become a rhetorical drumbeat in Chinese-language media; the United States and China needed to construct "a new model of great power relations."[77] Xi's address to the 19th National Congress describes China as a "great power" or a "strong power" twenty-six times.[78] He further proclaims that "It is time for us to take center stage in the world and to make a greater contribution to humankind."[79] This language marks a noteworthy departure from the foreign policy doctrine summed up in 1990 by Deng Xiaoping as "hide your strength and bide your time."[80] In Xi's view, China's time has come.[81]

Despite acknowledged concerns of rapidly slowing growth, on November 3, 2015, President Xi announced economic growth targets averaging 6.5 percent over the following five years.[82] This average matches the ceiling of analysts' estimates and suggested that at least 3 percent of growth from 2016 to 2020 must come from engineering- and science-based innovation.[83] And yet, by the CCP's own admissions, these innovation archetypes are where they are weakest.[84] Thus, Xi's bold growth forecast seems disconsonant with expressed concerns of a rapid growth slowdown. That said, over the prior decade, the CCP had adopted a multifaceted technology development strategy comprising licit and illicit methods to achieve China's innovation objectives. These included the legal and regulatory policies, science and technology research and development investments with international partners, mergers and acquisitions, joint ventures, nontraditional collectors (using individuals for whom science or business is their primary profession to target and acquire US technology), research partnerships (with government labs such as the Department of Energy labs), talent recruitment programs, academic collaborations, joint ventures, and economic espionage by the Ministry of State Security and military intelligence services.[85]

Although all facets could, over time, contribute to engineering- and science-based innovation growth, evidence supports the claim that an extensive cyber-enabled IP theft campaign from 2010 to 2013 was already bearing such fruit in 2015. It was producing growth with certainty, at scale, and within a fraction of the time experienced historically by other nations.[86] Evidence further supports the contention that it was doing so disproportionately relative to all other facets of the technology transfer program.[87]

The two primary sources on the extensiveness and centrality of China's cyber-enabled IP theft campaign are Mandiant's 2013 report on China's APT1 group and annual reports issued by the US DoD's Defense Counterintelligence and Security Agency (DCSA). Formerly known as the Defense Security Service, DCSA is responsible for providing annual statistical and trend analysis on the foreign entity cyber threat posed to the DoD's cleared contractor community (i.e., the firms most targeted by APT1).[88] The central sources of this campaign's early effectiveness are US Department of Justice indictments of China cyber operators.

Mandiant's 2013 report argued a strong case that APT1—2nd Bureau of the People's Liberation Army General Staff Department's 3rd Department, also known by its Military Unit Cover Designator as Unit 61398—was responsible for substantial cyber-enabled theft of US intellectual property in the 2010 to 2013 period.[89] As PLA reports directly to the CCP's Central Military Commission, Mandiant argues that CCP senior members centrally direct PLA enterprise cyber-enabled IP theft. US intelligence experts share this view.[90]

The report estimated that APT1 is staffed by hundreds or perhaps thousands based on the size of its physical infrastructure and that it has systematically stolen hundreds of terabytes of data from at least 115 US organizations (up through 2012), demonstrating the capability to steal from dozens of organizations simultaneously.[91] The report concludes these directly observed numbers represent only a small fraction of APT1 intrusions and, therefore, are the lower bounds of activity and capacity.[92] The conservatism regarding APT1's capacity is supported by a 2010 report from the FBI's former deputy director for counterintelligence. It notes that China sustains between 250,000–300,000 soldiers in the 3PLA dedicated to cyber espionage and that much of this capability can be deployed to support China's methods for stealing IP.[93]

Mandiant's analysis covers 2006 to 2012, but 76 percent of activity against US firms occurred from 2010 to 2012, suggesting a more concerted effort beginning in 2010. The industries targeted match China's Strategic Emerging Industries (SEIs), the central thrust of its engineering- and science-based indigenous innovation policy, suggesting a centrally directed effort.[94] A subsequent, more expansive analysis (up to 2015) concludes that all seven of China's SEIs were being served by numerous APTs' cyber operations—energy-efficient and environmental technologies, next-generation information technology, biotechnology, high-end equipment manufacturing, new energy, new materials, and new energy vehicles.[95] Data exfiltrated include a laundry list of IP: product development and use, including information on test results, system designs, product manuals, parts lists, and simulation technologies; manufacturing procedures, such as descriptions of proprietary processes, standards, and waste management processes; business plans, such as information on contract negotiation positions and product pricing, legal events, mergers, joint ventures, and acquisitions; policy positions and analysis, such as white papers and agendas and minutes from meetings involving high-ranking personnel; and emails of high-ranking employees.[96]

Additional, correlating evidence from DCSA supports the argument that, relative to other Chinese transfer technology methods, cyber-enabled IP theft was the central effort from 2010 to 2013. Although not identical, DCSA's annual report on the foreign entity cyber threat posed to the DoD's cleared contractor community uses Method of Operations (MO) categories aligning well with many of China's technology transfer methods.[97] Of all the MOs, the most important for this case study is Suspicious Network Activity (SNA), described as attempts to exfiltrate information via cyber intrusion, viruses, malware, backdoor attacks, acquisition of user names and passwords, and analogous targeting.

DCSA reporting discloses an increasing relative focus by China on SNA MO during from 2010 to 2013 period.[98] In 2009 and 2010, SNA reporting increased from third to second largest percentage. In 2011, SNA assumed the top MO position, and in 2012, they peaked at 42 percent of overall reported activity.[99] This jump in percentage further reflects a 245 percent increase of SNA reports, themselves, year-over-year, and a 1,443 percent increase in East Asia and the Pacific-attributed SNA reports from 2009 to 2012. In the same period, the number of confirmed (not merely reported) intrusions into cleared industries' unclassified networks grew by 1,138 percent.[100] In 2013, SNA was still the top MO, at 30 percent of all reported activity.[101]

Together, DCSA and Mandiant reporting supports a conclusion that around 2010, cyber-enabled IP theft from US companies became China's primary method for facilitating engineering- and science-based innovation growth necessary for avoiding the MIT. Not addressed yet in our case study, however, is an assessment of China's effectiveness in absorbing and actualizing acquired IP and speed in actualizing it against growth soon after acquisition. It has been argued that the absence of an ability to absorb and "re-innovate" IP explains why, after acquiring US fifth-generation jet fighter IP through cyber exploitation, China has been unable to develop a comparable fighter.[102] Thus, were Xi's 2015 confidence in continued 6.5 percent growth based on contributions from cyber-enabled IP theft, he would have needed evidence that re-innovation was, in fact, occurring. Such evidence existed.[103]

In May 2014, US federal prosecutors charged five PLA members with cyber intrusions into the computers of four US companies from 2010 to 2012 with the objective to "steal information from those entities that would be useful to their competitors in China."[104] The five charged are allegedly members of China's APT1 group. The companies targeted include SolarWorld, US Steel, and Westinghouse. In October 2018, prosecutors charged ten Chinese Ministry of State Security-affiliated persons (MSS) with conspiring to steal sensitive, commercial technological, aviation, and aerospace data from CFM International (and others) from 2010 to 2015 to support a Chinese State-owned, Enterprise-led effort (SOE) to build a turbofan engine of the same or similar design as developed by the targeted companies.[105] There is evidence that Chinese companies promptly re-innovated stolen IP from each of these companies to develop comparable, competitive, domestically produced products.[106]

In 2012, when SolarWorld was bringing to mass production Passivated Emitter Rear Contact (PERC) solar cells, PLA operatives allegedly conducted at least twelve intrusions into SolarWorld's computers, acquiring detailed PERC manufacturing metrics, technological innovations, and production line information.[107] By early 2014, a Chinese-based solar rival, JA Solar, announced it was converting to PERC technology and began mass production of PERC in May of that year. In early 2015, Chinese-based Trina announced its own PERC conversion and, later that year, brought to the market a comparable PERC technology.[108] In 2017, testimony before a special committee of the US Trade Representative, SolarWorld's AG CEO argued there was a clear connection between the IP theft and his Chinese rivals' rapid adoption of PERC technology, a technology that took SolarWorld eight years to develop.[109] Only two or three years after the IP theft, two Chinese SOEs were mass-producing the same technology.

Also in 2012, US Steel filed an International Trade Commission complaint that a company researcher's computer was breached in 2011 and that plans for a new steel technology that took a decade to develop had been stolen.[110] The plans included the chemistry for the alloy and its coating, the necessary temperature for heating and cooling the metal, and the layout of production lines—the product was known as Dual-Phase 980, one of US Steel's best performers. Two years after the alleged intrusion, Chinese SOE and steel giant Baosteel Group Corp. had a new line of products on the market—among them, Dual-Phase 980.[111]

In the period 2008–2009, Westinghouse signed a technology transfer agreement with China's State Nuclear Power Technology Corp to build four AP1000, third-generation reactors. This progressed in 2013 to a joint venture for building four Chinese nuclear power plants and getting Chinese nuclear scientists and technology up to speed.[112] From 2010 to 2013, PLA operatives allegedly exfiltrated from Westinghouse's computers proprietary and confidential technical and design specifications for pipes, pipe supports, and pipe routing associated with power plants Westinghouse contracted to build.[113] In December 2014, China's National Energy Administration approved a two-unit (HPR1000) construction plan for Fangchenggang Nuclear Power Plant (NPP) Project.[114] According to China Nuclear Power Group, the HPR1000, China's first third-generation reactor, "assimilates domestic and overseas experience on nuclear design, construction and operation."[115] One year later, Unit 3 of Fangchenggang NPP started construction, which was within four years of cyber-enabled theft of third-generation technical and design IP from Westinghouse.

Finally, in 2009, the State-owned Commercial Aircraft Corporation of China (COMAC) struck a deal with CFM International—a joint venture between US-based General Electric's aviation business and French aerospace company Safran—to develop a new commercial aircraft engine called LEAP-X. The deal called for CFM to develop LEAP-X1C, a variant of CFM's LEAP-1C engine, for China's C919 aircraft.[116] Around the same time, China's State-owned Assets Supervision and Administration Commission tasked both COMAC and the State-owned Aviation Industry Corporation of China (AVIC) with developing an "indigenously created" turbofan engine.[117] In June 2011, CFM International and COMAC signed a Memorandum of Understanding to study joint assembly—CFM and AVIC—of the LEAP-X1C engine in Shanghai; however, two years later, Chaker Chahrour, CFM's executive vice president, ruled out the joint assembly effort.[118]

Soon after COMAC signed the 2010 development deal, MSS cyber operators allegedly targeted Los Angeles–based Capstone Turbine, a manufacturer whose technology was key to the aircraft engine development.[119] Further, just over a year after Chahrour ruled out the assembly joint venture, MSS cyber operators allegedly intruded into networks in Safran's office in Suzhou, Jiangsu, China, and exfiltrated proprietary information on LEAP-X.[120] In August 2016, the State-owned Aero Engine Corporation of China (AECC) was established, with COMAC and AVIC as main shareholders, to domestically manufacture an "indigenously created" turbofan engine for the C919. Just over a year later, they completed an assembly process for the first CJ-1000AX demonstrator engine, an engine closely resembling both the LEAP-X and LEAP-1C engines.[121] A CrowdStrike analysis concludes it is highly likely that AECC benefited significantly from the cyber efforts of the MSS, knocking several years off CJ-1000AX's development time.[122]

Reported GDP figures support a conclusion that cyber-enabled IP theft and China's capacity to re-innovate the same has kept China, so far, from falling victim to the MIT and risking its great power status—the GDP reportedly averaged between 6.5 and 6.6 percent from 2015 to 2019.[123] When these data are coupled with a trend analysis of Global Innovation Index measurements for China, the conclusion gains additional support—from 2013 to 2019, China's innovation measure rose, relative to all other States, from thirty-fifth to fourteenth, thus establishing itself in the group of leading innovative nations.[124] This climb arguably reflects progress in engineering- and science-based innovation previously deemed as necessary for China to average 6.5 percent growth from 2015 to 2020.

All of that said, qualifications are prudent. An upward adjustment to China's 2018 GDP, announced in November 2019 by China's National Bureau of Statistics (NBS), feeds skepticism regarding the accuracy of recent NBS figures, as the adjustment aligns with Xi's declared 2020 growth target.[125] This adjustment may serve as evidence that US tariffs, first imposed in January 2018, are exerting downward pressure on growth—pressure Xi did not anticipate when declaring growth targets in 2015.[126] It may also be the case, as some argue, that China has *consistently* overreported its growth for many years.[127] At a minimum, then, a conservative conclusion is that cyber-enabled theft of IP is playing an important, independent role in posturing China to avoid the MIT and retain its great power status.

These cases illustrate that the advent of cyberspace has ushered in new ways and means for States to achieve strategic outcomes *absent the use of terrestrial force*, that is, an alternative to war. The DPRK "has developed a model that leverages the internet as a mechanism for sanctions circumvention that is distinctive, but not exceptional. This model is unique but repeatable, and disconcertingly can serve as an example for other financially isolated nations such as Venezuela, Iran, or Syria, for how to use the internet to circumvent sanctions."[128] Additionally, four years after the CCP adopted its asymmetric strategy, Xi held out China as a model for the new era, saying his country had developed its economy without imitating Western values—"It offers a new option for other countries and nations who want to speed up their development while preserving their independence."[129] These cases demonstrate that, not only can cyber *faits accomplis* campaigns short of armed conflict independently generate strategically significant outcomes, but the ways and means necessary to do so are available to many countries.

Competitive Interaction and Escalation within Cyberspace

We have argued that the primary and secondary manifestations of State exploitative behavior are cyber *fait accomplis* and direct cyber engagement, respectively, and that these behaviors result in the inter-State phenomenon of cyber agreed competition, a tacitly bounded strategic competitive space inclusive of operational restraint and exclusive of operations causing armed-attack equivalent effects. Cyber *faits accomplis* and direct cyber engagement also describe the mechanisms through which States seek to achieve strategic gains in and through cyber agreed competition. We describe the dynamic resulting from States leveraging these mechanisms as competitive interaction. In this section, we assess whether competitive interaction is the primary State dynamic in cyberspace and examine whether or not States have tacitly agreed to primarily compete in the cyber strategic environment via campaigns or operations whose effects are short of armed-attack equivalence. This assessment also addresses the likelihood of having the cyber escalation scenarios we offered in Chapter 3 coming to pass.

To evaluate the hypotheses, we adopt the "least likely" case study approach, often referred to as "hard" cases. "Least likely" cases pose difficult tests, in that one would not expect the hypotheses to be supported by a review of the case evidence. According to Harry Eckstein, "least likely" cases provide researchers with a great deal of inferential leverage, being "especially tailored" to disconfirming a set of hypotheses.[130] "Least likely" cases, in this instance, would be those including State dyads with a shared history of militarized crises or armed conflict predisposing them to breaching the upper bound of cyber agreed competition and further engaging in an escalation dynamic. In security studies parlance, such dyads are often referred to as *rivals*.

States are described as rivals when there is some degree of competitiveness, connection between issues, perception of the other as an enemy, and longstanding animosity.[131] Of the total population of States, rivals are the most likely to engage in crises, escalated conflicts, and wars.[132] There exists an open source dataset of cyber operations by rivals—the Dyadic Cyber Incident and Dispute Dataset—cataloguing rivals' behaviors from 2000 to 2014.[133] The 126 active rival dyads forming the foundation of the dataset are extracted from a preexisting enduring rival dataset and an additional strategic rival dataset.[134] For the purpose of this section, the most important coding category in the dataset is labeled "Severity," in which a cyber incident is coded against a 10-point scale, where coding values 6 through 10 represent attacks of armed-attack equivalent effects (to wit, 6 = single critical network widespread destruction, 7 = minimal death as a direct result of cyber incident, 8 = critical national economic disruption as a result of cyber incident, 9 = critical national infrastructure destruction as a result of cyber incident, and 10 = massive death as a direct result of cyber incident).[135] Of the 193 records in the database, none are coded as being in the severity range of 6 through 10—that is, *no State cyber action was coded as generating an armed-attack equivalent effect.*

Importantly, the dataset authors note that the 193 records do not represent every cyber incident initiated by a rival, as gathering and coding such data would be "unwieldy."[136] Each record, they say, could include thousands of incidents that they decided to bin into a single incident "as long as the goals of the perpetrator remained stable."[137] Thus, for example, a single incident coded in the dataset as a ShadyRat intrusion of US systems actually represents forty-nine ShadyRat intrusions of US-based targets.[138] In such cases, the incident with the highest severity score became the coded value.

What is the best way to classify and understand the dynamic comprising the hundreds if not thousands of cyber incidents between rival dyads from 2000 to 2014? The empirical record is not one of escalation or escalation dynamics. Those committed to a coercion frame might then argue that States are failing to coerce either because cyber means are poor coercive tools or because States have not yet developed the sophistication to apply them effectively. Cyber persistence theory offers an alternative explanation, and one that we find more compelling and supported by the range of empirical evidence we bring to bear in this volume. There is no evidence of States breaching the ceiling of cyber agreed competition and engaging in an escalation dynamic because they are playing a different strategic game entirely—they are exploiting to change security conditions, not coercing to shape behavior.

There is an important limited sub-category to this empirical record. As we noted previously, States have used cyber operations to create armed-attack equivalent effects *in contexts in which they were already in militarized crises or armed conflict.* This begs the important question of whether a State breaching cyber agreed competition's upper bound was the opening act creating those contexts, a phenomenon referred to as *cross-domain escalation.*

Cross-Domain Escalation

Martin Libicki has written about cyber escalation from the lens of coercion theory and presented empirical evidence on the topic. In 2019, he assessed a set of cyber-related escalation hypotheses against the empirical evidence available at that time.[139] One hypothesis from that set is particularly relevant to this section: whether cyberattacks whose effects are equivalent with armed conflict generate kinetic *responses* (i.e., cross-domain escalation). Libicki determines that there are no unambiguous examples of this phenomenon and that "Nothing *so far* suggests this as a plausible scenario."[140] Therefore, he concludes that "kinetic retaliation to a cyberattack is possible but cannot yet be deemed a likely consequence."[141]

Libicki notes an important case that has been cited by some as evidence of cross-domain escalation—Israel's May 2019 airstrike on a Gaza building said to house Hamas's cyber unit—but then dismisses it.[142] His dismissal cites arguments made by Robert Chesney that claims suggesting Israel's response crossed a Rubicon ignore the overall context in which the airstrike occurred.[143] Chesney clarifies that heavy fighting first broke out in Gaza and Israel, with Islamic Jihad and Hamas launching more than 690 rockets and mortars indiscriminately into Israel, and Israel countering with some 320 targeted airstrikes. During these kinetic attacks, according to the commander of IDF's cyber division, Hamas attempted to carry out a cyber operation aimed at "harming the quality of life of Israeli citizens."[144] Clearly, this kinetic response occurred within the larger context of an armed conflict and, therefore, should be viewed as unexceptional. It is activity that falls under the logic of the conventional strategic environment, not that of the cyber strategic environment.

Current Scarcity of Direct Cyber Engagement

We argue that direct cyber engagement will be present in a context of competition for control over key cyberspace terrain that States mutually perceive as strategically, operationally, or tactically significant for setting the conditions of security in their favor in and through cyberspace. Of all our hypotheses, this is perhaps the most difficult for which to find direct operational or tactical evidence in refutation or support because, as we argue, instances are currently scarce. That said, in the next section we present four open source reported cases directly illustrating the presence of the behavior. In this section, however, we adopt an indirect analytical approach focusing on a number of factors that, if present, would account for why direct cyber engagement is scarce today.

A factor constraining the prevalence of direct cyber engagement is the challenge States face in understanding how their national interests manifest in cyberspace, which limits their opportunities to respond through direct cyber engagement to exploitations of those manifestations. There is evidence this factor is widely present today. For example, a US government agency estimated it had about 250 "shadow IT" applications, only to learn it had 8,000.[145] The implication is that potential exploitation vectors available to adversaries was far more extensive than was understood. Moreover, by monitoring for adversary activity only those applications of which it was aware, the agency was excluding 97 percent of that potential exploitation surface. This case is not atypical—shadow, legacy, and abandoned applications are prevalent on information technology networks of the governments and the largest companies in the world, including critical infrastructure owners and operators, and defense contractors.[146]

Examples abound. A 2021 US government report noted that the US Department of Transportation lacks accurate IT system inventories. In a fiscal year

2020 assessment, the Department's Inspector General found the Department's hardware inventory failed to account for 14,935 assets, including 7,231 mobile devices, 4,824 servers, and 2,880 workstations.[147] Additionally, all eight agencies reviewed by their inspectors general—the Department of Homeland Security, State, Transportation, Housing and Urban Development, Agriculture, Health and Human Services, Education, and the Social Security Administration—were using legacy systems or applications that are no longer supported by the vendor with security updates.[148] An investigation of the cyber exploitation of a Florida water treatment plant in February 2021 revealed that the plant's computer systems were using Windows 7, for which Microsoft had ceased providing support or updates over a full year prior.[149] Similarly, the Russian APT 2021 compromise of French organizations was facilitated by the fact that they were using very outdated versions of open source IT monitoring software.[150]

A further factor constraining the prevalence of direct cyber exploitation is a lack of awareness of compromises. The previously presented cyber *fait accompli* evidence highlighting the prevalence of attacks that are initially unnoticed supports a claim that this factor is widely present today. The noted reticence (and legal limitations) to respond once aware by "hacking back" supports it further.

Other constraints flow from a State's domestic policies and laws. For example, States may limit the operational scope of their primary cyber forces to their own national defense department and agency systems, whereas adversaries see no such constraints on the systems they may seek to exploit. Evidence for the relevance of this factor can be found in States' varying cyber military doctrines. For example, in France, the commander of cyber defense (COMCYBER) is exclusively in charge of the defense of the Ministry of Defense's (MoD) networks. COMCYBER may use "offensive capabilities," presumably off-MoD networks, only if an attack targets operational military capabilities or the MoD's chains of command. Offensive cyber operations for purposes other than self-defense are the prerogative of the Directorate General for External Security (DGSE), the country's foreign intelligence service.[151] The French approach to cybersecurity and defense contrasts with that embraced by the United States or the United Kingdom.[152] Most notably, France assumes a clear separation between offensive and defensive cyber operations and actors.[153]

The situation in Germany differs from all three States mentioned above. In 2016, Germany created its military cyber command, the Cyber and Information Domain Service (KdoCIR, in German), and tasked it with cyber defense, limited offensive cyber operations, and defending against hybrid threats such as influence operations or disinformation. With the KdoCIR, the German Ministry of Defense claimed its stake in Germany's whole-of-government approach to cybersecurity. However, cybersecurity writ large in Germany remains the prerogative of the Department of the Interior. Due to constitutional constraints, the KdoCIR actually has little room to maneuver outside its defined lanes.[154]

Evidence shows that several constraining factors on direct cyber engagement are widely present today. Although the behavior is scarce, particularly outside armed conflict, it is not absent. The next section describes four cases of direct cyber engagement: one between the United States and Russia, and three involving State and non-State actors—a combined US and UK effort to disrupt a major botnet in 2015, US efforts to secure its 2020 presidential election process from ransomware, and the US cyber campaign against the Islamic State.[155]

Examples of Direct Cyber Engagement

In November 2014, APT29, a Russian State-sponsored group reportedly compromised US Department of State and White House systems.[156] After the compromise was recognized, National Security Agency (NSA) cyber operators engaged in a pitched competition over a twenty-four-hour period with Russian cyber operators who had breached the unclassified State Department computer system. Current and former US officials stated that whenever NSA operators cut APT29's link between their command and control server and the malware in the US system, the Russians set up a new one.[157] At one point, APT29 even gained access to the NetWitness Investigator tool that NSA operators were using to uproot Russian back doors, manipulating it in such a way that the Russian cyber operators continued to evade detection.[158] "It was hand-to-hand combat," said NSA Deputy Director Richard Ledgett. Ledgett also noted that the attackers' thrust-and-parry moves inside the network while defenders were trying to kick them out amounted to "a new level of interaction between a cyber attacker and a defender."[159] This reported case is a vivid example of two cyber peers competing for control over key cyber terrain that both perceive as strategically consequential, clear evidence of direct cyber engagement.

The most prolific and disruptive banking malware from 2011 to 2015 was known as Dridex (also referenced as Bugat). In 2015, an international disruption operation targeting Dridex's command and control servers was spearheaded by the US Federal Bureau of Investigation (FBI) and Britain's National Crime Agency (NCA). The law enforcement agencies seized the command-and-control servers being used by the Dridex administrators in an effort to disrupt the malware. As was reported on October 14, 2015, "The National Crime Agency is conducting activity to 'sinkhole' the malware, stopping infected computers—known as a botnet—from communicating with the cybercriminals controlling them," the NCA said. Moreover, "This activity is in conjunction with a U.S. sinkhole, currently being undertaken by the FBI."[160] The US Department of Justice

(DOJ) revealed this activity in an October 13 press release in which they stated the FBI took "measures to redirect automated requests by victim computers for additional instructions to substitute servers [sinkholing]."[161] The expectation was that this would be a temporary measure, essentially buying time for the users of infected computers to rid their systems of the malware—the DOJ encouraged users to do so when announcing the operation.[162] Numerous reports from security vendors revealed that expectation was warranted as Dridex administrators promptly contested the operation, and, in fact, Dridex was back in operation using the servers that were disrupted, albeit at a far lower capacity, less than forty-eight hours after the combined FBI-NCA operation.[163] A majority of security vendors commenting on the operation concluded that it certainly disrupted a portion of the Dridex operation, but it does not appear to have created a long-term cessation in activity.

In efforts to secure the US 2020 presidential election process from external influence, it was reported that in late September to early October 2020, US Cyber Command (USCYBERCOM) engaged in a campaign to temporarily disrupt what is described as the world's largest botnet—one used to drop ransomware, which US officials argued was one of the top threats to the 2020 election.[164] USCYBERCOM's campaign reportedly targeted the Trickbot botnet, a collection of more than two million malware-infected Windows PCs that are constantly being harvested for financial data and are often used as the entry point for deploying ransomware within compromised organizations.[165] On October 2, 2020, KrebsOnSecurity reported that twice in the preceding ten days, an unknown entity with inside access to the Trickbot botnet had sent all infected systems a command telling them to disconnect themselves from the Internet servers the Trickbot administrators used to control compromised computers.[166] Additionally, millions of bogus records about new victims were inserted into the Trickbot database—apparently to confuse or distract the botnet's operators.

Four US officials, who spoke on the condition of anonymity because of the matter's sensitivity, stated that the campaign was not expected to permanently dismantle the network.[167] Rather, it was viewed as one way to distract the botnet administrators, during the lead-up to the election, at least for a while as they restored operations. The day after the first operation (September 22), private-sector security researchers continuously monitoring Trickbot activity reported the Trickbot administrators were restoring control of the severed connections and continuing to operate using those still intact.[168] The same pattern evinced after the second operation (October 1). By mid-November, it was reported that the botnet's administrators had updated communication mechanisms and built a new command-and-control infrastructure based on a different router to better secure the infrastructure from exploitation.[169] "We believe that this shows a determination on the part of the actors behind Trickbot to defy the disruption activity against their operation," says Mark Arena.[170] Similarly, Alex Holden notes that "This was a punch in the gut for the bad guys, but not a knockout blow. . . . It was perceived as inconvenient, but most activity resumed within two to three days."[171] Other analysts remarked, "the endeavor proved to be more like a 'kneecapping' operation rather than cutting the hydra's heads."[172] If the disruption was primarily a defensive action to unbalance Trickbot operators in the run-up to the US election, the temporary nature of the disruption is not as important as whether its timing provided the security conditions necessary to sustain the integrity of electoral systems.

The Trickbot and Dridex cases are examples of competent cyber actors competing for control over the command-and-control servers of a substantial botnet in the context of strategic competition short of armed conflict. The back-and-forth, with no escalation by either party to cyber operations with armed-attack equivalent effects, is evidence of direct cyber engagement. The next case illustrates the presence of the same behavior within the context of armed conflict.

There is a growing consensus on the acceptable use of cyber capabilities in armed conflict in accordance with International Humanitarian Law.[173] A context of armed conflict mitigates factors constraining the prevalence of direct cyber engagement in several ways.[174] It down-scopes the range of operationally and tactically significant key cyber terrain (i.e., a focus on infrastructure, systems, or even user accounts that are highly relevant to the area of operations and/or warfighting), increasing the likelihood that mutual perceptions of key terrain will encourage States to compete for control. The context of armed conflict also likely loosens domestic constraints on the use of cyber capabilities and relaxes perceived international legal constraints (excepting International Humanitarian Law). Evidence of direct cyber engagement in recent conflicts is therefore not surprising.

The US cyber campaign focused on disrupting the Islamic State's ability to distribute propaganda—the opening engagements of Operation Glowing Symphony—is illustrative.[175] USCYBERCOM noted that the Islamic State's propaganda strategy was well supported by a group with varying levels of expertise to set up public and private internet infrastructure, and maintain the Islamic State's websites and mobile phone applications.[176] Glowing Symphony's lead planner and mission commander explained that nearly all of the Islamic State's propaganda was passing through the same ten nodes (servers) on the internet.[177] The mission team determined it was possible to target only the Islamic State's material on servers that also managed many others' commercial traffic. As part of the operation, USCYBERCOM reportedly obtained the passwords to a number of Islamic State administrator accounts and then used them to access the accounts, change the passwords, and delete folder directories and content such as battlefield video. It also denied the group's propaganda specialists access to their accounts.[178]

The Islamic State's response to these actions reflects direct cyber engagement—they contested them. US intelligence officers reportedly concluded about a month into the campaign that the impact was short-lived at best. The group either restored content or moved it to new servers.[179] Other assessments describe the impact differently, noting that Operation Glowing Symphony interrupted Islamic State's propaganda efforts and that although the Islamic State did its best to reconstitute and regain control, they could only do so at a lesser level of capability: "There would be a takedown, the Islamic State would recover slightly, but each time they were starting lower and lower on the ladder."[180] Based on remarks by Major General Matthew Glavy, then commander of the joint task force responsible for Glowing Symphony (JTF ARES), direct cyber engagement between these actors persists today. Glavy noted in a 2019 interview that "They've [the Islamic State] morphed a little bit. But let's face it, we got to be ever so diligent and vigilant about the media piece. And we cannot have—for them to gain the momentum that we saw in the past."[181] General Paul Nakasone, USCYBERCOM's commander, echoed the same sentiment, noting that before ARES, the fight against the Islamic State in cyberspace was episodic—now it is continuous.[182]

Data Representativeness and Counterfactual Exploration

Concerns that open source data are not representative of the totality of State cyber behaviors and dynamics—that much cyber activity is covert or clandestine and thus unavailable—have raised questions about the feasibility of empirically testing and validating cyber theories and hypotheses. While States rarely advertise their cyber activity, some, and perhaps most, activity and interactions can, in fact, be identified. This chapter references data from numerous diverse open sources in order to test cyber persistence theory's hypotheses. Our assessment is that these data are representative of the totality of State cyber behavior and interactions.

Given that cyber persistence theory's hypotheses are strongly supported by these data, we believe the theory should be a keystone for a State's developing cyber policy and strategies. That said, we are cognizant of concerns over deriving policy and strategy when data representativeness is in question. Therefore, we consider a counterfactual based on open source data representativeness and assess how such counterfactual evidence, were it true, would impact the value and risk of basing cyber policy and strategy on cyber persistence theory.

Consider the situation where open source reporting of State cyber behavior and dynamics is not fully representative of the totality of cyber strategic competition, that is, that competition also comprises discreet, covert, or clandestine behaviors and dynamics deviating from our hypotheses.[183] What alternative cyber strategic behavior short of armed-attack equivalence could this counterfactual data set comprise?

Three candidates come to mind—restraint, brute force, and coercion. The first two can be readily dismissed. Although the United States adopted broad restraint in its 2011 and 2015 cyber strategies, the voluminous State activity of Russia, China, North Korea, Iran, and others captured in a decade of open source materials undermines the claim that restraint describes primary State cyber strategic behavior. There is most certainly *more* behavior than is being reported, but there cannot be less. Additionally, in 2018, the United States adopted a new strategic posture of persistent engagement.

Brute force and coercive behavior in and through cyberspace must follow an exploitation, a point made earlier in this chapter. It is properly understood, then, as a potential *purpose* of exploitation, just as coercion, IP theft, and spreading misinformation may be. Thus, the remaining alternatives being considered are not that "brute force or coercive behavior is more prevalent than is being reported in open source materials." Rather, they are that "exploitation is being pursued for brute force or coercive purposes more than is being reported in open source materials." In the end, exploitation is the dominant behavior.

Brute force is described by Schelling as using capabilities to penetrate, to exhaust, or to collapse opposing military force—to achieve military victory—before those capabilities can be brought to bear on an enemy nation itself.[184] Given that such activity would generate armed-attack equivalent effects, it fails to satisfy our criterion of strategic activity that falls below that threshold.[185]

Coercive behavior requires both a threat and a demand, either explicit or implicit. Such threats and demands would most likely be revealed in open source materials as strategic threats are difficult to hide from the public eye, even those privately conveyed. Additionally, where demands were not met, punishments would be difficult to obscure from public view given their strategic character. The absence of coercive threats, demands, and strategic consequences in open source materials, therefore, strongly suggests their absence writ large in the total population of State cyber behavior in the competitive space short of armed conflict. What the open source materials referenced in this chapter show is that the few examples of such reported State behavior have manifested between States already engaged in ongoing militarized crises and armed conflict, as cyber persistence theory predicts. It is therefore unlikely that there is hidden coercive activity in an amount that would undermine the cyber persistence theory hypothesis that the cyber *fait accompli* is the primary mechanism used to advance interests in and through cyberspace.

While there is scant evidence of States using cyber operations to coerce other States, there is an abundance of reported coercive behavior by non-State actors, as evidenced in criminal gang's ransomware strategies where intruders clearly communicate they have compromised systems or data and make clear the demands that must be met for the defender to receive the key for file decryption. But this is not State behavior, the focus of cyber persistence theory. However, when executed at scale and over time against sub-State or private sector targets, such behavior could pose national security risks and thus should not be dismissed by States. And so States should consider the value and risks of pursuing a policy and strategy grounded in cyber persistence theory against non-State actor coercive cyber behaviors. From that perspective, such a policy would be sound, as the strategic principle of initiative persistence remains valid against non-State actors. The operations or campaigns targeting Dridex and Trickbot, albeit short-lived, support this argument.

An alternative angle for this counterfactual is considering the relative balance of cyber *faits accomplis* and direct cyber engagements. We have argued that cyber *faits accomplis* are the primary behavior in the cyber strategic environment and that direct cyber engagements are secondary. We have also noted that direct cyber engagements are scarcely reported. It may be the case, however, that direct cyber engagements are occurring with greater frequency but are underreported. Even were that the case, we do not think an increased frequency would be so significant as to elevate it to the primary State behavior because cyber persistence theory argues that States are incentivized to both avoid engagement and take advantage of abundant opportunities to unilaterally cumulate strategic gains. We have noted that as shared situational awareness of key cyber terrain improves across State actors, direct cyber engagements will likely increase in frequency. Even if direct cyber engagements were occurring with greater frequency than cyber *faits accomplis*, the relative balance of these behaviors has no bearing on the value or risk profile of using cyber persistence theory as a basis for cyber policy and strategy development.

What about State dynamics? We have argued it is unlikely that coercion short of armed conflict has been a prevalent but non-reported State cyber behavior. Thus, we logically conclude that a coercion dynamic is also not being underreported in the competitive space. Even if it were, policy prescriptions derived from cyber persistence theory would still be valuable—as noted in the discussion above—and no more risky than alternative prescriptions. A risk of inadvertent or accidental escalation into armed conflict from initiative persistent behavior may increase in a coercion dynamic, but that increase would be a consequence of the organic logic of the coercion dynamic itself (in which the mechanism for achieving strategic advantage is escalation), rather than from an initiative persistent posture, per se. If direct cyber engagements were occurring with greater frequency than cyber *faits accomplis*, the competitive interaction dynamic comprising those behaviors would still manifest. Thus, policy prescriptions derived from cyber persistence theory would still be sound from both a value and risk perspective.

In sum, no matter the specific representativeness of the data available through open source materials, policy prescriptions from cyber persistence theory are valuable and are no more risky than prescriptions that would flow from alternative theoretical orientations such as coercion theory, for example. Importantly, the same conclusion could not be made if considering coercion theory as the basis of cyber policy prescriptions because, if open source materials are representative of State behavior and dynamics (what we have shown in this chapter), coercion-based prescriptions would result in cumulative strategic losses, because they are not geared toward managing, mitigating, or deterring exploitation.

Conclusion

This chapter grounds cyber persistence theory's concepts in the empirical record. The record is actually robust as a whole and reveals the explanatory strength of the new theory. It illuminates the new dynamics associated with the cyber strategic environment in a manner that the application of coercion theory can and does not. Additionally, the empirical record provides confidence that prescriptions derived from the theory are supportable. Overall, we demonstrate the importance of viewing the cyber strategic environment as distinct from the nuclear and conventional environments through support for hypotheses derived from cyber persistence theory. Having introduced cyber persistence theory in Chapters 2 and 3 and evaluated its core concepts and logic empirically in this chapter, in Chapter 5 we turn to further theoretical development with an examination of the important concept of cyber stability through the lens of cyber persistence theory.

5

Cyber Stability

Is an environment of persistent action inherently unstable? Our conceptualization of the cyber agreed competition phenomenon suggests it is not. In this chapter, we turn to the notion of stability within the cyber strategic environment—what does it look like, what factors contribute to stability, and what factors may be destabilizing. We define cyber stability as a condition within the cyber strategic environment in which States are not incentivized to pursue armed-attack equivalent cyber operations or conventional/nuclear armed attack and thereby breach the tacit upper bound of the cyber agreed competition phenomenon. We argue that there are sufficient stabilizing factors that can guardrail State behavior so that even successful leveraging of cyber operations for strategic gain will remain within the context of exploitation and competition and not require or devolve into coercion and war.

Stability and Initiative Persistence

Cyber persistence theory argues that there is a structurally derived imperative for persistence in seizing and maintaining the initiative to set the conditions of security in one's favor. The pursuit of security thus necessitates continuous activity. Stability and continuous setting and resetting of the conditions in and through which further security-seeking action will take place seems logically disconnected at first blush. The notion of stability in the nuclear strategic environment was intuitively easier to understand—it was defined by factors that supported the absence of a prohibitively costly action, nuclear war. However, it was not, *ipso facto*, easy to produce.

Both the Soviet Union and United States were engaged in a global ideological, economic, political, and indirect military struggle. The conduct of their relations remained locked in the coercive conventional strategic environment and questions always remained as to whether activity would escalate to a nuclear

Cyber Persistence Theory. Michael P. Fischerkeller, Emily O. Goldman, and Richard J. Harknett, Oxford University Press.
 DOI: 10.1093/oso/9780197638255.003.0005

conflagration. The danger of escalation was, intriguingly, managed by actually linking the two strategic environments more tightly—the threat that such an unacceptable outcome was possible made it less likely (or so the logic of extended nuclear deterrence suggested). All of this rested on managing the prospective coercive threat of retaliation (and some denial) to shape the other side's decision calculus so that they did not engage in behaviors that might escalate inadvertently or explicitly toward the mutually unacceptable outcome of nuclear war.[1] When actions were taken (political, diplomatic, economic, military), much attention was directed at understanding them as signals to the other side that they should pause and think through their next steps. Coercion was primarily about communicating caution during the Cold War.

The cyber strategic environment is structured and incentivized differently. It does not appear at first glance to be a cautionary space; it is an environment of opportunity and reward due to the abundance of exploitable vulnerabilities. Cyber persistence theory expects States to seek rewards continually from parallel actions that reconfigure cyber terrain and the means to maneuver and cause effects in and through that terrain. Yet risk in this reward-abundant environment is containable because systemic resilience means that cumulative effects from cyber *faits accomplis* need not threaten the overall integrity of the system of network computing. Thus, cyber persistence theory predicts experimentation and strategic action that is not necessarily reckless.

Much can be achieved without "blowing up the Internet" or inviting coercive retaliation. Nevertheless, we examine three possibilities that could lead to cyberspace instability—winning/losing too much, unintended incidents, and spiraling complexity—and ways to bolster stability so that strategic action does not lead to sustained armed attack or its equivalence.

Destabilizing Elements

If stability is defined as a condition in which States are not incentivized to pursue armed attack or its cyber equivalent and are incentivized to bound their actions within a range of exploitative cyber *faits accomplis* or direct cyber engagements, the presence of those "bounded" actions is not instability. We should expect instability, the condition in which the cyber agreed competition phenomenon collapses, however, to follow from a particular form of cyber activity that advances destabilizing tendencies.

One possible instability action would be the deliberate decision of a State to arrest the loss of relative power due to cyber strategic competition. A loss that threatens national security may come from three related, but distinct destabilizing tendencies. First, a State might not respond to the structural imperative of the environment to persist and, thus, cede initiative to the other side to the point where relative power shifts and the State responds (potentially too late) with war. Failure to recognize the imperatives of strategic environments correctly can have catastrophic effects (e.g., both world wars). Second, a State might understand the imperative of initiative persistence but, for a variety of reasons, play this new game poorly and be bested continually by other actors across this interconnected space. And third, some cyber powers may recognize the imperative but respond to it ineffectively due to constraints such as self-imposed policy choices or resource limitations.[2] In such instances, an imbalance of sustained initiative in the cyber strategic environment may result in outcomes that produce cyber instability.

For example, some States may choose to adopt a cyber strategic approach of operational restraint and direct their policies and resources toward a strategic approach of defense and resilience. They would focus exclusively on internal-facing measures to secure their own government devices, systems, and networks or other devices, systems, and networks over which they exercise sovereign jurisdiction. Such an approach would focus on identifying and mitigating vulnerabilities or preparing to mitigate the consequences of potential adversary exploitation of those vulnerabilities.[3] Such a primary posture cedes the initiative and could lead to an imbalance of initiative that cumulatively undermines and overwhelms both defense and resilience as the initiative persistence of others succeeds.

Other States may adopt a cyber strategic approach of operational restraint that rests primarily on deterrence. This involves external-facing efforts to hold adversary targets at risk for future potential coercive purposes coupled with threats to impose costs in response. Such a primary posture also cedes the initiative allowing others to continuously set and reset conditions of security in their favor in ways that do not invite coercive retaliation—industrious actors can design around the deterrent threat and avoid its prospective consequences.

States may also adopt initiative persistence as their cyber strategic approach and conduct external-facing operations to continuously set the conditions of security in their favor. But, being willing to do so and doing so well are two very different things. Each of these three scenarios individually or in combinations can lead to a ceding of initiative that would create an imbalance of initiative to the point of significant degradation of relative power.

The concept of imbalance of initiative raises an important and fascinating area for further research into cyber foreign and military policy. For the State that benefits from an imbalance of initiative, when does pressing that initiative too far become counterproductive to the State's overall grand strategy? For a State generally content with the overall relative distribution of power (sometimes referred to as a status quo State), having well-affixed reins on "winning too much" in and though cyberspace should be part of a sophisticated grand strategy.[4] The same holds, however, for a State seeking revision of the distribution of power. A sophisticated grand strategy would advance its power and degrade others in relative terms short of war, if war is deemed to be an unacceptably high cost—a condition, at least with nuclear powers, that is likely to hold.[5]

Later in this chapter, we discuss how effective persistence through cyber countermeasures could mitigate the potential for imbalance to emerge. The point here is that stability can be reinforced through well-managed persistence that is cognizant of the upper boundaries of competition and avoids exacerbating the destabilizing tendencies that might emerge from others being "on the short end" all the time.

Whether a real imbalance exists or is perceived by one or multiple actors competing in the cyber strategic environment, there is no doubt that in the early part of the twenty-first century we remain in a situation that States are still feeling their way through the strategic consequences that may flow from their cyber actions. This lack of familiarity means that much of the contemporary (and foreseeable future) action is conducted without clear and defined understandings of acceptable and unacceptable behaviors among States acting through independent action and observation or mutually dependent interaction.

The absence of such understandings opens a door for the influence of factors generally accepted as undermining stability. These include uncertainty about another's strategy, motives, commitment, and utility functions (what actors value and how much they value it); and, a challenge amplified by cyberspace, uncertainty about who is actually responsible for observed behavior (the attribution challenge).[6] A greater number of such understandings, explicit or tacit, would not necessarily contribute to cyber stability—mutual understandings should not be construed as normative acceptance. However, tacit understandings in nuclear and conventional strategic environments have historically played important roles in increasing stability in great power competition.[7]

Uncertainty regarding the motives behind behaviors introduces a potential for inadvertent escalation. This condition presents a risk to cyber stability as States may seek to deny cumulative effects or outcomes or respond to uncertainties in ways that lead to unintended incidents and spirals of action and reaction that moves a State to a dynamic they did not intend to produce (i.e., inadvertent escalation). Such incidents are possible when a cyber operation creates or results in an unexpected armed-attack equivalent effect that did not result from escalating exchanges, necessarily.[8] This breach of the upper boundary of the cyber agreed competition phenomenon is possible, but if it emanates accidentally or inadvertently from operations, there are diplomatic conventions that have historically been used in the nuclear and conventional strategic environments to contain equivalent incidents and not worsen the situation. Such conventions were used to deal with several military shootdowns of commercial airliners, including the Soviet downing of KAL 007, US downing of Iran Air 655, Ukrainian separatist downing of Malaysia Air 17, and Iranian downing of Ukrainian Air 752.[9] This would require an admission of the operation or management of the diplomacy in the aftermath of a non-attributable incident. While such incidents would meet the definition of an instability action, whether there was sustained instability after a cyber incident of armed-attack equivalence would likely depend on the diplomatic and geopolitical context.

The more concerning destabilizing tendency would be if uncertainty around cyber operations leads to a spiraling of action-reaction cyber activity.[10] While no actor may be seeking to breach the upper boundary of the cyber agreed competition phenomenon, armed-attack equivalent effects could turn the interaction into a militarized crisis and push it into the coercive conventional or nuclear strategic environments. The possibility exists that parallel cyber *faits accomplis* from multiple actors might push States to react through traditional force. However, the dominant behaviors in the cyber strategic environment, chiefly their noncoercive exploitative nature, should reduce the likelihood of spiral conditions.

We now turn to an examination of stabilizing mechanisms that can mitigate these destabilizing paths.

Stabilizing Mechanisms in an Initiative Persistent Environment

The logic of the cyber strategic environment is distinctive. That does not mean that the dynamics and processes used to maintain stability in the conventional and nuclear environments are irrelevant. States in the twenty-first century must manage all three strategic environments simultaneously, which adds complexity; yet it also adds further management tools.[11] Moreover, the influence of these tools is additive, and perhaps multiplicative, in forming and reinforcing the tacit upper bound of the cyber agreed competition phenomenon, and consequently, in contributing to cyber stability.

The risk to cyber stability can be addressed through two macro-courses of action: explicit, formal bargaining efforts and additional tacit coordination and/or increased tacit bargaining efforts to, respectively, establish or construct more explicit and/or tacit, mutual understandings of acceptable and unacceptable behaviors within the bounds of cyber agreed competition.[12] Some international relations scholars have argued that explicit, formal agreements or conventions provide greater contributions to stability than do tacit agreements.[13] Formal agreements reduce uncertainty by allowing greater attention to detail and explicit consideration of contingencies that might arise. They permit the parties to set the boundaries of their commitments, to control them more precisely, or to create deliberate ambiguity and omissions on controversial matters. Formal agreements also raise the political costs of flagrant or deliberate violations, thereby discouraging unilateral violations and, consequently, engendering stability. For these reasons, explicit, formal bargaining arguably contributes to stability more than tacit approaches because formal agreements reflect precise (or intentionally ambiguous), mutually dependent, mutual understandings among the parties.

Tacit coordination and tacit bargaining also have advantages. Tacit coordination describes common but independent behaviors around a shared strategic interest. Tacit bargaining describes a process resulting in mutually dependent, mutual understandings. Tacit coordination and tacit bargaining are more flexible than explicit, formal bargaining—they are willows, not oaks.[14] They can be adapted to meet uncertain circumstances and unanticipated shocks. This flexibility is useful if there is considerable uncertainty about the distribution of future benefits under a particular agreement. The parties need not try to predict all future states and comprehensively assess or contract for them.[15] Their informal character also reduces the reputational costs incurred when abandoning explicit, formal agreements. And, tacit bargaining's iterative character permits progress to occur in small increments with low transaction costs.[16] All of these attributes of tacit coordination and tacit bargaining are particularly germane when seeking agreements for or within complex, rapidly changing environments like the cyber strategic environment. Finally, tacit agreements can serve as foundations for future formal agreements.

The approach adopted to enhance cyber stability by increasing mutual understandings of acceptable and unacceptable behaviors within the bounds of the cyber agreed competition phenomenon should not be biased by traditions for arriving at agreements or by bureaucratic, functional silos.[17] Rather, it should be informed by whether explicit or tacit approaches, or some combination thereof, is best suited to the strategic environment—both its geopolitical and technological (or sociotechnological) features.

One's approach should also be informed by the vast knowledge the social sciences have accumulated on how "norms" are constructed.[18] It is through these lenses that we consider explicit and tacit approaches to construct a larger (and more granular) set of mutual understandings of acceptable and unacceptable behaviors within the tacit boundaries of the cyber agreed competition phenomenon. We examine evidence of each approach's contribution to date where available and highlight challenges to each in the current strategic environment.

Explicit, formal approaches encompass recent and ongoing State efforts to establish a global, explicit consensus on the relevance of existing international law or new international law (or other formal agreements) and efforts to establish voluntary, non-binding norms of responsible behavior. There are similar ongoing efforts seeking international group, but not global, explicit consensus.

Examination of potential focal points for tacit coordination looks beyond observations of the ongoing unilateral, independent operational behaviors currently dominating the cyber strategic environment (i.e., cyber *faits accomplis*) to explore States' unilateral, independent efforts to declare acceptable and unacceptable behaviors through expressions of *opinio juris*.[19] We also assess the prospects of tacit bargaining in cyberspace, a process through which States could construct mutually dependent, mutual understandings of acceptable and unacceptable behaviors through direct cyber engagement. Since expressions of *opinio juris* aligned with cyber *faits accomplis* and/or direct cyber engagement represent two key elements to forming new rules of binding, customary international law—an additional approach to reinforcing stability—we address the potential contributions of tacit approaches to cyber stability from that orientation.[20]

Seeking Consensus on International Law (or Other International Formal Agreements)

Although a long-standing international convention aligns with the upper bound of the cyber agreed competition phenomenon, several policymakers have suggested that international law might not be up to the task of governing cyberspace.[21] Their concerns are that the complexity, dynamism, and novelty of the cyber strategic environment present insurmountable challenges to international laws established among States in strategic environments of a far different character. There have, nevertheless, been important, explicit international efforts to establish a global consensus on if and how international law might inform the establishment of explicit, mutual understandings of acceptable and unacceptable behaviors that would contribute to the substance of the cyber agreed competition phenomenon.

The most notable global effort is the United Nations (UN) Group of Governmental Experts on Developments in the Field of Information and Telecommunications in the Context of International Security (GGE) process. It began in 2004 with a group of fifteen countries and to date has consisted of five such groups, with the 2016 group comprising twenty-five members. The five permanent UN Security Council members have always been group members.

In 2013, the third GGE reached consensus that international law, and in particular the UN Charter, is applicable to cyberspace. The report included recommendations on norms, rules, and principles of responsible behavior by

States, which were seen as deriving from existing international law.[22] Signatories concurred that (1) State sovereignty and international norms and principles that flow from sovereignty apply to State conduct of activities related to information communication technology (ICT); (2) State efforts to address the security of ICTs must go hand in hand with respect for human rights and fundamental freedoms; and (3) States must meet their international obligations regarding internationally wrongful acts attributable to them, must not use proxies to commit internationally wrongful acts, and should seek to ensure that their territories are not used by non-State actors for unlawful use of ICTs.[23]

The 2015 iteration of the GGE was tasked with analyzing the specific application of international law principles elaborated in the 2013 report, that is, how international law applied. This was a contested discussion because States' understanding and interpretations of international law frequently vary and did so in the context of cyberspace.[24] To get past contestation and make progress, the GGE adopted a new construct: general non-binding, voluntary norms, rules, and principles for the responsible behavior of States.[25] The fourth GGE report, published in 2015, moved the discussion forward cautiously by noting consensus on a set of "voluntary and non-binding" norms of acceptable behavior.[26]

Still, the question of *how* international law and, now, norms derived from international law applied in the context of cyberspace remained. And so in late 2015, the General Assembly tasked a fifth GGE "to study, with a view to promoting common understandings, . . . how international law applies to the use of information and communications technologies by States, as well as norms, rules and principles of responsible behavior of States, confidence-building measures and capacity-building."[27] The resulting 2017 GGE process represented significant regression. Several members—most notably Russia and China—objected to including references in the 2017 report to "self-defense" and "international humanitarian law" even though they had indicated their acceptance of both in the prior 2015 report.[28]

Some scholars argued that this "collapse"[29] signaled the improbability in the current geopolitical and cyber strategic environments of reaching any explicit, formal global agreement governing cyberspace that details acceptable and unacceptable behaviors.[30] States' viewpoints seemed to be diverging rather than converging. To that point, in November 2018, the General Assembly's first committee adopted two separate (and, some say, competing) resolutions on the actions of States in cyberspace—a new Open Ended Working Group (OEWG) resolution sponsored by Russia and a US-sponsored resolution looking to further the GGE 2015 framework.[31] Further evidence of divergence were Russian efforts to create new international laws on cybercrime.[32] This Russian resolution has influential detractors, convincing some scholars that "the prospects for new [international] laws applicable to cyberspace are slim."[33]

The reports from the 2021 OEWG and GGE groups offer no evidence that a detailed, formal agreement on acceptable and unacceptable behaviors is likely anytime soon. Although the OEWG report, by virtue of the 193 States participating in the process, increases the number of States who concur that international law, particularly the UN Charter, applies in the context of cyberspace, it does not address the specific question of *how* it applies.[34] Some scholars describe the report as being "new without bringing much new" and largely failing "to deliver on the OEWG's key objectives, namely, to address the root causes of global cyber instability today."[35]

Likewise, although the 2021 GGE report includes a substantive step forward by acknowledging that international humanitarian law (IHL) applies to cyber operations during an armed conflict, "some disagreement remains about *how* IHL governs cyber operations during armed conflicts."[36] The report continues to kick the can down the road on specifics by calling for States to make voluntary national contributions (in the original language of submission without translation) to an official compendium on the subject of how international law applies to the use of ICTs by States.[37]

The United States posited after the 2017 GGE meeting that the realization of global cyber norms may not be achievable through a UN effort and that it is time to consider other approaches.[38] This conclusion will not surprise social scientists studying norms. Martha Finnemore and Duncan Hollis argued that, despite their [norms'] newfound popularity, the discourse on cyber norms is sorely underdeveloped and the GGE process illustrates that immaturity. "Those calling for cyber norms," they say, "have largely focused on the desired products—the particular behaviors that any new cyber norms may mandate. Efforts like the GGE's pronouncement of 'peacetime' norms focus on what norms ought to say, as if dictating the contours of a norm makes it a reality . . . it is not enough to know what cyber norms we want; we must know more about the processes for cultivating them."[39] In the end, there is a difference between aspirations and a convergence of expectations about behavior around a particular activity (a norm). Cultivating such convergence is the challenge, even among allies and partners.

Seeking International Group, but Not Global, Consensus

Alongside the GGE process, the United States and "like-minded" States have engaged in a complementary explicit, formal effort in pursuit of consensus on explicit principles of responsible State behavior in cyberspace.[40] The 2015 G20 Leaders' Communiqué and 2017 G7 declaration on "responsible" behavior are outcomes of that effort. The 2015 G20 Communiqué notes, among other things, that "no country should conduct or support ICT-enabled theft of intellectual property, including trade secrets or other confidential business information, with the intent of providing competitive advantages to companies or commercial sectors"; that "All States in ensuring the secure use of ICTs, should respect and protect the principles of freedom from unlawful and arbitrary interference of privacy, including in the context of digital communications"; and, in reference to the 2015 GGE report, that "international law, and in particular the U.N. Charter, is applicable to State conduct in the use of ICTs and commit ourselves to the view that all States should abide by norms of responsible state behavior in the use of ICTs in accordance with U.N. resolution A/C.1/70/L.45."[41] The 2017 G7 Declaration on Responsible State Behavior in Cyberspace was similar in content, and it also referenced the GGE 2015 report.[42]

Though not a global consensus, the G7 and G20 outcomes suggest an explicit, formal approach to establishing more expansive, explicit, mutual understandings of acceptable and unacceptable behaviors can bear fruit. However, as the agreements are among like-minded States, this conclusion rests on tautology.[43] The G7 and G20 cases represent a most likely sample of cases to result in consensus given that members are like-minded.[44] This introduces selection bias, which increases the likelihood of consensus.[45] Consequently, the outcomes of these cases represent the weakest support for making a generalization that an explicit, formal approach will be equally successful at the global level.[46]

An additional, shared attribute of the 2015 (and 2021) GGE report and G7 declaration deserves further scrutiny. The former declares as an objective the identification of "voluntary, non-binding norms for responsible State behavior and to strengthen common understandings to increase stability and security in the global ICT environment." The latter notes the members "support the promotion of voluntary, non-binding norms of responsible State behavior in cyberspace during peacetime, which can reduce risks to international peace, security and stability." As these agreements are voluntary and non-binding, they do not represent mutually dependent, mutual understandings among the group members. While the veneer of the explicit, diplomatic processes behind them suggests these are formal agreements, *their voluntary and non-binding character is representative of a tacit coordination outcome*, where States have a shared strategic interest (avoiding instability) but act independently (voluntarily) on common but independent (not mutually dependent) understandings of behaviors they do and do not consider acceptable.

Arguably, then, the substance of the GGE and G7 agreements offers no greater contribution to the substance of the cyber agreed competition phenomenon than has tacit coordination among cyber powers incentivized primarily by opportunities for strategic rewards by operating short of armed-attack equivalence and secondarily by a few core principles and rules of international law. This claim leaves advocates of an explicit, formal approach in a position of having to argue either that the explicitness of the approach, rather than the outcome of the approach, makes it superior for arriving at mutual understandings; or that the precision in the GGE and G7 agreements reduces uncertainty among the parties more so than does unilateral, independent State cyber activity. Both of these positions are untenable.

Explicitness raises the political costs of flagrant or deliberate violations of formal agreements, but only if those agreements are structured in a way where political costs can be incurred.[47] A voluntary, non-binding structure only weakly, if at all, satisfies this requirement.[48]

Regarding uncertainty, Michael Schmitt argues that "while these [GGE] efforts largely settled the issue of international law's applicability to cyberspace and confirmed the relevance of its core principles and rules," they did not achieve the granularity required to reduce the susceptibility of those principles and rules, or the norms derived from them, to exploitation from cyber operations.[49] The G7 agreement recognized the same, noting that "To increase predictability and stability in cyberspace, we call on States to publicly explain their views on *how* existing international law applies to States' activities in cyberspace to the greatest extent possible in order to improve transparency and give rise to more settled expectations of State behavior."[50,51] Thus, these agreements do not appreciably reduce uncertainty because they lack greater precision above and beyond the existing global set of tacit, mutual understandings of acceptable and unacceptable cyber behaviors.

Schmitt extends his assessment into a forecast, arguing that "Nor, is establishing such granularity likely to occur [following this approach] given the typically slow pace of progress in multinational fora dealing with international law [and norms derived from the law]."[52] This sentiment has been echoed by others arguing that "[w]ith the accelerating pace of change in cyberspace and the glacial speed at which conventional law develops, new international law [and norms] will likely come through State practice."[53] Thus, the explicit, formal approach, when considered solely as a process, is also not responsive to the dynamism of the cyber strategic environment.

In sum, the current geopolitical and cyber strategic environments dampen the potential contribution of explicit, formal approaches to establishing granular, global or limited-group, explicit, mutual understandings of acceptable and unacceptable cyber behaviors. This is not to suggest that granular, mutual understandings are unachievable. It is useful to recall Kenneth Adelman's remarks regarding the failure of explicit, formal US-Soviet arms control efforts: "To assign arms control talks responsibility for eliminating or even diminishing geostrategic competition is to burden them with much more than they can conceivably carry. To laden arms control with such unrealistic expectations is inevitably to cause it to break down. Arms control can best be considered one single element in a full panoply of political, economic and defense efforts. But, frankly, such modesty has been lost since arms control has been thrust forward as the barometer by which superpower relations (indeed, global tranquility) are gauged."[54]

In 1960, Thomas Schelling recognized that even in geopolitical environments characterized by deep mutual mistrust and uncertainty concerning new technology, States nonetheless have a common interest in avoiding the kind of false alarm, panic, misunderstanding, or loss of control that may lead to unintended or non-deliberate escalation.[55] Although Schelling's focus at the time was on US-Soviet relations and nuclear weapons, his recognition applies equally well today to relations between the United States, Russia, China, DPRK, Iran (and others) in the cyber strategic environment. In such environments, Schelling argued that States have a common interest in not getting drawn or provoked or panicked into war by the actions of other parties (whether a party intends that result or not),[56] and they may have an interest in saving some resources by not doing things that tend to cancel out. Importantly, these common interests do not depend on trust or good faith. "In fact," he argued, "it seems likely that unless thoroughgoing distrust can be acknowledged on all sides, it may be hard to reach any real understanding on the subject."[57] Further, "[t]he intellectual clarity required to recognize the nature of the common interest may be incompatible with the pretense that all parties trust each other, or that there is any sequence of activities in the short run by which any side could demonstrate its good faith to the other."[58] These strategic realities motivated Schelling to propose both an alternative concept and a process to explicit, formal bargaining in order to arrive at and sustain strategic stability between States that fundamentally distrusted one another. He thus introduced and developed the concept of tacit coordination and the process of tacit bargaining.[59]

Understanding the current geopolitical context as one of great power distrust is important for two reasons. First, to Schelling's point, it suggests that informal, tacit approaches to arriving at mutual understandings may be more fruitful than formal, explicit approaches. Second, as is discussed below, it suggests that granular, mutual understandings rather than general norms should be the desired outcome of tacit approaches. This suggestion hints that the GGE and OEWG processes are off track.

As Adamson notes, the norms pursued in these processes are best understood as general standards, norms that are often goal-oriented, allow discretion for interpretation, and do not prescribe or proscribe the specific action needed to conform to or violate the standard.[60] Since standards are open-ended and allow for discretion, their power depends upon trust and solidarity among the community. Moreover, when the issue to be regulated occurs rarely or episodically, that is, single isolated incidents, general standards alongside trust ensure that given the circumstances, actors will balance all relevant interests when deciding how to act.[61]

Alternatively, granular, mutual understandings are better suited to opposite circumstances. They allow for very limited discretion by setting expectations for behavior in order to convey an obligation to achieve a certain outcome through certain means and measures. Specific norms are best suited for circumstances where there is no solidarity or there is limited trust among a community and the issue arises frequently. These circumstances describe well the current geopolitical environment and State behaviors within the cyber strategic environment.

We argue that tacit coordination and the tacit bargaining process are key to the construction of specific norms in cyberspace and, therefore, to cyber stability itself. From this perspective, we consider two avenues for arriving at granular, mutual understandings and also, potentially, granular, mutually dependent understandings of acceptable and unacceptable behaviors that fall within the bounds of the cyber agreed competition phenomenon: State unilateral and independent declarations of interpretations of relevant international law that could be accepted as evidence of expressions of *opinio juris*; and tacit bargaining through direct cyber engagement. As noted previously, *opinio juris* coupled with consistent State practice—cyber *faits accomplis* and/or direct cyber engagement—are two key elements forming binding, customary international law.

Tacit Coordination—Mutual Understandings through Independent Expressions of *Opinio Juris*

Opinio juris is an independent, general statement recognizing a State's obligation under international law to act or refrain from acting in a particular manner.[62] Even though the evidence of expressions of *opinio juris* presented in this section comprises explicit State declarations, we argue that their independent and general character (not directed at any specific State) are consistent with actions associated with tacit coordination (vice explicit, formal agreements).[63] Schelling notes, for example, that tacit coordination may coalesce around "statements and declarations that are not direct communication to the enemy."[64]

The Tallinn process is a notable effort that perhaps set the foundation for States to begin declaring interpretations of international law's applicability to cyberspace that could be accepted as evidence of expressions of *opinio juris*. The Tallinn process is a multi-year program examining how to interpret existing international law in the contexts of both cyber conflict and competition.[65] *Tallinn Manual 2.0*, published in 2017, specifically included a focus on "legal analysis of the more common cyber incidents that states encounter on a day-to-day basis and that fall below the thresholds of the use of force or armed conflict." It thus aligns with the competitive space described by the bounds of the cyber agreed competition phenomenon.[66] The manual highlights areas of consensus and non-consensus among nineteen international law scholars, with its contents serving an early indicator of the common and differing views that States have begun to express in the years since its publication.

We offer brief summaries of three areas; two roughly align with a notion of unacceptable behaviors: the international law principle of sovereignty and the rule of non-intervention, which are described as areas "ripe for exploitation [by cyber operations] and in need of clarification by States."[67] The third speaks to the principle of countermeasures, a potential remedy for States seeking to reinforce their views of what cyber behaviors are and are not acceptable. Some States have since offered interpretations addressing all three areas in relation to cyberspace. Those are reviewed following this summary.

The *Tallinn Manual* begins with a discussion of sovereignty and makes the point in its first rule that "[t]he Principle of Sovereignty applies to cyberspace" and, further, that "[a] State must not conduct cyber operations that violate the sovereignty of another State."[68] The assumption underlying the latter quote is that sovereignty is a rule of international law, the violation of which is an internationally wrongful act.[69] However, the participating experts held differing views on what constitutes a violation of sovereignty. Among the many possibilities offered by the experts on which there was not consensus was a cyber operation causing cyber infrastructure or programs to operate differently; altering or deleting data stored in cyber infrastructure without causing physical or functional consequences; emplacing malware into a system; installing backdoors; and causing a temporary, but significant, loss of functionality, as in the case of a major DDoS operation.[70] Moreover, the assumption that sovereignty is a rule is not a universally held view. Gary Corn notes that "An opposing view holds that sovereignty is a baseline principle of the Westphalian international order undergirding binding norms such as the prohibition against the use of force in Article 2(4) of the UN Charter, or the customary international law rule of non-intervention, which States have assented to as an exercise of their sovereign equality."[71] And Eric Jensen offers a view that "sovereignty is a principle that depends on the domain [air, land, sea, and space] and the practical imperatives of states and is subject to adjustment in interstate application."[72]

The Tallinn process and other international legal analyses are wedded to law premised on coercion.[73] Their starting point is that forceful intervention into the "internal or external affairs" of other States is a prohibited internationally wrongful act.[74] Two conditions must be met to determine a violation of the prohibition—experts agreed that both are vague and open to varying interpretations by States. First, the prohibition only applies to matters that fall within another

State's *domaine réservé*. These are matters that international law leaves to the sole discretion of the State concerned, such as the "choice of a political, economic, social and cultural system, and the formulation of foreign policy."[75] For example, elections and an exclusive right to regulate online communication in the exercise of sovereignty are both understood by many to fall within the *domaine réservé*, so using cyber means coercively to disrupt them would raise issues of intervention.[76]

The second condition to finding a violation of the prohibition on intervention is that an act must involve coercion. In the simplest terms, a coercive act is one designed to compel another State to take action it would otherwise not take or to refrain from taking action it would otherwise engage in.[77] However, things are not so simple because "coercive" is, again, up to State interpretation. The government of the Netherlands notes that "The precise definition of coercion, and thus of unauthorized intervention, has not yet fully crystallized in international laws."[78] International experts participating in the Tallinn process described cyber coercion as referring "to an affirmative act designed to deprive another State of its freedom of choice, that is, to force that State to act in an involuntary manner or involuntarily refrain from acting in a particular way."[79] Harriet Moynihan argued that States should instead understand coercive behavior "as pressure applied by one State to deprive the target State of its free will in relation to the exercise of its sovereign rights in an attempt to compel an outcome in, or conduct with respect to, a matter reserved to the target State."[80] Others advocate for lowering the threshold at which mere influence becomes unlawful coercion, claiming that a hostile cyber operation should not necessarily have to deprive a State of all reasonable choice, so long as it renders making the choice difficult.[81] Still others have argued that the scope of coercion "must be understood to encompass actions involving some level of subversion or usurpation of a victim State's protected prerogatives, such as the delivery of covert effects and deception actions that, like criminal fraud provisions in domestic legal regimes, are designed to achieve unlawful gain or to deprive a victim state of a legal right."[82] Finally, it has also been argued that, for the rule of non-intervention to be more relevant in the cyber context, "coercion" as a condition should be supplemented with "exploitation" given that exploitation, not coercion, is the dominant behavior through which States are seeking gains in and through cyberspace that endanger international peace and security.[83]

A consequence of these differing views is that Tallinn's participating experts' opinions varied as to whether, for example, the 2016 Russian cyber campaign including the compromise of Democratic National Committee servers was coercive. Some have argued that as the subsequently released e-mails had not been altered, the campaign amounted to mere espionage that, "without more," is not a breach of international law.[84] Other participating experts argued that the cyber campaign manipulated the process of elections and therefore caused them to unfold in a way that they otherwise might not have. In this sense, the Russian campaign was coercive.[85] An alternative view based on exploitation rather than coercion is that the campaign was a violation of the rule of non-intervention.[86]

The general principle of countermeasures is that a State may be entitled to take countermeasures (cyber or otherwise) that might otherwise be considered an internationally wrongful act (i.e., a breach of international law), in response to a breach of an international legal obligation owed by another State.[87] The purpose of countermeasures is to cause the responsible State to cease its unlawful action (or omission) and to provide assurances or guarantees and make reparations where appropriate. Countermeasures (as described in the law) are reactive, not prospective. Moreover, the injured State must notify the other State of its intention to take countermeasures. Additionally, countermeasures are generally characterized as temporary measures and therefore, according to the International Law Commission, "must be as far as possible reversible in their effects in terms of future legal relations between the two States."[88]

The participating experts in the Tallinn process were divided on the issue of whether there is a requirement to attempt lesser means of convincing another State to desist in its internationally wrongful conduct (i.e., retorsion) before turning to countermeasures.[89] Nor was there consensus on the duty to notify in advance of a countermeasure and whether States need to select the cyber countermeasure option that is most easily reversed or simply one that is, in fact, reversible at all. The participating experts agreed that when a cyber operation in question is but one in a series of ongoing actions that, for the purposes of State responsibility, constitute a single internationally wrongful act, countermeasures remain available.[90] From the perspective of cyber persistence theory, this is a particularly relevant opinion given that the theory argues that States are incentivized to engage in cyber campaigns (a series of ongoing actions) in pursuit of strategic ends.

Perhaps motivated by the output of the Tallinn process, State declarations of their views of how international law applies in the cyber context have been trickling out since 2018; these declarations serve as evidence of expressions of *opinio juris*.[91] Regarding sovereignty, the Attorney General of the UK declared in June 2018, "Sovereignty is of course fundamental to the international rules-based system. But I am not persuaded that we can currently extrapolate from that general principle a specific rule or additional prohibition for cyber activity beyond that of a prohibited intervention. The U.K. Government's position is therefore that there is no such rule as a matter of current international law."[92] Some claim the French government's position is diametrically opposed to the British position, referencing the Ministry of the Armies' 2019 publication of *International Law Applicable to Operations in Cyberspace*[93] as the basis for their argument.[94] Those holding this view claim that France affirms that "State sovereignty and international norms and principles that flow from sovereignty apply to the conduct by States of ICT-related activities."[95] Therefore, it exercises sovereignty over information systems located within its territory. Thus, they conclude that *any* cyberattack (i.e., any operation that breaches the confidentiality, integrity, or availability of the targeted system) constitutes, at a minimum, a violation of French sovereignty, if attributable to another State. It is particularly noteworthy that, according to this view, a violation of sovereignty occurs not only when effects are produced on French territory, but already when there is an intrusion into French computer systems. This has been described as a "pure sovereignty" perspective.[96]

Additional interpretations of sovereignty have been offered by the governments of the Netherlands, Austria, the Czech Republic, New Zealand, and US government legal advisors.[97] The views on sovereignty of US government legal advisors is close to that of the UK.[98] The US DoD General Counsel states that "the DoD view, which we have applied in legal reviews of military cyber operations to date, shares similarities with the view expressed by the U.K. Government in 2018."[99] He further suggests, as has the UK attorney general, that "many States' public silence in the face of countless publicly known cyber intrusions into foreign networks precludes a conclusion that States have coalesced around a common view that there is an international prohibition against all such operations (regardless of whatever penalties may be imposed under domestic law)."[100] Finally, he notes that "The implications of sovereignty for cyberspace are complex, and we continue to study this issue and how State practice evolves in this area, even if it does not appear that there exists a rule that all infringements on sovereignty in cyberspace necessarily involve violations of international law."[101] Similarly, the government of New Zealand argues that "the standalone rule of territorial sovereignty also applies in the cyber context but acknowledges that further state practice is required for the precise boundaries of its application to crystallise."[102]

The governments of the Netherlands, Austria, and the Czech Republic have expressed views close to, but not as "pure" as, the aforementioned understanding of the French position. The Netherlands asserted that it "believes that respect for the sovereignty of other countries is an obligation in its own right, the violation of which may in turn constitute an internationally wrongful act."[103] Austria argued, "A violation of the principle of State sovereignty constitutes an internationally wrongful act—if attributable to a State—for which a target State may seek reparation under the law of State responsibility."[104] And the Czech Republic "concurs with those considering the principle of sovereignty as an independent right and the respect to sovereignty as an independent obligation."[105]

Regarding the prohibition on intervention, the French declaration is brief, stating only that interference by cyber means with the internal or external affairs of France—that is, with its political system, economy, in social or cultural matters, or foreign policy—may constitute a violation of the principle of non-intervention.[106] Although New Zealand offers a similar perspective,[107] the statement of the UK attorney general and the UK government's 2021 policy paper on international law offer more granularity by detailing examples of violations, including cyber operations that manipulate the electoral system to alter the results of an election in another State, intervene in the fundamental operation of Parliament, target essential medical services, or impact the stability of the UK's financial system.[108] The US DoD general counsel states that "a cyber operation by a State that interferes with another country's ability to hold an election" or that tampers with "another country's election results would be a clear violation of the rule of non-intervention." He also acknowledges but does not commit to the UK's position on intervention into the operations of a legislative body or a financial system.[109] Israel's Deputy Attorney General (International Law) concurs with the DoD general counsel's position on non-intervention and elections.[110] Australia, in its 2019 International Cyber Engagement Strategy, acknowledges and commits to the UK's views on what would constitute violations: the use of cyber operations by a hostile State to manipulate the electoral system to alter the results of an election in another State, intervene in the fundamental operation of Parliament, or affect the stability of States' financial systems.[111]

Regarding countermeasures, the government of Japan declared, "A State that is the victim of an internationally wrongful act may, in certain circumstances, resort to proportionate countermeasures against the State responsible for the wrongful act."[112] In so doing, it adopted several strict requirements: the injured State must establish a violation of an obligation under international law that applies between the injured State and the responsible State; the cyber operation must be attributed to the responsible State; countermeasures must be temporary and proportionate, may not violate any fundamental human rights, and may not amount to the threat or use of force; and the injured State must notify the other State of its intention to take countermeasures.

Other States addressing countermeasures have taken differing views on the requirement of "notice." The Dutch Minister of Foreign Affairs affirmed the general notification requirement "in principle," but emphasized that it may be dispensed with "if immediate action is required in order to enforce the rights of the injured state and prevent further damage."[113] The aforementioned French document also rejected an absolute duty of prior notice before taking countermeasures, taking a position that a State could derogate from this rule where there is a "need to protect its rights" in urgent cases.[114] In 2016, then US State Department legal advisor Brian Egan did not dispute the "prior demand" requirement but noted that "[t]he sufficiency of a prior demand should be evaluated on a case-by-case basis in light of the particular circumstances of the situation at hand and the purpose of the requirement."[115] The more recent 2018 comments by US DoD general counsel are equally noncommittal to the duty, noting that "In the traditional view, the use of countermeasures must be preceded by notice to the offending State, though we note that there are varying State views on whether notice would be necessary in all cases in the cyber context because of secrecy or urgency."[116] Australia is equally noncommittal in declaring that "if a state is a victim of malicious cyber activity which is attributable to a perpetrator state, the victim state may be able to take countermeasures (whether in cyberspace or through another means) against the perpetrator state, under certain circumstances."[117] The UK has stated that States are always legally obliged to give prior notice before taking countermeasures against wrongdoing States and that it would "not be right for international law to require a countermeasure to expose highly sensitive" defense capabilities.[118] Finally, the Israeli Deputy Attorney General (International Law) noted that "[w]ith respect to the issue of countermeasures, I would like to echo the positions taken by the UK, the US and other States, to the effect that there is no absolute duty under international law to notify the responsible State in advance of a cyber-countermeasure."[119]

Two States commenting on countermeasures have directly addressed the consensus opinion of the Tallinn experts that, under certain circumstances, a series of actions that individually would not violate international law can rise to the level of a violation for which countermeasures would be an internationally lawful remedy. The UK government states that adversarial cyber activities that "cease almost instantaneously or within a short timeframe" may nevertheless be part of "a wider pattern of cyber activities [that] might collectively constitute an internationally wrongful act justifying a response."[120] France has adopted the same perspective.[121] The North Atlantic Treaty Organization has also taken a position on the issue. With the publication of the NATO Brussels Summit Communique on June 14, 2021, the alliance re-conceptualized how and what kind of adversarial activities can lead to crossing the threshold of an armed attack, including a reference to cumulative behavior. According to paragraph 32 of the Communiqué, allies now recognize that "the impact of significant malicious cumulative cyber activities might, in certain circumstances, be considered as amounting to an armed attack."[122] When asked to clarify the insertion of the term "cumulative," the NATO press office responded that (a) the term was indeed used deliberately, and (b) the reason for using it is because the alliance has recognized that the cyber threat landscape is evolving, and that several low-impact cyber incidents by the same threat actor can have the same impact as a single destructive cyberattack.[123] The Estonian Ministry of Defense added via email that "it is paramount that we would also take into account long-term cyber operations and attacks that might cause cumulative damage equal to what a single cyber-attack could cause."[124]

Notably, Estonia became the first State to publicly endorse the idea of *collective* countermeasures in the cyber context. In a May 2019 speech, Estonian president Kersti Kaljulaid cited the inherent right to self-defense and noted that "[a]mong other options for collective response [attribution], Estonia is furthering the position that states which are not directly injured may apply countermeasures to support the state directly affected by the malicious cyber operation."[125] New Zealand shares a similar perspective, arguing, "Given the collective interest in the observance of international law in cyberspace, and the potential asymmetry between malicious and victim states, New Zealand is open to the proposition that victim states, in limited circumstances, may request assistance from other states in applying proportionate countermeasures to induce compliance by the state acting in breach of international law."[126] Jeff Kosseff argued that collective countermeasures would allow States to better address "the persistent nature of the threats that they face in cyberspace . . . more likely to consist of constant adversarial actions, rather than the discrete events that shaped the debate over collective countermeasures in the non-cyber context."[127] Others sharing this view encourage States to go even further. For example, to account for the fact that States can achieve strategic outcomes through cyber campaigns short of armed-attack equivalence that were previously attainable only through armed attack, Gary Corn and Eric Jensen argued that the rule of countermeasures should be as robust against such operations as the rule of self-defense is against armed attack. The robustness of the self-defense rule follows from its associated exceptions, including allowing collective action on behalf of a victim State, allowing action against non-State actors, and allowing actions in anticipation of an armed attack. The concept of anticipatory countermeasures, arguably, is a recognition of cyber persistence theory's strategic prescription for security (i.e., persistence in seizing and maintaining the initiative to set the conditions of security in and through cyberspace in one's favor). Such exceptions for countermeasures, however, have yet to be endorsed by States.[128]

What do these unilateral, independent State expressions of *opinio juris* offer to the construction of granular, mutual understandings of acceptable and unacceptable behaviors? While not being explicit in the sense of being indicative of an explicit, formal approach, these expressions of *opinio juris*, nonetheless, offer some precision for which explicit, formal approaches are prized (because precision reduces uncertainty). Where all signatories of the GGE, OEWG, G7, and G20 agreements concurred that State sovereignty and international norms and principles that flow from sovereignty apply to State conduct of ICT-related activities, these unilateral expressions of *opinio juris* provide clarity on State's interpretations of *how* some apply. In so doing, the expressions highlight differences among significant cyber powers on their interpretations of sovereignty but also reveal mutual understandings among some States regarding specific cyber behaviors that would represent violations of the non-intervention principle, understandings that are more granular and expand on the set put forth in the GGE reports.

Where differences in interpretation are present, they could nonetheless serve as focal points around which States could tacitly bargain to create granular, mutually dependent understandings acceptable and unacceptable behaviors. This notion is explored in this chapter's following section on tacit bargaining. Areas where States' expressions of *opinio juris* currently align, however, such as common views of prohibited behaviors among American, British, Israeli, and Australian authorities, serve as important bases for continued development of additional granular, mutual understandings.

Accordingly, it can be argued that a unilateral, independent approach pursuing granular expressions of *opinio juris* is at least equivalent to, and in several respects more promising for, cyber stability than an explicit, formal approach to pursuing granular, mutual understandings of acceptable and unacceptable behaviors.[129] Indeed, Schmitt argued that this approach is where most of the normative activity regarding cyberspace will take place over the middle term.[130]

Several additional pathways to further reduce uncertainty have been proposed by scholars. Quoting Jens David Ohlin, Gary Corn and Eric Jensen noted, "Despite the patina of precision in its French rendering, the concept [of *domaine réservé* in the rule of intervention] has little internally generated content."[131] Thus, they argued that a more precise articulation of the boundaries between protected and unprotected interests—"domains and activities"—comprising *domaine réservé* would better serve international peace and security by placing States on notice of the areas of interference most likely to generate legal consequence and potentially escalatory responses.[132] Schmitt took a slightly different tack, noting that States could further relax the interpretation of the scope of the term "internal or external affairs" to more broadly include the *target* of the cyber operation and not just the "domains or activities" of the State.[133] In addition, Moynihan argued, "There is perhaps likely to be more commonality between states about whether particular state behavior constitutes an internationally wrongful act [a violation of the intervention rule] and why, then there is about whether sovereignty is a rule or a principle and how it relates to intervention."[134]

All of these *prima facie* seem useful prescriptions for further reducing uncertainty and introducing granular content to mutual understandings of acceptable and unacceptable behaviors. However, it should be acknowledged that there are significant, practical limits to pursuing these prescriptions in the current environment.[135] These limits, imposed more by the cyber strategic than the geopolitical environment, are discussed below and accompanied by notional, potential mitigations.

If States are more specific regarding the systems, policies, functions, and/or activities comprising their *domaine réservé*, it follows that they should also have an obligation to make clear how those manifest technically in and through cyberspace. Otherwise, significant uncertainty will still plague States' understandings. Therein lies a challenge. Most, if not all, States have limited understandings of such matters. Additionally, the processes they might use to develop those understandings may differ. For example, when considering "economic matters," a key question States should ask is, "What are the key functions of the economy that rely on an operational Internet?" Some might stop there. Others might go a level deeper and ask, "What are the industrial processes, sectors, and entities (physical and logical layers) that directly support those functions that also rely on an operational Internet?" Yet other States might go another level deeper and ask, "What are the data central to those processes, sectors, and entities, and where are they stored in cyberspace?"[136] Finally, what personnel participate in the operations of all of the above (i.e., the cyber persona layer)? This line of inquiry also speaks to the practical limits of Schmitt's recommendation to include targets in an expanded interpretation of "internal and external affairs." A major State's financial or electoral systems, for example, likely includes thousands if not hundreds of thousands of targets.[137] This challenge is illustrated in the final report of the Global Commission on the Stability of Cyberspace.[138] In calling for a prohibition on intrusions into the "public core" of the Internet, the report describes the core in two levels of detail, including:

> such critical elements of the infrastructure of the Internet as packet routing and forwarding, naming and numbering systems, the cryptographic mechanisms of security and identity, transmission media, software, and data centers.... Packet routing and forwarding elements include, but are not limited to, (1) the equipment, facilities, information, protocols, and systems that facilitate the transmission of packetized communications from their sources to their destinations; (2) Internet Exchange Points (the physical sites where Internet bandwidth is produced); (3) the peering and core routers of major networks which transport that bandwidth to users; (4) systems needed to assure routing authenticity and defend the network from abusive behavior; (5) the design, production, and supply-chain of equipment used for the above purposes; and (6) the integrity of the routing protocols themselves and their development, standardization, and maintenance processes.[139]

Such enumeration efforts suggest quite a bit of homework for States, but, arguably, work that will have to be done to some level not only to support more granularity in expressions of *opinio juris* but also, we argue later, to support tacit bargaining through direct cyber engagement. Indeed, we contend that enumerating States' *domaine réservé* is a necessary condition for creating granular, mutual understandings of acceptable and unacceptable behaviors.

Given the above, even under a scenario where two or more States declare "narratively" similar, more precise *domaine réservé* (absent details on technical manifestation), tacit coordination around that precision should not be a foregone conclusion as their understandings may not be "technically" similar, that is, an understanding grounded in mutual understandings of the hardware-software-processes those functions comprise. In the expressions of *opinio juris* referenced above, for example, several States offered views that intervention into the operations of a legislative body, the electoral system, and the financial system is an internationally wrongful act. Thus, the narrative appearance of a potential focal point for tacit coordination is present. However, it is not obvious that those States have comparable understandings of how those State functions and their associated activities manifest technically in their own States, let alone in the other States.

Another challenge comes in the form of States having dissimilar views over what *domaine réservé* comprises (an issue raised previously). For example, China, Russia, and others consider the management of all ICTs, their content, and the content they may receive or share in or through their sovereign territory to be within their *domaine réservé*, a view not widely shared among States. Tacit coordination among States holding different views of *domaine réservé* is far less likely, but not impossible. History shows that many tacit understandings are not composed of "symmetric exchanges," because great power strategic cultures are often idiosyncratic in what they value. After World War II, for example, the United States recognized the security interests of the Soviet Union in Eastern Europe but continued to press Moscow for an "open" rather than a "closed" sphere of security interest, one in which Eastern European States would enjoy a measure of political freedom and access to the West.[140]

Finally, it is likely that the technical manifestations of *domaine réservé* will routinely change for States, both as a result of discovery (i.e., revelations regarding how *domaine réservé* manifests in and through cyberspace) and through choice (as States respond to continuous changes within cyberspace itself). Both speak to the dynamism of the cyber strategic environment. This raises an issue of how frequently States should or would have to update expressions of *opinio juris* to ensure they are keeping customary international law in the context of cyberspace up to date in a manner relevant for and consistent with States' uses thereof and cyberspace's inherent dynamism. A mitigation for this issue may be found in the second element of customary international law: State practice. It is likely that once a State has discovered or determined that a new or existing system, policy, function, and/or activity is cyber-enabled, it will likely first act to secure its cyber manifestations through State practice before (or if) expressing in *opinio juris* that it should be considered part of its sovereign zone of protected interests.[141] State practice, then, is a reasonable manner through which updates could be communicated. In fact, it is necessary if States want to consider customary international law as a potential regulatory regime for cyber behaviors.

As averred previously, *opinio juris* coupled with consistent State practice—cyber *faits accomplis* and/or direct cyber engagement—would serve as the constituent elements of new, binding rules of customary international law addressing the cyber strategic environment. The presence of only one constituent element does not suffice for the identification of a rule of customary international law.[142] A belief that something is (or ought to be) the law (*opinio juris*) unsupported by practice is mere aspiration; practice without acceptance as law, even if widespread and consistent, can be no more than a non-binding usage; it is the two together that establish the existence of a rule of customary international law.[143] Importantly, the existence of *opinio juris* and State practice need not follow a particular order for the identification of a new rule of customary international law. This flexibility aligns with the dynamism of the cyberspace strategic environment and argues for a regime of cyberspace customary international law that is dynamic, at times and perhaps often updated by expressions of *opinio juris* or by State practice.[144] This recommendation is consistent with social science observations regarding agreements in areas where rapid obsolescence is likely.[145] Likewise, it aligns with observations on the dynamism of norms. Finnemore notes, "Part of the utility, and the challenge, of norms is that their meanings are dynamic. Every new application of a norm to a new situation refines understandings of exactly what the norm entails. These accumulations of shared understanding can give norms depth and make them robust, but these processes can also be contested and messy. Contestation of cyber norms is to be expected, particularly because changing technology constantly creates new situations."[146] Constructing robust regimes and processes through which to have these debates is one way to manage these challenges.

Only when State practice is a *general* practice, meaning that it is sufficiently widespread and representative (as well as consistent), can it be the basis of a rule of customary international law. There are several potential evidentiary sources of State practice, the most relevant for this discussion being operational conduct "on the ground," described by the International Law Commission as including battlefield or other military activity as well as law enforcement and seizure of property. The Commission, importantly, explicitly requires that "Practice must be publicly available or at least known to other States," further noting that it is difficult to see how confidential conduct by a State could serve such a purpose unless and until it is revealed.[147] Some reference this requirement to argue that "State cyber practice is mostly classified" or "otherwise shielded from observation by other states . . . [making it] difficult to definitively identify any cyber-specific customary international law."[148] This argument, however, ignores the bounty of publicly available data on State cyber activity (APTs) that we referenced in Chapter 4. That said, absent States' concomitant expressions of *opinio juris*, these data are evidence only of "non-binding usage" (practice without acceptance in law) and therefore are not evidence of general practice. Nonetheless, that State cyber behavioral data are widely available in open source materials counters the argument that due to classification issues, it will be difficult to identify general practice.

In sum, the emergence of expressions of *opinio juris*, in and of itself, represents a valuable contribution to a process of constructing granular, mutual understandings of acceptable and unacceptable behaviors, as well as an important and necessary step for constructing a binding, adaptive customary international law regime for cyberspace. Further, by establishing focal points, their emergence sets the stage for tacit coordination and unilateral State practice around those focal points. If State practice manifests broadly and consistently, general practice will be established. Indeed, there are some areas where this arguably has already occurred absent expressions of *opinio juris*, for example, areas highlighted in the 2015 GGE agreement. However, there are also ongoing unilateral behaviors (cyber *faits accomplis*) upon which States do not concur and which are not representative of shared strategic interests that could serve as focal points for tacit coordination. In the absence of concurrence and presence of divergent interests, States could instead pursue tacit bargaining through direct cyber engagement with the objective of arriving at mutually dependent, mutual understandings of acceptable and unacceptable behaviors. Tacit bargaining agreements would also aid in establishing general practice. In the next section we take up the prospects for tacit bargaining in cyberspace and its potential contributions to cyber stability.

Tacit Bargaining—Seeking Mutually Dependent, Mutual Understandings through Cyber Interactions

A State bargains tacitly when it attempts to set a context for another State's policy choices through behavior rather than formal or informal diplomatic exchanges. The process is tacit because actions (excluding directed rhetoric) constitute the critical medium of communication; it is bargaining for stability and not coercion because the actions aim to influence an outcome that can only be achieved through some measure of joint, *voluntary* behavior.[149] It is important to note that the pursuit of security through independent parallel reconfiguring of the terrain and the means to maneuver and cause effects in and through that terrain is the primary assumed goal of State action in the cyber strategic environment. The choice to emphasize stabilizing tendencies through tacit bargaining is a secondary objective in an initiative persistent environment. More precisely, then, tacit bargaining in and through cyberspace describes mutually dependent State actions, and thus denotes a shift away from unilateral, independent State actions that underpin tacit coordination (including cyber *faits accomplis* and expressions of *opinio juris*) and toward direct cyber engagement. In this section, we discuss the complementarity of tacit bargaining with the cyber strategic environment.[150] We then address a general criticism of tacit bargaining, as well as specific concerns about its potential to construct new rules of customary international law in cyberspace. We conclude with practical limitations on States engaging in tacit bargaining in cyberspace which constrain direct cyber engagement and suggest mitigations to address those factors.

The cyber strategic environment is one of persistence and interaction, and therefore is structurally and strategically supportive of tacit bargaining, which, itself, is rooted in action. Should States adopt the strategic prescription for initiative persistence, there could be substantial opportunities for tacit bargaining through direct cyber engagement to reach mutually dependent, mutual understandings of acceptable and unacceptable behaviors. Tacit bargaining, moreover, has a "natural affinity" for managing relations in complex and dynamic environments, a challenge that some policymakers feel is an insurmountable hurdle for international law governing in cyberspace.[151] Finnemore and Hollis state that the "value of cyber norms comes in the processes by which they operate as much as the contents (or products) that such processes generate . . . in important ways, the process is the product when it comes to cyber norms."[152] For all of the reasons just listed, the value of a tacit bargaining process to granular norms construction is clear.

However, a tacit bargaining approach also faces challenges.

A general criticism of tacit bargaining is that, due to its tacit character, resulting agreements cannot be substantively weighty. A stronger criticism, and one particularly salient for the cyber strategic environment, is that tacit bargaining is sensitive to "noise and interpretation."[153] The challenge of attribution may be the most significant "noise" factor impacting efforts to assess or monitor a particular State's behavior against a sensitive system (e.g., is a State trying to bargain or holding to a bargain) given that multiple State or non-State actors may be targeting that system.[154] In addition, "interpretation" of behavior in any tacit approach is challenging given the absence of direct, formal communication.[155] The general critique is not supported empirically, as Karin Koch and Charles Lipson make clear in their research on the role tacit bargains in the functioning of the Organization for European Economic Cooperation (OEEC) and the Organization of the Petroleum Exporting Countries (OPEC), respectively.[156] The Missile Technology Control Regime serves as another example.[157] Regarding noise, attribution is less of a challenge than it once was. Profiles of APTs have revealed relatively consistent preferences in motivations, types of targets, techniques, tactics, and procedures that increase confidence in conclusions regarding the source of behaviors.[158] Regarding interpretation, a State's confidence that a tacit understanding is being sought or sustained is entirely informed by observation. Should States follow cyber persistence theory's prescription of initiative persistence, they could have a rich set of observations to inform interpretations of each other's behaviors. Noise and interpretation challenges are formidable, but not insurmountable. Moreover, their impact on tacit bargaining will likely vary over time with changes in State policy and technology. This suggests that these challenges should inform tacit bargaining policy rather than be referenced as reasons for dismissing tacit bargaining as a pathway toward cyber stability.

Noise and interpretation have implications for the simplicity and initial comprehensiveness of tacit bargaining agreements that could be achieved and their potential relation to the establishment of general practice in support of the identification of new rules of customary international law. Because problems of noise and interpretation work against the construction of complex tacit agreements, we would generally expect successful cyber tacit bargaining to involve a specific focus. Historically, this seems to have been the case—tacit bargaining in the Anglo-German and Anglo-French naval races, for example, each involved a single type of ship.[159] In cyberspace, for example, given the scope and technical complexity of the "public core" of the Internet, realizing an all at-once comprehensive tacit agreement to prohibit cyber operations targeting the same, while preferable, seems unlikely. Policymakers, therefore, should consider an iterative tacit bargaining process that focuses on pursuing specific agreements in series, each contributing to the larger objective. An initial effort could focus on prohibiting intrusions into the peering and core routers of major networks that transport that bandwidth to users, a follow-on effort could focus on systems needed to assure routing authenticity and defend the network from abusive behavior, and so on. By slightly altering a State's expectations about the strategy and motivations of its opponent, every successful tacit bargain increases the likelihood that another may occur. Over time, tacit, mutually dependent, mutual understandings of acceptable and unacceptable behaviors would expand, creating a broader base for establishing general practice that would support new rules of customary international law. In time, the climate could improve to a point where explicit, formal bargaining to address the most difficult and complex issues could occur.

The natural affinity of tacit bargaining for breaking complex problems into smaller constituent parts, resolving them, and moving forward can motivate decision makers to bargain in specific areas even while most of their unilateral cyber behaviors or mutually dependent cyber interactions continue to be at odds.[160] Additionally, noise and interpretation demand continuous, consistent, and observable behavior (e.g., general practice) to create and sustain tacit bargains. Absent sustained, consistent, predictable behavior, tacit bargains collapse. States adopting cyber persistence theory's prescription of initiative persistence, therefore, are well postured to both create and sustain a tacit bargain.

Applying these observations to tacit bargaining over *domaine réservé* requires that we revisit the challenges States face in understanding how their *domaine réservé,* and national interests more broadly, manifest in and through cyberspace. States must do their homework before they can communicate to each other any potential opportunities for tacit bargaining that may result in mutual understandings. There is no escaping this requirement. In addition, institutional constraints—operational capacity, domestic laws, and national cyber policies—may further restrain tacit bargaining. Some aspects of *domaine réservé* may be outside the legal purview or resources of a State's primary cyber forces. And States may experience human and/or technological constraints on their ability to attribute and respond in a timely and repetitive manner to mitigate "noise" and "interpretation."

The tacit bargaining discussion up to this point presumes that bargaining occurs through direct cyber engagement at the "logical location" of contested key cyber terrain, which reduces the effect of "noise." Yet so long as effects are proportionate, countermeasures in cyberspace need not be confined to the logical location of the internationally wrongful act or to the specific entity that authored the wrongful act.[161] Thus, a State could employ a cyber *fait accompli* or direct cyber engagement to communicate the unacceptability of the adversary's actions. This could exacerbate the deleterious effects of "noise" and "interpretation" on achieving a bargain, but the trade-off may be acceptable for operationally constrained States.

Additional mitigations would be available to States facing capacity and/or institutional constraints to securing key cyber terrain if two countermeasure exceptions recommended by more forward-leaning States and scholars appeared more broadly in State expressions of *opinio juris*: collective action on behalf of a victim State; and "no notice," anticipatory countermeasures. Estonia's president addressed the progress made toward collective countermeasures: "Our ability and readiness to effectively cooperate among allies and partners in exchanging information and attributing malicious cyber activities has improved. The opportunities for malicious actors to walk away from their harmful actions with plausible deniability are clearly shrinking. Last year [2018] demonstrated that states are able to attribute harmful cyber operations both individually and in a coordinated manner."[162]

Anticipatory countermeasures would allow the offended State to preclude competition over its key cyber terrain from an adversary's set of potential strategic choices. This exception would require high confidence that an adversary will likely and opportunistically act to gain control over a State's specified key cyber terrain. The countermeasure itself, preferably, would represent a minimum necessary, or an estimated proportional action to preclude the adversary wrongful act(s).[163] Assessments in support of this exception are achievable and need not be endlessly complex.[164] The criterion and recipe offered by Estonia's president for *ex post* attribution are equally appropriate for supporting this *ex ante* exception. President Kaljulaid noted, "At the end of the day what is required from the attributing state, is not absolute certainty but what is reasonable. When assessing malicious cyber operations we can consider technical information, political context, established behavioural patterns and other relevant indicators."[165] This perspective is consistent with that offered by Brian Egan in 2015, then legal advisor to the US Department of State, who noted, "Absolute certainty is not—and cannot be—required. Instead, international law generally requires that States act reasonably under the circumstances when they gather information and draw conclusions based on that information."[166]

In sum, the concept of tacit bargaining aligns well with the cyberspace strategic environment and could serve as a key vehicle for constructing granular, mutual understandings of acceptable and unacceptable behaviors. This, in turn, could come to represent general practice, inform new rules of customary international law, and help mitigate the destabilizing tendencies that might come from an imbalance of initiative, unintended incidents, or spirals of action and reaction. Although the cyber strategic environment enables this possibility, States must overcome a range of constraints to fully capitalize on the opportunity. These include limited understandings of how their national interests manifest in cyberspace, lack of capacity to fully secure those interests, and institutional restrictions on the operational reach of their primary cyber forces. New interpretation of the principle of countermeasures could overcome some of these constraints. Advances in cyber technologies offer another route, albeit one that raises potential risks for cyber stability, which we address in the next section.

Ever-Changing Technology

High mutability is a feature of the cyber strategic environment and must be taken into account in any consideration of cyber stability. We focus on two projected more immediate evolutions: the transition to 5G technology and the prospect of a marked increase in AI-enabled cyber operations.

Although we and others argue that States are not incentivized to disable or degrade large segments of the Internet, States are nonetheless incentivized to position themselves in ways that facilitate exploitation at times, locations, scales, and for durations of their choosing. The development of 5G networks presents a strategic opportunity for States seeking a foundational advantage. There are several ways in which 5G networks themselves are more vulnerable to cyber exploitation than their predecessors.[167] First, they move away from centralized, hardware-based switching to distributed, software-defined digital routing, thereby denying the potential for chokepoint inspection and control (yet another example of the ongoing default of building more and easier access for efficiency even if it introduces more security challenges). Second, 5G networks now execute through software higher-level network functions formerly performed by physical devices. These functions are supported by a common language of Internet Protocol and well-known operating systems that have proven to be valuable tools for those seeking to exploit networks.[168] Thus, were a State-sponsored manufacturer of 5G technology platforms to dominate a local, regional, or global infrastructure, they could potentially hold the "keys to the kingdom."[169]

Third, the rollout of 5G portends the introduction of tens of billions of "insecure smart devices" attached to the network colloquially referred to as the Internet of Things (IoT).[170] In 2018, there were 7 billion IoT devices; in 2019, the number of active IoT devices reached 26.66 billion; and in 2020, experts estimated the installation of 31 billion IoT devices. Every second, 127 new IoT devices are connected to the web. By 2025, it is estimated that more than 75 billion IoT devices will be connected,[171] broadly binned into five types of applications: consumer, commercial, industrial, infrastructure, and military.[172]

A future cyber strategic environment including billions more "insecure smart devices" further enriches the abundance of organic opportunities for exploitation by States.[173] The Mirai botnet DDoS attack on Dyn in October 2016, an effort led by a college student, serves as an early indication of what is possible. The botnet, comprising exploited commercial IoT devices, at its peak comprised 600,000 devices with only forty-six IoT devices being central to its growth—primarily video security cameras and digital video recorders.[174] The attack disrupted Internet traffic for most of the US East Coast.[175] As we have argued, the cyber strategic environment does not incentivize States to cause such disruption, but they have been opportunistic nonetheless. On May 23, 2018, Cisco Talos published an alert regarding its discovery of "VPNFilter" malware on over 500,000 small and home offices routers and storage devices spread across at least fifty-four countries.[176] The malware, attributed by the US Federal Bureau of Investigation to APT28 (a Russia State-sponsored APT), was designed to conduct surveillance on its targets and gather intelligence, interfere with Internet communications, monitor industrial control systems (such as those used in electric grids, factories, and other infrastructures), and conduct destructive operations.[177] In April 2019, Microsoft reported that APT28 exploited commercial IoT—a voice-over-IP phone, an office printer, and a video decoder—in multiple customer locations.[178] The ingenuity States exhibited in the last decade of cyber campaigns and operations is on display in operations integrating IoT devices—not simply as bots or to gain accesses to more valuable devices, systems, and networks, but also as part of nontraditional command-and-control server infrastructure.[179]

Reporting on the APT28 incident, Microsoft noted, "These simple attacks taking advantage of weak device management are likely to expand as more IoT devices are deployed in corporate environments."[180] This speaks to the problem of shadow, legacy, and abandoned applications on networks—a problem exacerbated by commercial IoT devices and further compounded by the now prevalent "bring your own device to work" culture (consumer IoT).[181] Consumer IoT devices are purposefully designed to connect to a network and thus may connect to an organization's networks with little management or oversight. In large complex enterprises, IT operation centers often do not know IoT devices are connected to their networks.[182]

Concomitant with an increase in IoT devices, we should expect State cyber *faits accomplis* targeting those devices to increase and outpace direct cyber engagements. To the degree this increase contributes to imbalances in initiative, it introduces a risk to cyber stability. Additionally, to the degree it contributes to the gap between State behaviors and mutual understandings of acceptable and unacceptable behaviors, it could undermine cyber stability. States' cyber capabilities and capacities could have a strong influence on the validity of this expectation and conclusions, however. We next consider how advances in the development of AI platforms could influence States' abilities to secure or exploit this ever-expanding virtualscape.

Potential Impact of AI

Advances in machine learning and deep learning, heretofore referenced as artificial intelligence or AI, will impact cyber security and, we argue, cyber stability.

There are valid competing arguments for whether AI platforms will provide more advantage to those wanting to secure the international status quo vice those seeking to alter it.[183]

Schneier believes that "AI has the capability to tip the scales more toward defense" yet offers arguments supportive of both offense and defense.[184] AI will support discovering new vulnerabilities for offensive operations to exploit along with new types of vulnerabilities for defensive operations to patch, enabling automatic exploitation and patching.[185] AI will support reacting and adapting to an adversary's actions, both offensively and defensively. AI will support abstracting lessons from individual incidents, generalizing them across devices, systems, and networks, and applying those lessons to increase overall attack and defense effectiveness. And AI will support identifying strategic and tactical trends from large datasets and using those trends for both offense and defense.[186]

Others emphasize how AI could advantage attackers by automating tasks and thus alleviating the existing trade-off between scale and efficacy of attacks.[187] AI could support rapid and wide-scale identification of vulnerabilities and their automated exploitation. AI may expand the threat of labor-intensive cyberattacks like spear phishing. AI could enable novel attacks to exploit human vulnerabilities, such as using speech synthesis for impersonation. AI could also identify and exploit the vulnerabilities of others' AI systems through adversarial inputs, data poisoning, and model extraction.[188] Absent a contribution from AI that is more qualitative than quantitative in character, some believe the defense will eventually gain the upper hand.[189]

At the time of this writing, the technical community has not reached consensus on whether AI will advantage the offense or defense. From the perspective of cyber persistence theory, this lack of consensus is immaterial because the analytical measure itself (offense-defense balance) is not applicable to the cyber strategic environment, except in limited tactics. However, in the logic and lexicon of cyber persistence theory, a couple of observations can be made about the impact of AI on cyber stability.

Greater adoption of AI has the prospect of further compressing time to observe, orient, decide, and act. The literature on stability suggests that time compression exacerbates destabilizing tendencies. Additionally, AI holds out the prospect of automating the exploitation of vulnerabilities as well as their prevention and mitigation. Given that testing of AI defensive systems has demonstrated that most are vulnerable to exploitation through a discovery process of repeated intrusion attempts,[190] Wyatt Hoffmann proposes that "unlike other domains where engagements between attackers and defenders might be episodic, e.g. autonomous weapon systems in kinetic warfare," competitors will be inclined, instead, toward persistent behavior to seek out vulnerabilities in adversary defenses.[191] Although Hoffmann argues this will introduce instability, we argue at the outset of this chapter that a strategic environment of persistent action, that is, the cyber strategic environment, is not inherently unstable.[192]

A question worthy of critical examination is "Where must the human decision maker sit on the loop to make consequential decisions over introducing destabilizing tendencies or controlling the choice to breach cyber agreed competition?" Policymakers also should be concerned with the global rates of diffusion and adoption of AI platforms.[193] Should diffusion be slow or adoption be limited, imbalances of initiative may emerge and incentivize States on the "losing ends" to resort to armed attack to remedy their perceived loss of relative power.

Intriguingly, the process of machine learning could, itself, be leveraged as a stabilizing reinforcement. If core security algorithms leveraged by States for cyber *faits accomplis* or direct cyber engagements emphasize tactics, operations, and campaigns that cumulatively advance national interests short of actions and outcomes that would produce instability, the algorithms can be taught and trained to promote cyber stability.[194]

Conclusion

The voluminous activity comprising cyber agreed competition does not preclude the coexistence of stability and fluidity. For this to occur, States must align with the structural imperative of the environment and effectively execute strategies that map to the prescriptions and expectations of cyber persistence theory. The GGE, OEWG, and other explicit, formal approaches to establishing norms of acceptable behavior in support of cyber stability are not aligned with key features of either the current geostrategic or the cyber strategic environments. To mitigate the consequences of this misalignment, tacit coordination and tacit bargaining should play more prominent roles in States' efforts to construct granular, mutual understandings of acceptable and unacceptable behaviors. Pivoting toward this new approach as well as other better-aligned strategies is the subject of our remaining two chapters.

6

The Cyber Aligned Nexus of Theory and Policy

Our purpose in writing this book was to align theory with the reality of the cyber strategic environment and create a foundation upon which to derive strategy and policy. This required us to: (1) articulate the theory of cyber persistence and position it for further development by scholars and practitioners; (2) demonstrate the theory's explanatory and prescriptive power and juxtapose these findings with the dominant paradigm of deterrence; and (3) offer policy recommendations based on the new theory. Here we summarize our efforts at theory development and offer recommendations for the academic community of cyber security studies scholars and the policy community managing national security strategy in the digital age.

The logic of cyber persistence theory derives from the fundamental features of the networked computing environment. Cyberspace is "a global domain within the information environment consisting of the interdependent network of information technology infrastructures and resident data, including the Internet, telecommunications networks, computer systems, and embedded processors and controllers."[1]

The theory of cyber persistence adopts a systems-level analysis. It starts from the premise that cyberspace is a sociotechnical environment, and interconnectedness its central feature. Interconnectedness means that States are in a structurally imposed condition of constant contact with all other actors in this global system.[2] Constant contact is not a policy choice; it is the core condition that logically follows from having to operate in an interconnected world. This approach differs from a unit-level analysis of State behavior that assumes contact may also be imminent, potential, or episodic, but not constant.

In cyber persistence theory, the interconnectedness of cyberspace carves out a distinct strategic environment that is defined by the prospect that at every minute of every day some actor somewhere has both the capacity and will to

Cyber Persistence Theory. Michael P. Fischerkeller, Emily O. Goldman, and Richard J. Harknett, Oxford University Press.
 DOI: 10.1093/oso/9780197638255.003.0006

exploit some vulnerability that allows access to one's national sources of power directly or indirectly. The terrestrial organizing principle of segmentation (the opposite of interconnectedness) assumes that contact with others has some degree of separation salient enough to give time and or distance to protect, so that the contact can be deemed episodic (potential, imminent contact with sources of national power that typically require crossing some demarcated line that is organized, seen, and actuated). Constant contact with sources of national power raises a fundamentally different security situation. Just as nuclear weapons overwhelmed defense and raised a new fundamental security question—How can I secure when I cannot defend?—global networked computing overwhelms deterrence, raising yet another new fundamental security question—How do I secure in an interconnected environment of constant contact?[3] This question requires a redefining of security itself, where reactive avoidance of action (deterrence) and reactive mitigation of action (defense) are not sufficient. Where deterrence moved the locus of security out of one's own hand and placed it in the mind (decision calculus) of the adversary, interconnectedness and constant contact place the locus of security in a continual struggle over initiative—who has it will be more secure.

The global networked computing environment is a warehouse for and gateway to troves of sensitive, strategic assets that translate into wealth and power, and the capacity to organize for the pursuit of both. In line with its original architecture, this global networked computing environment is at once macro-resilient and micro-vulnerable, with very low entry costs for acquiring cyber capabilities. The combination of resilience and vulnerability creates a distinct dynamic—States can seek to exploit vulnerabilities at scale with little fear of destabilizing the overall technical environment. As the potential for exploitation is ever-present and States are in constant contact due to interconnectedness, they should assume their sources of national power—economic, political, social, and military—are vulnerable and can be undermined in and through cyberspace. Cyber activity, thus, must be a strategic concern.

Cyber persistence theory argues that this resilience-vulnerability convergence produces a structural imperative for States to persist in seizing and maintaining the initiative to set the conditions of security in and through cyberspace in one's favor. States set conditions by exploiting adversary vulnerabilities and reducing the potential for exploitation of their own.

Thus, cyber persistence theory argues that the dominant form of State behavior in cyberspace will be exploitation short of armed-attack equivalence, not coercion or brute force. This primarily takes the form of cyber *faits accomplis*—a limited unilateral gain at a target's expense where that gain is retained when the target is unaware of the loss or is unable or unwilling to respond. China's illicit cyber-enabled acquisition of IP and North Korea's exploitation of international financial systems to circumvent international sanctions are examples. A less prevalent type of exploitative action is direct cyber engagement, where a State directly engages with another actor for control over key cyberspace terrain. Examples are the US grappling with Trickbot and with ISIS network administrators and the November 2014 competition between the United States and Russia's APT29 for control over Department of State and White House IT systems.

The cyber strategic environment can reward States engaging in these behaviors with cumulative strategic gains over time. These behaviors, in turn, produce an inter-State dynamic of competitive interaction short of armed conflict rather than cyber escalation into armed conflict. The result is cyber agreed competition, a tacitly bounded, self-limited strategic phenomenon whose lower and upper bounds are inclusive of operational restraint and exclusive of operations causing armed-attack equivalent effects.

Set against the empirical record, the principles of cyber persistence theory provide a more robust explanation for the behavior we are witnessing in the early twenty-first century. The abundance of open source evidence available to researchers and policymakers aligns more clearly and comprehensively with cyber persistence theory than with the expectations and assumptions of theories of coercion. Cyber persistence theory presents a stark contrast to the explanations, predictions, and prescriptions of deterrence theorists, who rely on a coercion frame to explain cyberspace behavior and dynamics. A focus on unilateral behavior, rather than mutually dependent behavior, yields greater explanatory power. Competitive interaction, rather than escalation, is the dominant dynamic in this space below armed attack equivalence.

This alignment between theory and reality goes beyond greater explanatory strength and carries forward prescriptively into the policy and strategy realm. States whose behavior aligns with the expectations of cyber persistence have been gaining relative to States whose policies and strategies align with legacy theories of coercion. What these States have found is that persistence in seizing the initiative, rather than restraint, is central to defending and advancing one's interests and values in and through cyberspace.

Policy Implications

As noted above, cyber persistence theory has explanatory power and provides prescriptive direction. It opens further research avenues for the academic community and informs policy development for practitioners managing national security strategy in the digital age. Throughout this book we have offered a new strategic vocabulary better aligned with the realities of the cyber strategic environment that we hope postures the cyber security studies community for future academic progress. Here, we highlight the most important prescriptive implications of the theory for the policy community.

Planning and Posturing for the Cyber Strategic Environment

Given that exploitative campaigns below the threshold of armed attack are the dominant form of cyber behavior, policymakers seeking enhanced security in and through cyberspace should adopt a "campaign mindset" when they prepare, plan, and posture. They should view adversary behavior through the lens of campaigns rather than individual intrusions, hacks, or incidents. Planning and preparing for "significant" incidents or catastrophic attacks,[4] while important, omit how ongoing campaigns comprising activities whose individual effects never rise to the level of a significant incident, and therefore rarely elicit a timely response, cumulatively produce strategic gains. In strategic cyber competition, the campaign is the relevant unit of analysis, and cumulative effects,[5] rather than use-of-force or armed-attack equivalence, is the primary metric of consequential behavior.

This reorientation does not supersede planning for crisis and conflict scenarios that approach or breach the armed attack threshold. Rather, contingency planning for militarized crises and armed conflict must coexist with planning and executing continuous cyber campaigns short of armed conflict. A reflexive focus on low probability–high impact events, therefore, must be resisted in the cyber strategic environment where campaigns in competition are strategically impactful not just as a potential step toward war or a pause between crises. Policymakers need to shift their focus to where the preponderance of consequential action is now and where cyber persistence theory expects it to remain, to evolve in unanticipated ways, and to accelerate in frequency.

Adopting the strategic framework of exploitation as the dominant State behavior with competitive interaction as the central dynamic and cyber agreed competition as the prevailing strategic phenomenon means pivoting away from a focus on deterrence, escalation, and armed attack. This prescription is counterintuitive for practitioners who hold a taken-for-granted coercive bargaining mindset. From that perspective, States can influence each other's strategic cost-benefit calculations to act (or not act); they must impose costs that exceed benefits, credibly signal redlines, hold targets at risk, and have at the ready on-the-shelf response options. Cyber persistence theory explains why there is an imperative to be active in cyberspace. Therefore, resources and planning efforts should focus on precluding adversary options for exploitation (action) rather than trying to alter the adversary's cost-benefit calculus to act no matter how hard it is to break from that nuclear strategic environment structured mindset. It is through initiative persistence that security can be obtained, not by ceding decisions and action to the other side.

In support of this more active orientation, domestic policy and legal frameworks should be crafted or amended to enable cyber operational persistence, agility, and initiative rather than reinforcing inaction and restraint. For example, in the case of the United States, the 2019 National Defense Authorization Act included new and clarified authorities enabling an increased tempo of USCYBERCOM's operations outside the DoD networks.[6]

Interconnectedness, the core structural feature of the cyber strategic environment, requires continuous integrated campaigning, supported by ongoing collaboration, integration, and synchronization, across all relevant cyber planning and operational players and all instruments of national power. A quick test policymakers should use in cyber-related policy decisions is to always ask, "Does this policy create greater synergy or segmentation?" and choose solutions that favor synergy. The challenges of interconnectedness cannot be solved through separation.

Operating in and through the Cyber Strategic Environment

Cyberspace is continuously changing through both intentional action and organic evolution. Myriad vulnerabilities exist, as do opportunities to alter the terrain (physical, logical, persona) and interactions within it. If a State is not controlling the tempo of activity with respect to an adversary, then that State is reacting and the odds are that it is operating under less than favorable conditions set by more active States, who have gained initiative.

Cyber persistence theory argues that security rests in initiative persistence in setting the conditions of security within the virtualscape. This is accomplished by anticipating how one can be exploited in and through cyberspace and taking away—or precluding—exploitation opportunities that could produce strategically consequential effects, either at once or cumulatively over time. States would also be wise to exploit adversary vulnerabilities (for example, to expose, disrupt, or degrade) as a hedge against potential aggressors. The cyberspace strategic environment is flush with vulnerabilities that are ripe for exploitation and offers the potential for cumulative gains from operating in the competition space below armed conflict that exceed any cumulative costs incurred through retorsion, the "go-to" playbook for States reacting to cyber exploitations. In this competitive space the cyber *fait accompli* is the primary form of State cyber behavior used to define the security contours of the cyber strategic environment (e.g., the vulnerability landscape).[7] Operators, therefore, should persist in seizing and maintaining initiative, continuously setting, and resetting, the conditions of security so that their leaders can defend and advance national interests and values in and through cyberspace.

Cyberspace campaigns are ongoing across space and time; and so too must approaches be to thwarting them. There is no operational pause in cyberspace, and for this reason operational restraint cedes initiative. This does not mean being everywhere all the time; it does mean that the struggle to sustain the initiative and set favorable security conditions in cyberspace is enduring. This requires a continuous operational tempo supporting preclusion efforts with a campaign-based, rules-of-engagement mindset that exploits adversary cyber targets of opportunity now, rather than trying to hold targets at risk for some future contingency or episodically reacting to events. Security and stability in cyberspace flow from deliberate, cumulative action, not the threat of prospective action.

There will be those who wave the escalation and instability flags. As we presented in Chapter 5, cyber persistence theory argues that the cyber strategic environment offers a strategic incentive (the potential to cumulate gains short of armed conflict to strategic effect) that, when coupled with a core technical feature (macro-resilience and micro-vulnerability), discourages escalatory behavior. Policymakers should understand the conditions that might incentivize escalation, for example, a gross and sustained imbalance of initiative or an accumulation by an adversary of extraordinary gains. However, after more than a decade of competitive activity in the cyber strategic environment, escalation out of the cyber competition space into armed conflict has not occurred. States have been emboldened over time, engaging in malicious behavior routinely when unimpeded. Still, emboldened behavior within the cyber competition space is not escalation out of competition and into crisis or armed conflict.

Maturing the Cyber Strategic Environment

Cyber persistence theory views cyber activity below armed conflict not as an anomaly but rather as a manifestation of the emergence of a new competitive space where agreement over the substantive character of acceptable and unacceptable behaviors is immature. The past decade has witnessed the emergence of de facto norms defined by massive theft of intellectual property, expanding control of Internet content, attacks on data confidentiality and availability, violations of privacy, and interference in democratic debates and processes. Unless actively contested by smartly using all instruments of national (and international) power, these activities will continue to be normalized.[8] Policymakers should therefore be actively working to mature this space by applying cyber persistence theory's core prescription of seizing and maintaining initiative.

One line of effort should be to evolve international law and diplomacy to address strategic cyber competition. Policymakers and diplomats can leverage existing international law or seek to establish customary international law to manage instability that may result from cyber strategic campaigns short of armed conflict. For example, States should support the creation of exceptions for countermeasures that parallel those for self-defense—including an ability to target non-State actors and utilize collective and anticipatory countermeasures.

A second line of effort is for States to articulate how their national interests, including *domaine réservé,* manifest in and through cyberspace; and then declare interpretations of existing principles and rules (*opinio juris*) of international law in the context of cyberspace, as well as manifestations of national interests and *domaine réservé* that are subject to those rules. For example, if a State declares that holding free and fair elections falls under its *domaine réservé,* it should describe the infrastructure enabling that function and declare that actions against the infrastructure that disrupt free and fair elections are considered violations of the rule of non-intervention.

This would support a third line of effort to employ tacit bargaining through cyber campaigns that fall within the bounds of cyber agreed competition in order to construct and reinforce mutual understandings of acceptable and unacceptable behaviors around and about that function or infrastructure. This activity would serve as evidence of State practice and, as such, could contribute to the establishment of binding, customary international law in the context of cyberspace.

Paradigm Change

The logic of cyber persistence theory does not amount to just a competing explanation of cyber security dynamics in the early twenty-first century—it represents a new paradigm premised on new assumptions, key concepts, lexicon, and methodology that explains and describes the overwhelming majority of State activity in and through cyberspace. Consequently, cyber persistence theory should supplant deterrence theory as States' touchstone for developing cyberspace strategy and policy.

Paradigm changes mark new beginnings. For scholars, this book represents a starting point for building a robust, comprehensive body of cross-disciplinary theory on cyber security. For practitioners, the book can anchor new strategies and policies that marry cyber persistence theory's strategic prescriptions with

States' capacities, institutional constraints, and interpretations of international law.

Given the enormous creativity that it took to create cyberspace, we remain hopeful that such positive creativity will prevail in securing it.

In the next, final chapter of this book, we examine to what degree and when US policymakers came to understand that the cyber strategic environment warranted a new strategic approach and how that recognition led to adjustments in strategy and policy.

7

United States Case Study

Previous chapters demonstrated the explanatory power of cyber persistence theory. Here we turn to the theory's prescriptive potential. To do so, we examine the evolution of US cyber strategic thinking and cyber activities from the establishment of US Cyber Command in 2010 up to 2021 in order to bridge the gap between theory and practice. The case study reveals a story about a status quo State wedded to legacy strategies and structures struggling with the dynamics of the cyber strategic environment—a struggle that leads to a significant pivot in approach aligned with the prescriptions of cyber persistence theory. Adopting a Kuhnian frame, we argue that this pivot is by no means complete or irreversible and that additional policy adaptation consistent with cyber persistence theory's policy prescriptions is necessary if security interests are to be advanced within the cyber strategic environment. While seated in an examination of the United States, we conclude with the larger observation that these prescriptions are generalizable to all States.

What makes the United States a useful case from which to generalize is that it represents a "least likely" case for the prescriptions of cyber persistence theory. Relative to other countries, the United States is most likely to adhere to legacy security strategies that have served it well in sustaining its leading status in the global distribution of power. Evidence presented in preceding chapters shows that the behaviors of other cyber powers such as Russia, China, and North Korea align with the expectations of cyber persistence theory. Evidence that the United States is adopting strategies and behaviors also aligned with the core tenets of cyber persistence indicates that the prescriptive potential of the theory may also match its explanatory capacity. In the end, the theory expects that not adopting the policy prescriptions logically derived from it will be costly to States. Whether States adjust accordingly is another matter; evidence that the United States is adapting is, therefore, interesting for theory development and promising for policy outcomes.

Cyber Persistence Theory. Michael P. Fischerkeller, Emily O. Goldman, and Richard J. Harknett, Oxford University Press.
 DOI: 10.1093/oso/9780197638255.003.0007

Specifically, the case study traces to what degree and when US policymakers came to understand that the cyber strategic environment warranted a new strategic approach.[1] It shows an evolution in threat assessments leading initially to gradual adjustments in strategy and policy and in 2018, to a more dramatic and public shift aligned with cyber persistence theory's arguments and prescriptions. Evidence suggests that cyber persistence theory's core strategic principle of persistence in seizing the initiative to set and maintain the conditions of security in and through cyberspace has been applied across the US government since 2018, but unevenly and in an ad hoc manner.

We adopt a Kuhnian framework on paradigm change to explain why US policymakers have struggled to recognize and adapt. The legacy of deterrence as the US central security strategy continues to cast a long shadow over cyber strategy, limiting opportunities to better protect and advance US national interests in and through cyberspace and to promote cyber stability. We identify specific areas where adaptation has not occurred, propose ways to align activity to the policy prescriptions of cyber persistence theory, and advance a whole-of-nation-plus framework to achieve the political and organizational synergy the interconnected environment of cyberspace requires.

The case study unfolds over three periods: 2010–2014, 2015–2017, and 2018–2021. From 2010 to 2014, the White House promoted law enforcement as a deterrent and the establishment of responsible cyber norms; DoD strategy focused on defense and resilience.[2] By 2015, strategic guidance reflected an expanded threat assessment with significant concern over catastrophic attacks or attacks of "significant consequence." The White House's strategy of deterrence expanded beyond law enforcement to embrace all instruments of national power. Diplomatic efforts continued to pursue international norms, while deterrence joined defense and resilience as core tenets of DoD's cyber strategy.[3] By 2018, the threat assessment had shifted to great power strategic competition, with equal concerns expressed for adversary cyber campaigns short of armed conflict and catastrophic attacks. The White House's cyber strategic guidance continued prescribing deterrence using all instruments of national power and pursuing international norms through diplomacy.[4]

However, in 2018, DoD made a sharp pivot with its "defend forward" cyber strategy. The strategy to persistently contest adversary cyber activity in day-to-day competition short of armed conflict now stood side by side with deterrence of significant cyber incidents.[5] The defend forward strategy reinforced ideas from US Cyber Command's (USCYBERCOM) 2018 *Command Vision,* which introduced the operational approach of *persistent engagement.*[6] New presidential policy as well as cyber-specific statutory provisions enacted by Congress in the FY19 National Defense Authorization Act provided vital support to this pivot. Taken together, these guidance documents and actions embody and support many of the strategic prescriptions and policy recommendations of cyber persistence theory[7] although, as we assess, to enhance security in the digital age these pivots are necessary but not sufficient actions consistent with prescriptions of cyber persistence theory as a full US Government paradigm change had not occurred.

2010 to 2014

The White House and DoD

In October 2011, US Deputy Defense Secretary William Lynn III announced that "as the scale of cyberwarfare's threat" to US national security had come into view, "the Pentagon has formally recognized cyberspace as a new domain of warfare."[8] This declaration followed the formation in May 2010 of USCYBERCOM as a sub-unified command under US Strategic Command and the May 2011 release of the White House's *International Strategy for Cyberspace*.[9] The *International Strategy* declares, "The United States will defend its networks, whether the threat comes from terrorists, cybercriminals, or states and their proxies. Just as importantly, we will seek to encourage good actors and dissuade and deter those who threaten peace and stability through actions in cyberspace. We will do so with overlapping policies that combine national and international network resilience with vigilance and a range of credible response options." Moreover, it argues, "Risk reduction on a global scale will require effective law enforcement; internationally agreed norms of state behavior; measures that build confidence and enhance transparency; active, informed diplomacy; and appropriate deterrence."[10]

The roughly similar ordering in both statements reflects the weight of effort assigned to various tools of national power to address cyber threats: defense/resilience, law enforcement as a deterrent, norms establishment, and credible response options as a deterrent. An unclassified fact sheet on Presidential Policy Directive 20, signed October 2012, reinforces this argument by declaring, "It is our policy that we shall undertake the least action necessary to mitigate threats and that we will prioritize network defense and law enforcement as preferred courses of action."[11] Indeed, in 2014, Admiral Michael Rogers, then commander of USCYBERCOM and director of the National Security Agency (NSA)/Central Security Service (CSS), testified that the United States tends to view highly methodical and systematic adversary cyber operations short of armed-attack equivalence "as a law enforcement issue."[12]

The 2011 DoD cyber strategy aligned with the White House's *International Strategy* and made clear that defense/resilience is DoD's priority. To wit, "The implementation of constantly evolving defense operating concepts is required to achieve DoD's cyberspace mission today and in the future. As a first step, DoD is enhancing its cyber hygiene best practices to improve its cybersecurity. Second, to deter and mitigate insider threats, DoD will strengthen its workforce communications, workforce accountability, internal monitoring, and information management capabilities. Third, DoD will employ an active cyber defense capability to prevent intrusions onto DoD networks and systems.[13] Fourth, DoD is developing new defense operating concepts and computing architectures. All of these components combine to form an adaptive and dynamic defense of DoD networks and systems."[14] The only mention of deterrence addressing non-insider threats is a repetition of the sentence from the White House's *International Strategy* regarding the need to develop credible response options.

In October 2012, then US Secretary of Defense Leon Panetta elevated the specter of a "cyber Pearl Harbor," a scenario he described as involving cyber actors launching several attacks on US critical infrastructure at once in combination with a physical attack. The collective results of such an attack would cause physical destruction and loss of life and would also paralyze and shock the country.[15] This concern, however, had no noticeable effect at that time on how the US government intended to leverage its most potent cyber instrument of national power, namely DoD. When General Keith Alexander retired in 2014 as the dual-hatted commander of USCYBERCOM and director of NSA/CSS, the tenor of his farewell ceremony makes clear that policymakers wanted the US military to refrain from employing its operational capability and capacity in cyberspace. "DoD will maintain an approach of restraint to any cyber operations outside the U.S. government networks. We are urging other nations to do the same," said Defense Secretary Chuck Hagel at Alexander's retirement.[16] Hagel's statement portended the increasingly central role that a strategy of deterrence would play in DoD's next cyber strategy.

US Department of State and US Department of Justice

During this period, the US Department of State (DOS) did not produce a cyber strategy, but its chief cyber diplomat, Chris Painter, had a lead role in writing the White House's 2011 *International Strategy for Cyberspace*. Accordingly, diplomats participated in several explicit, formal diplomatic fora seeking to establish international norms of responsible behavior in cyberspace: the UNGGE, G20, and G7 efforts described in Chapter 5.[17] Also, in May 2014, the US Department of Justice (DOJ) released its first unsealed indictment against State-sponsored cyber operators for cyber-enabled intellectual property theft, an effort supporting the White House's emphasis on law enforcement-based deterrence.[18]

Although the *International Strategy* opens with a quote from President Barack Obama describing cyberspace as "interconnected,"[19] there is no apparent recognition of the deeper implications associated with that structural feature and its attendant logic of cyber persistence. US adversaries, on the other hand, were drawing a very different set of lessons about the nature of cyberspace based on the Arab Spring, which began in December 2010. Global interconnectedness fostered by the Internet, they learned, could facilitate uprisings and coups, and even topple regimes. Cyberspace could have strategic consequences *outside* of armed conflict. Autocrats, worried that their political hold on power would be undermined by digital-age capabilities empowering civil society, soon began to increase repression, surveillance, and active social manipulation of their own populations in and through cyberspace. They then turned those techniques outward on the rest of the world.

The year 2013 was a strategic inflection point as regimes that felt threatened by Internet-based subversion began launching offensive cyberspace operations to harass Western governments and corporations.[20] More capable adversaries began operating against US and allied corporate and government systems, illicitly acquiring intellectual property and personally identifiable information at scale, and targeting critical infrastructure. Where once espionage was the major concern, there was a shift to disruptive (e.g., the 2012–2013 DDoS attacks conducted by the Iranians against financial networks in New York), and then destructive (e.g., 2014 data deletion attack by the Iranians against a US casino and the North Korean attack against Sony Pictures) attacks.

Despite this activity, the United States adhered to its "doctrine of restraint," even as allowing adversary activity to go largely unchallenged appeared to embolden and incentivize aggressors to continue experimenting and operating with impunity. The reality of State behavior and interaction in cyberspace was beginning to diverge markedly from the model of war, catastrophic attack, and coercion upon which US cyber strategy and policy were based.

When assessing the 2010–2014 period against the arguments and prescriptions of cyber persistence theory, we observe the following:

- No recognition of a structural imperative to operate persistently;
- No recognition that cyber *faits accomplis* short of armed-attack equivalence are the primary behavior in cyberspace;
- No recognition that States are able to achieve strategic outcomes through cyber competitive interaction short of armed-attack equivalence (via cyber *faits accomplis* or direct cyber engagement operations/campaigns);
- No recognition or acceptance that a strategy of deterrence is ineffective against such campaigns;

- No recognition or acceptance that tacit, rather than explicit bargaining may be a more fruitful approach to constructing a binding regime of international norms for the cyber strategic environment.

2015 to 2017

The White House and DoD

The 2015 DoD cyber strategy differs from the 2011 DoD strategy primarily in perspective—it reflects an aspiration for leveraging and maturing US cyber capabilities in the service of national security. The 2015 strategy detailed three missions: DoD must defend its own networks, systems, and information; DoD must be prepared to defend the United States and its interests against cyberattacks of significant consequence; and DoD must support operational and contingency plans.[21] These also represent the defense, deterrence, and warfighting missions assigned to USCYBERCOM at the time. The first mission—defense—is a carryover from the 2011 DoD strategy and was the Command's priority mission.[22] The third mission represents a recognition and acceptance that maturing cyber capabilities should be considered in the deliberate contingency planning process and in ongoing operations.[23] The second mission reflects a recognition that DoD has a role in defending the nation in cyberspace, but it was a distant third priority and tightly circumscribed in terms of responding to a catastrophic attack. The strategy makes clear DoD's and USCYBERCOM's role in defending the nation from such attacks (i.e., they would establish a deterrence posture of operational restraint): "As a matter of principle, the United States will seek to exhaust all network defense and law enforcement options to mitigate any potential cyber risk to the U.S. homeland or U.S. interests before conducting a cyberspace operation."[24] Additionally, the strategy declares, "The United States will always conduct cyber operations under a *doctrine of restraint*, as required to protect human lives and to prevent the destruction of property."[25]

Shortly after publication of the 2015 DoD strategy, the White House submitted to Congress a report describing the nation's cyber deterrence strategy. The report notes that in seeking "to counter malicious cyber activity that poses significant threats to the nation, and to deter nation-states and non-state actors seeking to harm the United States through cyber-enabled means" the "United States' cyber deterrence policy relies on all instruments of national power—diplomatic, information, military, economic, intelligence, and law enforcement—as well as public-private partnerships that enhance information security for U.S. citizens, industry, and the government."[26] The report's section entitled "What the United States Will Seek to Deter" declares that the US government seeks to deter cyberattacks "that pose a significant threat to the national or economic security of the United States or its vital interests. Specifically, this includes cyber threats that threaten loss of life via the disruption of critical infrastructures and the essential services they provide; or that disrupt or undermine the confidence in or trustworthiness of systems that support critical functions, including military command and control and the orderly operation of financial markets or that pose national-level threats to core values like privacy and freedom of expression."[27] The focus on catastrophic attacks is clear.

DOS, DOJ, and US Department of Treasury (USDT)

As in the previous period, the US DOS supported the White House's strategy to establish norms by participating in several explicit, formal diplomatic fora, with the 2015 UNGGE agreement setting the high mark for such efforts and the failure of the UNGGE to sign a 2017 agreement setting the low mark. Additionally and notably, in 2016, Brian Egan, a DOS legal advisor, hinted at a complementary approach to constructing norms. Egan noted that States must "do more work" to clarify their interpretations of the principles of sovereignty and non-intervention and that mutually agreed-upon behaviors regarding both "ultimately will be resolved through the practice and *opinio juris* of States."[28]

The US DOJ continued supporting deterrence by law enforcement. It released seven new unsealed indictments and joined with the UK's National Crime Agency in a combined operation that disrupted the operational infrastructure of the Dridex malware administrators.[29,30] The capabilities of the US Department of Treasury (USDT) were employed in support of a cyber deterrence strategy through Executive Orders 13694 (April 2015) and 13757 (December 2016), enabling the use of economic sanctions against malicious cyber actors.[31]

By 2016, in the wake of Russian attempts to influence the presidential elections and in spite of all of these government-wide efforts, frustration in Congress and the broader policy community with ongoing adversary cyber operations and campaigns was palpable.[32] Political leaders, particularly in Congress, clamored for *more cyber deterrence*, including a strategy and options, to halt the barrage of intrusions and attacks across government, industry, and academia. One of the most vocal and persistent critics was then-Chairman of the Senate Armed Services Committee, Senator John McCain. He pushed relentlessly for the government to develop a comprehensive cyber deterrence strategy and equally chastised the Obama and Trump administrations for not delivering a cyber deterrence policy.[33] During a March 2017 hearing, McCain complained that the United States was still "treating every [cyber] attack on a case-by-case basis" and projecting weakness in cyberspace that "has emboldened our adversaries."[34] He continued, "As America's enemies seized the initiative in cyberspace, the last administration offered no serious cyber deterrence policy and strategy."[35] This statement is particularly pertinent for two reasons. First, it is indicative of the predisposition of many US policymakers to presume that deterrence is the appropriate strategy for the cyber strategic environment—it will work if we try harder, many believed.[36] Second, characterizing adversary behavior as "seizing the initiative" and yet calling for more deterrence actually highlights the continuing misalignment between traditional security thinking from the nuclear strategic environment and the core strategic principles of cyber persistence theory. As of yet, the United States had not considered that what other States were doing was in fact strategically grounded in a different view of the cyber strategic environment.

In sum, when assessing the 2015–2017 period against the arguments and prescriptions of cyber persistence theory, we observe the following:

- Still no recognition of a structural imperative to operate persistently;
- A recognition and acceptance that a cyber *fait accompli* of armed-attack equivalence could occur, but (still) no recognition that the cyber *fait accompli* short of armed-attack equivalence is the primary behavior in cyberspace;
- Still no recognition that States are able to achieve strategic outcomes through cyber competitive interaction short of armed-attack equivalence;
- Still no recognition or acceptance that a strategy of deterrence is ineffective against such campaigns;
- A recognition (the DOS comment on State practice and *opinio juris*), but no acceptance, that tacit, rather than explicit, bargaining may be a more fruitful approach to constructing a binding regime of international norms for the cyber strategic environment.

2018 to 2021

The White House and DoD

The 2018 US national cyber strategy marked a notable shift in cyber threat assessment, arguing "that the United States is engaged in a continuous competition against strategic adversaries . . . [who] use cyber tools to undermine our economy and democracy, steal our intellectual property, and sow discord in our democratic processes."[37] It concludes that these "[n]ew threats and a new era of strategic competition demand a new cyber strategy that responds to new realities."[38] This assessment is consistent with the US *2017 National Security Strategy*, which notes that "adversaries and competitors became adept at operating below the threshold of open military conflict and at the edges of international law" thus requiring the United States to "rethink the policies of the past two decades" and develop "new operational concepts and capabilities to win without assured dominance in air, maritime, land, space, and cyberspace domains, including against those operating below the level of conventional military conflict."[39] However, the *2018 National Cyber Strategy* indicates no such shift in policy or operational concepts; rather, it continues promoting defending the homeland by protecting networks, systems, functions, and data and preserving peace and security by strengthening the United States' ability—in concert with allies and partners—to deter and, if necessary, punish those who use cyber tools for malicious purposes.[40]

DoD's 2018 "defend forward" cyber strategy, however, took up the call for change, arguing that, in addition to deterring significant cyber events, it is necessary to "persistently contest malicious cyber activity in day-to-day competition" short of armed conflict.[41] The central mechanism for this change was described in USCYBERCOM's 2018 *Command Vision*: persist in seizing and maintaining the initiative in and through cyberspace based on new cyber operational concepts of anticipatory resilience, defend forward, and contest.[42] General Paul Nakasone, who assumed command of USCYBERCOM in May 2018, recognized that the *Command Vision* aligned with his experience in 2016 leading Joint Task Force Ares to combat ISIS, and he embraced it.[43] He called for USCYBERCOM's operational teams to shift from being a "response force" to a "persistence force," and argued that "[u]nlike the nuclear realm, where our strategic advantage or power comes from possessing a capability or weapons system, in cyberspace it's the use of cyber capabilities that is strategically consequential. The threat of using something in cyberspace is not as powerful as actually using it."[44,45]

General Nakasone also made the following arguments:

- Persistent engagement "cannot be successful if our actions are limited to DOD networks. To defend critical military and national interests, our forces must operate against our enemies on their virtual territory as well . . . adopting a posture that matches the cyberspace operational environment."[46]
- USCYBERCOM has "shifted away from the earlier emphasis on holding targets 'at risk' for operations at a time and place of our choosing," and instead "will operate continuously to present our decision makers with up-to-date options."
- USCYBERCOM is "building relationships with U.S. institutions that are likely to be targets of foreign hacking campaigns—particularly in the Nation's critical infrastructure—before crises develop, replacing transactional relationships with continuous operational collaboration among other departments, agencies, and the private sector. These relationships are crucial to thwarting attackers before they strike."
- Finally, "we are ensuring our capabilities, operational tempo, decision making processes, and authorities enable continuous, persistent operations."[47]

According to then-deputy assistant secretary of defense for cyber policy Ed Wilson, the 2016 Russian cyberattacks on the Democratic National Committee network helped drive a consensus in the US government to "take the next steps with active cyber defense, including indictments and the [2018 DoD] Defend Forward strategy."[48]

After the release of the 2018 DoD cyber strategy and USCYBERCOM *Command Vision*, Rob Joyce, then White House cybersecurity coordinator, commented:

> We've decided that we've got to have one element of our national power be cyber capabilities. . . . Looking at a strategy that just says: "We're going to wait until the attacks come to us, and then we'll defend them at the boundary, we'll clean up and remediate and try to push them back out after there's been a compromise, we'll recognize that we lost information"; that's not a winning strategy.[49]

Joyce further noted that "It's about making it harder for them [US adversaries] to succeed. . . . Some of that will be taking away the infrastructure they're using. Some of it [is] exposing their tools."[50] These statements from the White House are evidence that, although the cyber strategic prescriptions in the 2018 *National Cyber Strategy* did not represent the strategic shift the document called for, the White House nevertheless embraced the strategic shifts put forth in the 2018 DoD cyber strategy and USCYBERCOM *Command Vision*. Further evidence is found in the 2018 drafting of National Security Presidential Memorandum 13 to accompany the 2018 National Cyber Strategy. In discussing the new strategy, then-National Security Advisor John Bolton noted that the administration had also "repealed what is known as PPD-20, an Obama administration PD on offensive cyber operations," and replaced it with a new presidential directive that "effectively reversed those restraints." "We're not just on defense as we have been," stated Bolton.[51]

Congress also embraced this shift in the 2019 National Defense Authorization Act (NDAA). The 2019 NDAA includes new/clarified authorities enabling an increased tempo of USCYBERCOM's operations outside the DoD networks.[52] Under 10 U.S.C. § 394 (a) and (b), specifically, it is stated that the Secretary of Defense shall "develop, prepare, and coordinate; make ready all armed forces for purposes of; and, when appropriately authorized to do so, conduct, military cyber activities or operations in cyberspace, including clandestine military activities or operations in cyberspace, to defend the United States and its allies, including in response to malicious cyber activity carried out against the United States or a United

States person by a foreign power," and that Congress affirms that "the activities or operations referred to in subsection (a), when appropriately authorized, *include the conduct of military activities or operations in cyberspace short of hostilities* . . . or in areas in which hostilities are not occurring."[53] These subsections are highlighted because such language is a necessary condition for cyber operations in the competitive space short of armed conflict. As such, it supports execution of the 2018 DoD cyber strategy guidance calling for the United States to "assertively defend our interests in cyberspace below the level of armed conflict."[54]

As noted above, when appropriately authorized to do so, the law also states that the US Secretary of Defense shall conduct clandestine military activities or operations in cyberspace. 10 U.S.C. § 394 (c) further clarifies that "[a] clandestine military activity or operation in cyberspace shall be considered a traditional military activity."[55] This designation ensures that military cyber operations do not trigger the statutory covert action framework even when conducted on a deniable basis, thereby supporting cyber actions at the speed of relevance that cyber persistence theory argues is essential for managing the cyber strategic environment.[56]

Finally, the DoD general counsel's March 2020 remarks noting that "many States' public silence in the face of countless publicly known cyber intrusions into foreign networks precludes a conclusion that States have coalesced around a common view that there is an international prohibition against all such operations" marked an interpretation of international law that aligned with the defend forward/persistent engagement strategic approach.[57]

The 2018 DoD cyber strategy and USCYBERCOM *Command Vision* represent a strategic shift in how the US government uses its cyber capabilities to secure national interests in and through cyberspace and, as such, answers the 2018 *National Cyber Strategy's* call for a new policy and operational concepts.

It is no coincidence that in October 2019, General Paul Nakasone, as dual-hatted Commander of USCYBERCOM and Director of NSA/CSS, stood up the National Security Agency Cybersecurity Directorate to "prevent and eradicate cyber threats in national security systems and critical infrastructure" with an initial focus on the companies that build and maintain defense and weapons infrastructure and accompanying capabilities.[58] The largest agency in the US intelligence community shows signs of persistence in seizing the initiative to set the conditions of security in and through cyberspace.

These changes, however, represent a departure from DOS, DOJ, and USDT approaches that continue to vigorously pursue activities aligned to a deterrence policy framework, despite the fact that a failure of deterrence policy is a primary source of ongoing US policymaker frustrations.[59]

DOS, DOJ, USDT

The US DOS, still absent its own cyber strategy, successfully advocated for the inclusion in the 2018 *National Cyber Strategy* of a "Cyber Deterrence Initiative (CDI) to build such a coalition [of like-minded States] and develop tailored strategies to ensure adversaries understand the consequences of their malicious cyber behavior."[60] In this initiative, the *Strategy* declares the United States "will work with like-minded states to coordinate and support each other's responses to significant malicious cyber incidents, including through intelligence sharing, buttressing of attribution claims, public statements of support for responsive actions taken, and joint imposition of consequences against malign actors."[61] Chris Ford, US Assistant Secretary of State for International Security and Nonproliferation, in speaking about the CDI noted, "[W]e have worked within the U.S. government and with international partners to build a shared capacity to swiftly impose consequences when our adversaries transgress this framework [of responsible behavior]. Working with interagency colleagues, we have developed policies, processes, and response options that allow us to act quickly."[62] It is unclear what those response options are, but their intent is clear: deterrence. From the DOS perspective, options would include acts of retorsion expressed as, for example, withdrawing or expelling diplomatic missions or staff, a formal written démarche, limiting high-level visits, or "attribution diplomacy."[63]

Ford argues that "attribution diplomacy" is a critical part of the CDI, a line of effort consistent with the 2018 *National Cyber Strategy*'s claim that deterrence "includes diplomatic, information, military (both kinetic and cyber), financial, intelligence, *public attribution*, and law enforcement capabilities."[64] He asserts, "Our policy and our actions are clear in this respect, and they contribute both to reinforcing norms of responsible behavior and to deterring irresponsible actions. We will 'name and shame' foreign adversaries who conduct disruptive, destabilizing, or otherwise malicious cyber activity against the United States or our partners. And we do."[65]

US DOJ indictments of foreign cyber operators picked up pace in this period (law-enforcement based deterrence), with ten unsealed indictments in 2018 alone. For example, in September 2018, an individual associated with the North Korean regime was charged with malicious activities including the creation of the malware used in the 2017 WannaCry 2.0 global ransomware attack; the 2016 theft of $81 million from Bangladesh Bank; the 2014 attack on Sony Pictures Entertainment (SPE); and numerous other attacks or intrusions on the entertainment, financial services, defense, technology, and virtual currency industries, academia, and electric utilities.[66] The most publicized unsealed indictment, perhaps, was of twelve Russians associated with the Russian Main

Intelligence Directorate (GRU) who were accused of interference in the 2016 US presidential election.[67] Shortly after the release of the Russian indictment, the DOJ's Cyber Digital Task Force published its 2018 strategy. In announcing the publication, Deputy Attorney General Rod Rosenstein argued that public indictments achieve deterrence.[68] The Task Force's strategy also notes that "the State Department often uses information from our investigations and criminal indictments in diplomatic efforts to attribute malign conduct to foreign adversaries to build consensus with other nations to condemn such activities, and to build coalitions to counter such activities"[69]

Attributions behind DOJ indictments are also used to support USDT actions. Following the aforementioned indictment of twelve Russian GRU officers, the Secretary of the Treasury imposed financial sanctions against those individuals under an executive order authorizing sanctions for malicious cyber-enabled activity targeting US elections.[70] The USDT's actions blocked all property and interests in property of the designated persons subject to US jurisdiction and prohibited US persons from engaging in transactions with the sanctioned individuals. Similarly, following the aforementioned September 2018 indictment of an individual associated with the North Korean regime, the USDT announced sanctions for "significant activities undermining cybersecurity through the use of computer networks or systems against targets outside of North Korea on behalf of the Government of North Korea or the Workers' Party of Korea."[71]

When assessing the 2018–2021 period against the arguments and prescriptions of cyber persistence theory, we observe the following in official US documents and pronouncements:

- A recognition and acceptance (by DoD, the White House, and some members of Congress) of a structural imperative to operate persistently;
- A recognition and acceptance that a cyber *fait accompli* of armed-attack equivalence could occur and that cyber *faits accomplis* short of armed-attack equivalence are the primary behavior in cyberspace;
- A recognition and acceptance that States are able to achieve strategic outcomes through cyber competitive interaction short of armed-attack equivalence;
- A recognition and acceptance (by DoD) that a strategy of deterrence is ineffective against such campaigns;
- A recognition (from the 2016 DOS legal counsel and 2020 DoD general counsel comments on State practice and *opinio juris*), but no acceptance, that tacit, rather than explicit bargaining may be a more fruitful approach to constructing a binding regime of international norms for the cyber strategic environment.

Viewing US Cyber Strategy through the Lens of Kuhnian Paradigm Change

The US cyber strategy evolution covered by this chapter bears the hallmarks of a Kuhnian paradigm crisis and change moment.[72] Until 2018, the United States applied as its central approach to cyberspace a strategy of deterrence, defaulting toward restraint in the face of provocations. Deterrence had become a synonym for security in cyberspace and thus came to mean all things to all people. Rather than ask how to increase security in cyberspace, policymakers and their congressional overseers asked how to deter in cyberspace—assuming that key features of the physical domains supporting a deterrence strategy were also present in the virtual domain.[73]

This framing, which presumed the answer, became commonplace in US government open discussions and during congressional testimony.[74] Deterrence was going to stop everything: strategic cyber war, ISIS's virtual caliphate, the theft of intellectual property, as well as attacks against US government systems and critical infrastructure. Every approach to security was either called deterrence or framed in the deterrence language of imposing costs and denying benefits. How did this come to be?

Several factors contributed to the reflexive reliance on deterrence theory and the iron grip that strategies of coercion hold on US policymakers and strategic thinkers. For the DoD, it reflected a legitimate preoccupation with the consequences of cyber capabilities for war. Prevailing in war is the military's preeminent no-fail mission, and cyberspace was considered strategically significant if it had bearing on armed conflict—either winning the conflict or deterring it. By the 1980s, cyberspace was already a military concern and viewed through the lens of war.[75] National Security Decision Directive (NSDD)-145 in 1984 identified national security risks from converging telecommunications and automated information systems. It anticipated that national security data could be not only exploited by foreign adversaries but also corrupted or destroyed, with strategic implications.[76]

After the swift victory of US-led coalition forces in the 1991 Persian Gulf War, cyberspace quickly became intertwined with the concept of information war and integrated into joint warfighting doctrine. The battlefield now extended beyond geographic terrain to the electromagnetic spectrum. Prominent defense intellectuals John Arquilla and David Ronfeldt warned that "Cyberwar Is Coming!"[77] In 1996, joint doctrine subsumed computer network attack under information operations.[78] The integration of information operations—including computer network defense and attack—into joint warfighting doctrine was bolstered by military conflicts in the Balkans (1999) and Iraq (2003).

By the time the precursor organization to USCYBERCOM—Joint Task Force-Computer Network Operations (JTF-CNO)—was stood up in 2002 and then transferred to Joint Functional Component Command for Network Warfare (JFCC-NW) in 2005, cyber operations were comfortably ensconced in the strategic space of armed conflict. With cyberspace formally recognized as a domain of warfare in 2010, as critical to military operations as land, sea, air, and space,[79] it was but a short step to apply the strategic approach of deterrence to cyberspace.

The "deterrence default" was reinforced by a national security enterprise dominated for nearly two generations by deterrence thinking. The vast majority of contemporary national security practitioners and senior academics were schooled during or immediately after the golden years of Cold War scholarship, which produced a trove of classics focused on coercion theory.[80] Unsurprisingly, fixation on coercion, militarized crisis, and war in cyberspace led to a "high-and-right" bias in the cyber literature. For over two decades, practitioners and academics have been debating if, when, and how cyberwar will occur.[81]

Deterrence theory, thus, had attained "paradigm" status. During the Cold War, it was firmly ensconced as the logic that drove the strategies of the superpowers (whether it did in reality or not), and anyone who wanted to advocate for different policy had to address the theory's core logic.[82] Unsurprisingly, in the first decades of the twenty-first century, the United States and other Western democracies assumed cyberspace was a deterrence strategic environment and that prospective response and operational restraint would produce positive norms and stability.

The reality of State behavior and interaction in cyberspace, however, proved to be quite different from the model of war, catastrophic attack, and coercion underlying US cyber strategy and policy. Most adversary State-sponsored cyber activity occurred outside armed conflict and was not primarily coercive in application.[83] Despite all evidence to the contrary, practitioners and scholars succumbed to paradigm lock—the inability to change from a deterrence mindset, instead tinkering with the paradigm to explain away cyber anomalies.[84] These are hallmarks of what Kuhn called a "paradigm crisis."[85]

As friction between reality and theory became more pronounced with adversaries making strategic gains in and through cyberspace outside of armed conflict, the first condition for paradigm change—misalignment between theory and observed reality—was in place. But as Kuhn notes, this is a necessary, but not sufficient condition for paradigm change.

> The decision to reject one paradigm is always simultaneously the decision to accept another, and the judgement leading to that decision

> involves comparison of both paradigms with nature *and* with each other.[86]

By 2018, an alternative way of thinking emerged in the logic of cyber persistence. The validity of cyber persistence theory's core arguments and strategic prescriptions was recognized and accepted by key US policymakers, leading to a new DoD strategic approach of defend forward/persistent engagement and a host of supporting actions by the White House and Congress.

The strategic principle of seizing the initiative to set and maintain the conditions of security in and through cyberspace also took hold beyond the Pentagon in a few places, albeit in an ad hoc manner and within the wider context of the strategic tit-for-tat reaction dynamic characterizing US-China relations since 2018.[87] The June 2018 US declaration of increased tariffs to halt Chinese "unfair trade practices" was, in part, a reaction to China's cyber-enabled IP theft.[88] However, other US actions reflect initiative persistence. For example, after the defeat of its nominee to China's preferred candidate for leading the United Nation's (UN) Food and Agriculture Organization, the DOS seized the diplomatic initiative in early 2020 and launched an assertive diplomatic campaign to defeat China's nominee to lead the UN's World Intellectual Property Organization.[89] In so doing, the DOS denied China the opportunity to legitimize in a global forum its practice of cyber-enabled IP theft.

Additionally, recognizing that China could potentially accrue strategic advantage if its 5G equipment supported US 5G infrastructure, the 2019 National Defense Authorization Act prohibited US government use of Huawei, ZTE, Xiaomi, and other untrusted vendors from China. In April 2020, the DOS announced the associated program of "Clean Path" 5G to ensure DOS networks and supply chains remain free of dependencies on untrusted foreign vendors like Huawei and ZTE.[90] Later that year, the DOS announced the Global Clean Network program, which rejects Chinese government affiliated vendors and sets the requirements for trusted vendors.[91] To inhibit Huawei's 5G infrastructure component production capacity, in May and August 2020, the US Department of Commerce (DOC) announced restrictions on Huawei's ability to acquire electronic components developed using US technology.[92] In September 2020, the DOC also announced it was prohibiting WeChat and TikTok transactions based on a national security concern that China would harvest personally identifiable information (PII) of the users of those applications.[93]

That said, deterrence theory has displayed extraordinary staying power and Kuhn's framework suggests it would be premature to write its epitaph. Cyber deterrence is hardly a straw man or fringe idea in either policy or academic circles. To wit, the DOS persists in its argument that "attribution diplomacy" contributes both to reinforcing norms of responsible behavior and to deterring irresponsible actions, even though this view is not supported by empirical evidence. Florian Egloff's study of the effects of public attribution on future State cyber behavior shows that States' concerns about public attribution typically lead to "a more careful tailoring of their offensive operations." For States still building offensive capabilities, Egloff surmises the effect will be efforts to "adapt their policies and procedures to prevent indiscriminately delivering effects."[94] Neither of these "consequences" is indicative of deterrence or of reinforcing norms, unless adversary efficiency is considered a desirable norm. Of course, it may well be the case that some States are not at all concerned with public attribution. When discussing the issue of Russian cyber operations seeking to influence the US presidential election, for example, President Obama stated, "[T]he idea that somehow public shaming is going to be effective, I think doesn't read the thought process in Russia very well."[95]

Likewise, the DOJ's claim that public indictments achieve deterrence is equally tenuous. Garret Hinck and Tim Maurer show that "Based on the existing record, bringing criminal charges against foreign hackers and online influence operators does not appear to impose enough costs on adversaries to convince them to cease from further malicious activity."[96] Both Jack Goldsmith and Peter Machtiger further argue that after five years of high-profile indictments of foreign cyber operators, there is little evidence to support Rosenstein's argument that DOJ indictments "stop or even slow these activities."[97] US Attorney David Hickton, on the other hand, considers the law enforcement effort against foreign cyber operators to be a long-term one, likening it to indictments issued in Florida against South American drug kingpins during the height of the drug war. Then, as now, skeptics wondered what the point was of bringing cases against individuals who seemed all but certainly beyond the reach of US law enforcement. Today, Hickton says, US prisons are filled with drug traffickers. Left unsaid, however, to Goldsmith's and Machtiger's point, is that drugs continue to flow across the border in significant volume.[98]

The Cyber Digital Task Force's note that DOS often uses information from their investigations and criminal indictments to build coalitions to counter malicious cyber activities implies that the indictments support deterrence. But this circles back to the fragile "attribution supports deterrence" claim.[99] Moreover, there is no guarantee that US allies will join in a chorus of public attribution as then-Assistant Attorney General John Demers learned after a July 7, 2020, indictment charging cyber operators supported and directed by the Chinese Ministry of State Security with cyber-enabled intellectual property.[100] The indictment specifically mentions twelve companies affected that were located in Australia, Belgium, Germany, Lithuania, the Netherlands, South Korea, Sweden and the UK, a clear case of an international campaign. Demers noted in the DOJ's July 21 press conference that the case was "another example of how like-minded countries can stand together to counter malicious state-sponsored cyber activities," and, he said, "[W]e are appreciative of the statements that are going to be made by several of these countries in the coming hours."[101] Yet few of the promised statements actually arrived from the countries listed in the indictment, and statements from the handful of countries that did weigh in were noticeably soft toward China.[102]

Attributions behind DOJ indictments are also used to support the application of USDT sanctions; however, there is scant evidence that the punitive impact caused by sanctions against these individuals will deter State cyber behavior.[103] Would there be a higher likelihood of deterrence if sanctions targeted groups instead of individuals? The USDT has, in fact, enacted group sanctions. In September 2019, Sigal Mandelker, Under Secretary of the Treasury for Terrorism and Financial Intelligence, announced, "Treasury is taking action against North Korean hacking groups that have been perpetrating cyberattacks to support illicit weapon and missile programs."[104] To extend this line of thinking to its logical end, might the likelihood increase even further if States themselves were the targets of USDT sanctions?[105] When these questions are examined in light of evidence, the answer is consistently no. Russia, North Korea, and Iran, all subject to significant USDT and other international sanctions for cyber and non-cyber transgressions, have shown no consequent inclination to curb their ongoing strategic cyber campaigns.

In fact, our earlier discussion of North Korea (Chapter 4) makes clear that the regime leans heavily on cyber operations to circumvent the impact of sanctions. North Korea demonstrates that the cyber strategic environment rewards operational persistence, not operational restraint, and so sanctioned States do not sit idly by as time passes. Consequently, the cumulative effects of sanctions simply do not, and arguably cannot, offset the cumulative gains adversaries accrue through continuous cyber operations short of armed conflict. Indeed, the trajectories of the two over time are mirror images. With the strategic value of imposed costs diminishing over time and the strategic value of cumulative gains increasing, a prohibitive cost threshold that might deter future behavior will never be reached.[106]

In sum, policymaker arguments that these DOS, DOJ, and USDT activities "achieve" deterrence or deter "irresponsible actions" find no support when subject to rigorous analysis *whether considered in isolation or in combination.*[107] Some policymakers admit as much. Then-White House National Security Advisor Robert C. O'Brien reflected in August 2020, "We've sanctioned the heck out of the Russians—individuals, companies, the government. . . . We've kicked out literally scores of spies. We've closed down all of their consulates on the West Coast. We closed down diplomatic facilities. There's not a lot left we can do with the Russians." In October 2020, he further noted, "One of the problems we have with both Iran and Russia is that we have so many sanctions on those countries right now that there's very little left for us to do."[108] This is not to dismiss the important roles and capabilities these departments could bring to a national cyber strategy better aligned with the cyber strategic environment than is a strategy of deterrence; rather, it is to suggest they have been suboptimized in support of deterrence.

Early evidence from the Biden administration reveals that deterrence may, again, play a central role in US national cyber strategy.[109] The administration promotes "integrated deterrence" as the DoD strategy, although its meaning has not been fully fleshed out.[110] The role of cyber mission forces in contingency operations and warfighting once again is a focus of discussion in the DoD, even as the White House remains equally concerned with cyber's strategic impact outside of armed conflict (e.g., ransomware, SolarWinds, Microsoft Exchange, Colonial Pipeline).[111] Support for a tacit bargaining approach to complement explicit bargaining efforts to construct a more granular set of mutual understandings of acceptable and unacceptable cyber behaviors, while publicly recognized by a few as a fruitful if not inevitable path forward, has not been widely embraced by policymakers. The DOS continues to rely on explicit bargaining, defining redlines, and promoting the Cyber Deterrence Initiative.[112]

Even those who applaud reported successes of defend forward/persistent engagement operations continue to call for deterrence as the central cyberspace strategy and describe these successes in deterrence terms.[113] This is not merely a matter of semantics. Paradigms provide basic assumptions, key concepts, mechanisms, and methodologies. Evoking deterrence implies a set of assumptions, concepts, mechanisms, and dynamics that are wholly distinct from those comprising cyber persistence theory. Strategic vocabulary should be consistent with strategic action. In a national security context, disconnects stall the acceptance of a new paradigm, which leads to a mis- or underutilization of department and agency authorities and capabilities, contradictory strategic communications, and an inconsistent and noncohesive national strategy.

All of the above supports an argument that it is premature to declare a paradigm change. The US shift from a central reliance on deterrence to a more comprehensive adoption of cyber persistence is still underway, with the groundwork for a paradigm change firmly in place. Kuhn's observations serve as a cautionary note for anyone expecting new paradigms to be accepted quickly and evenly. We expect no less with cyber persistence theory. Existing paradigms are sticky. Some States or entities within States will adopt new precepts more readily than will others. Consistent with Kuhn's observations, outside the DoD, a few US departments and agencies have begun to adopt the new paradigm's strategic prescriptions in an ad hoc manner. However, deterrence precepts remain influential because of policymakers' familiarity and comfort with deterrence thinking, despite overwhelming evidence that the approach is ineffective against adversaries' primary (strategic) cyber behaviors, which occur below the level of armed conflict. In this regard, we are reminded of the comment by military historian Liddell Hart, "There are over two thousand years of experience to tell us that the only thing harder than getting a new idea into the military mind is to get an old one out."[114]

Actions to Further Institutionalize a Kuhnian Change

US policymakers can go much further to align policy and strategy to the logic of the cyberspace strategic environment.[115] The potential contributions of US DOS and DOJ efforts to seize and maintain initiative have, until now (end of 2021), been underutilized because of an emphasis on setting redlines, signaling, and imposing costs. DOS's singular focus on explicit bargaining and the Cyber Deterrence Initiative to establish and reinforce norms of responsible behavior also neglects opportunities to expand tacit bargaining, as well as lay the groundwork for defend forward/persistent engagement at scale.

By altering the framing of public attributions, DOS (and like-minded States) could move beyond a reliance on voluntary, *non-binding* norms to a *binding* regime of customary international law. Declaring the "irresponsible behaviors" being called out as internationally wrongful acts in violation of a principle or rule of international law would begin to amass evidence of State practice necessary for the establishment of new rules of customary international law for cyberspace.[116]

The international public attribution of Russia as the State actor behind the large-scale cyberattack against Georgia on October 28, 2019, was a lost opportunity toward this end.[117] In late February 2020, Georgia, the UK, the United States, the European Union, and many other States, in a series of generally coordinated statements, publicly attributed the cyberattack to the Russian GRU. The attributions are most notable for what they omit: a clear reference to a rule of international law the attributed cyber conduct allegedly breached. The US statement called out Russia's irresponsible, reckless, and destabilizing activities. The UK and the Netherlands decried Russia's reckless, destabilizing, disruptive behavior aimed at undermining Georgian sovereignty. New Zealand condemned Russian attempts to interfere in Georgia's political and economic freedom. Australia, Canada, and Latvia also reference international law *without, however, pointing to a specific obligation.*[118] Only Georgia comes close to invoking international legal parlance when describing the cyberattack as "infringing

Georgia's sovereignty."[119] By not specifying in legal parlance their views of which international rules the October 2018 Russian cyberattack actually violated and how it did so, the United States and its allies missed an opportunity to begin constructing new rules of customary international law.[120]

The statement by US Secretary of State Mike Pompeo affirmed that the GRU action "contradicts Russia's attempts to claim it is a responsible actor in cyberspace and demonstrates a continuing pattern of reckless Russian GRU cyber operations against a number of countries. These operations aim to sow division, create insecurity, and undermine democratic institutions."[121] The UK's statement noted, "These cyber-attacks are part of Russia's long-running campaign of hostile and destabilising activity against Georgia" and that "[t]he UK is clear that the GRU conducted these cyber-attacks in an attempt to undermine Georgia's sovereignty, to sow discord and disrupt the lives of ordinary Georgian people." Foreign Secretary Dominic Raab called the "GRU's reckless and brazen campaign of cyber-attacks against Georgia, a sovereign and independent nation . . . totally unacceptable."[122] The Netherlands declared, "The GRU undermined Georgia's sovereignty and disrupted the lives of ordinary Georgian people."[123] New Zealand asserted that "These malicious cyber activities . . . were designed to interfere in Georgia's political and economic freedom. Activities which seek to undermine democratic processes are unacceptable."[124]

In other ways, the United States—in an ad hoc manner—is laying the groundwork for a tacit bargaining approach. In 2019, Secretary of Defense Mark Esper declared that "Moving forward, I consider election security an enduring mission for the Department of Defense."[125] Similarly, Lt. Gen. Charles Moore, Deputy Commander USCYBERCOM, declared in November 2020 that "We are not stopping or thinking about our operations slacking off on Nov. 3 [2020]. . . . Defending the elections is now a persistent and ongoing campaign for Cyber Command."[126] Coupled with the DoD general counsel's March 2020 remarks, these statements could be construed as a declaration that elections are a component of the United States' *domaine réservé*.[127]

The hard work of identifying and communicating to other States the cyber manifestations of their *domaine réservé* remains, but this represents a step toward defining the ground rules and guardrails for cyber strategic competition rather than establishing redlines for a deterrence framework that is untenable. The US DOS and the foreign ministries of US allies can use their considerable diplomatic portfolios to start defining what acceptable competition is as much as proscribing what it is not; otherwise, their efforts will remain peripheral to the cyber *fait accompli* environment they seek to influence.

As with DOS public attributions, the framing of the US DOJ's unsealed public indictments only as violations of US domestic law to support law-enforcement-based deterrence also represents a lost opportunity to establish a binding regime of cyber international law.[128] For example, in the case *United States of America v. Internet Research Agency*, the unsealed public indictment detailing Russian interference in the US 2016 presidential election included the charges of conspiracy to defraud the United States, conspiracy to commit wire fraud and bank fraud, and aggravated identity theft.[129] Some international law scholars have argued that Russia's cyber behavior also violated the international law rule on non-intervention (into an election process) or the basic right of self-determination, a legal concept that captures the right of a people to decide, for themselves, both their political arrangements (at a systemic level) and their future destiny (at a more granular level of policy).[130] No such claims were made by agents of the US government, however.

Were adversary behaviors described in unsealed public indictments framed as internationally wrongful acts, the extraordinary detail in the indictments should make policymakers comfortable with pursuing countermeasures, if the behavior identified in the indictment is ongoing. Although international law does not prescribe a prerequisite evidentiary burden with respect to undertaking countermeasures,[131] a State is responsible for countermeasures that are later proved undertaken based on flawed or mistaken evidence.[132] Additionally, the accumulation of evidence within or across multiple indictments addressing the same State could serve as evidence in support of a call for anticipatory countermeasures if it identifies consistent patterns in, for example, opportunism, exploitation approach (vulnerability exploited, malware used, tactics, techniques, and procedures), intentions, and targets.

By describing State-sponsored cyber actions as internationally wrongful acts, unsealed public indictments could leverage tacit bargaining to help establish new rules of customary international law for cyberspace.[133] Martha Finnemore and Duncan Hollis contend that an accusation can serve as an opening bid in a bargaining process, aimed at a particular community, indicating not just the accuser's disapproval of the cited cyber behavior(s), but often, too, its proposal (perhaps implicit) that all such conduct should be barred (i.e., that there should be an international rule against such conduct).[134] Accusations could thus lay out the contours of "bad behavior" along with an argument about why, exactly, the behavior is undesirable.[135] Finnemore and Hollis continue: "Other States may then respond to the accusation. They may accept some of it; they may accept all of it; they may accept it in some situations but not others; or they may reject it entirely. It is these interactions between the accuser, the accused, and third-party audiences that—over time—could result in the creation of a new norm (or its failure)."[136]

In sum, viewing State cyber behaviors through the lens of deterrence strategy serves to blind US policymakers to opportunities in international law (i.e., countermeasures) to lawfully respond to adversary campaigns short of armed-attack equivalence (and to make arguments for anticipatory responses).[137] Absent a unified, concerted effort to consign the deterrence paradigm to a still important, but nonetheless secondary, role in cyber strategy, the shadow of deterrence will limit opportunities to best leverage the instruments of national power and international law to achieve security in the cyber strategic environment.[138] The US case is indicative of the inertia, but not permanence, of established thinking as well as the organizational rigidity that sits behind established roles and responsibilities across institutional arrangements.

Organizing to Seize the Initiative—From Whole-of-Government to Whole-of-Nation-Plus

The strategic principle of cyber persistence theory—persistence in seizing and maintaining the initiative to set the conditions of security in and through cyberspace in one's favor—should serve as the core strategic principle of any State's approach to the cyber strategic environment.[139] The challenge of cyber security is not merely a technical one. It is a political, economic, social, organizational, and behavioral challenge in a technically fluid environment. The consequences of interconnectedness and the condition of constant contact that flows from it require a framework that can position a State comprehensively.

The solution to interconnectedness cannot be greater segmentation. Cyber persistence theory would expect States to begin to adopt greater coordination of their societies combining three major elements with a fourth line of effort: intergovernmental coordination (sometimes referred to as whole-of-government approaches), alignment between the private sector and their government, and some engagement with their populace along with some international effort, minimally with allies, to align their intergovernmental, private sector, and citizenry efforts with those of other States at all three levels. The whole-of-nation-plus (WON+) thus stands in contrast to stand-alone whole-of-government efforts, separated public-private partnerships, and basic cyber awareness education approaches. WON+ recognizes the implications of interconnectedness and, as such, orients toward not leaving any flank open. In a strategic environment in which the weakest link is potentially consequential, all other approaches will fall short.

Between 2010 and 2021, a number of States began to move toward greater organizational coherence by shifting formerly disparate national components into more coordinated organizational forms. Examples include the UK's National Cyber Security Centre, Australia's National Cyber Security Centre, and Israel's Cyber Security Directorate.[140]

Although the last decade certainly witnessed an increasing amount of US government activity focused on securing and advancing US national interests in and through cyberspace, an uptick in activity in and of itself is not evidence of persistence in seizing and maintaining the initiative to set the conditions of security. Much of non-DoD US government strategic cyber activity remains reactive to adversary behaviors (i.e., responses to deterrence failure). The DOJ China Initiative was only stood up many years after evidence of China's cyber-enabled theft campaigns dating back to 2010 was revealed.[141] The DOS CDI, by intent, is a reactive approach to adversary cyber behavior. Still, there is evidence that the United States is moving toward initiative persistence and also bridging segmentation.

Whole of Government

In the US case, the strategic approach of defend forward/persistent engagement recognizes that security relies on reducing or eliminating seams between external- and internal-facing cyber security operations and activities, the departments and agencies responsible for the same, and the relationships those departments and agencies have with the private sector and international partners. It provides the strategic rationale for operating with less bureaucratic and legal segmentation.

Persistent engagement's operational concepts of anticipatory resilience and defend forward address these issues. Although USCYBERCOM can, and has, contributed independently to the US government's internal-facing defensive measures by posting on the VirusTotal website unclassified malware discovered through "defend forward" operations or campaigns, there are statutory constraints limiting USCYBERCOMs internal-facing contributions.[142] Anticipatory resilience acknowledges those constraints and seeks to overcome them through, as General Nakasone noted, "building relationships with U.S. institutions that are likely to be targets of foreign hacking campaigns . . . before crises develop, replacing transactional relationships with continuous operational collaboration among other departments, agencies, and the private sector."[143]

The 2018 establishment of the Russia Small Group marks an important event contributing to the maturation of the Command's operational approach. The Group—with members from NSA and USCYBERCOM—worked with the DOJ, the Intelligence Community, and the Department of Homeland Security (DHS) to eliminate seams in a coordinated effort to secure the US 2018 midterm elections from foreign interference.[144] In the course of doing so, Brig. Gen. Timothy Haugh, the commander of Cyber Command's cyber national mission force at that time, noted, "We had to build new relationships, whether that was with DHS, whether that was with FBI, or our teammates in the Department of State. . . . All of those we had to build together in a really short timeframe."[145]

Those efforts continue to pay dividends and have expanded to include additional domestic partners, including the USDT. In a September 2019 announcement of sanctions targeting three North Korean State-sponsored malicious cyber groups, USDT's Office of Foreign Assets Control (OFAC) highlighted that DHS's Cybersecurity and Infrastructure Security Agency (CISA) and USCYBERCOM "have in recent months worked in tandem to disclose malware samples to the private cybersecurity industry, several of which were later attributed to North Korean cyber actors, as part of an ongoing effort to protect the U.S. financial system and other critical infrastructure."[146] "This, along with today's OFAC action," the announcement noted, "is an example of a government-wide approach to defending and protecting against an increasing North Korean cyber threat and is one more step in the persistent engagement vision set forth by USCYBERCOM."[147]

Similar coordination supported a 2020 filing of a civil forfeiture complaint in a US federal court, identifying 280 accounts that "North Korean actors" allegedly used to steal funds in two cyber operations with proceeds ultimately flowing through Chinese cryptocurrency laundering networks. "Department of Defense cyber operations do not occur in isolation," Brig. Gen. Joe Hartman, commander of the USCYBERCOM's Cyber National Mission Force (CNMF), said in a statement announcing the law enforcement filing with the FBI, the Internal Revenue Service, and officials from the DHS and DOJ. "Persistent engagement includes acting through cyber-enabled operations as much as it does sharing information with our interagency partners to do the same."[148]

Beginning in July 2019, numerous joint advisories and alerts have been published by the US government, indicating the continued success and value of this coordination in supporting national defense/resilience.[149] In early July 2019, CISA and USCYBERCOM coordinated on a limited-disclosure advisory with USCYBERCOM posting on Twitter that a threat group was actively using a Microsoft Outlook vulnerability previously leveraged by an Iran-linked malware campaign, and CISA sharing an associated TLP: Amber-designated advisory with industry.[150] On February 14, 2020, in an effort to "enable network defense and reduce exposure to North Korean government malicious cyber activity," six "unlimited disclosure" Malware Analysis Reports (MARs) coauthored by CISA, the FBI, and USCYBERCOM's CNMF identified Trojan malware variants used by the North Korean government.[151] On July 23, 2020, CISA and NSA issued a public joint advisory warning that "over recent months, cyber actors have demonstrated their continued willingness to conduct malicious cyber activity against critical infrastructure (CI) by exploiting internet-accessible operational technology (OT) assets."[152] On October 27, 2020, a

TLP: White-designated advisory coauthored by CISA, the FBI, and CNMF described the tactics, techniques, and procedures used by a North Korean APT group.[153] Finally, on October 29, 2020, CISA, the FBI, and CNMF coauthored a TLP: White-designated MARs of Zebrocy, malware associated with Russia's Turla APT group.[154]

USCYBERCOM also supports DoD's Energy Sector Pathfinder initiative with the DHS and the US Department of Energy to "advance information sharing, enhance training and education to understand risks and develop joint operational preparedness and response activities to cybersecurity threats."[155]

Since 2018, the interagency Committee on Foreign Investment in the US (CFIUS) has also been extraordinarily active compared to years prior.[156] Led by the Secretary of the Treasury, with other members from the State, Defense, Justice, Commerce, Energy, and Homeland Security departments, CFIUS is charged with reviewing acquisitions of, and some investments in, American businesses by foreign buyers to determine if the deals pose risks to national security.[157] Since 2018, CFIUS findings disrupted Broadcom Ltd.'s proposed $117 billion takeover of Qualcomm Inc., arguing the deal could curtail US investments in chip and wireless technologies, and could handleadership to Huawei Technologies Company; disrupted the planned acquisition of MoneyGram International Inc. by Ant Financial, a Chinese financial-services form, due to concerns of Chinese access to PII; and reached an agreement with the Chinese company Beijing Kunlun Tech Company to unwind its purchase of Grindr and to refrain from accessing information about Grindr's users.[158]

A further step toward reducing seams across the US government interagency and with the private sector is the August 6, 2021, announcement by CISA to launch a Joint Cyber Defense Collaborative (JCDC). This agency effort will focus on developing whole-of-nation cyber defense plans and coordinating the execution of defensive cyber operations across the interagency, private sector, and State, local, tribal, territorial governments. CISA Director Jen Easterly described the JCDC as "a unique planning capability to be proactive vice reactive in our collective approach to dealing with the most serious cyber threats to our nation."[159]

Public-Private Alignment

There are a number of US government–private sector collaboration and coordination efforts indicative of a whole-of-nation approach to persistence in seizing the initiative to set and maintain the conditions of security in and through cyberspace.[160,161]

In response to the potential threat posed by VPNFilter (discussed in Chapter 5), Cisco Talos was sharing information about the malware with the FBI, and on May 23, 2018, three things happened simultaneously: the FBI seized infected Web domains it suspected the Russian hackers would exploit, Cisco published its findings in a blog post, and all members of the Cyber Threat Alliance (a nonprofit forum) were sent simultaneous urgent notices describing how to protect against the Russian attack.[162]

In September 2019, representatives from Facebook, Google, Microsoft, and Twitter met US government officials from the FBI, the DHS, and the Office of the Director of National Intelligence to discuss their preparations for the presidential election. "Participants discussed their respective work, explored potential threats, and identified further steps to improve planning and coordination," said Facebook's head of cybersecurity policy Nathaniel Gleicher.[163] Similarly, Richard Salgado, Google's director of law enforcement and information security stated, "We will continue to monitor our platforms while sharing relevant information with law enforcement and industry peers. . . . It is crucial that industry, law enforcement and others collaborate to prevent any threats to the integrity of our elections."[164]

The benefit of this collaboration manifested when the FBI seized ninety-two domain names that were unlawfully used by Iran's Islamic Revolutionary Guard Corps (IRGC) to engage in a global disinformation campaign centered on the US 2020 election. The investigation supporting the seizure was initiated by intelligence the FBI received from Google, and "was a collaborative effort between the FBI and social media companies Google, Facebook, and Twitter," said FBI Special Agent in Charge Bennett.[165] Similarly, the October 2020, Trickbot infrastructure disruption reviewed in Chapter 4 is a prominently reported example, with USCYBERCOM and Microsoft Corporation both disrupting Trickbot's infrastructure in an effort to secure the 2020 US presidential election.

In June 2021, the NSA stood up the Cybersecurity Collaboration Center for engagement with the private sector. The partnerships aimed to help NSA to prevent and eradicate foreign cyber threats to National Security Systems (NSS), the DoD, and the Defense Industrial Base (DIB) by creating an environment for information sharing and to combine respective expertise, techniques, and capabilities to secure the United States' most critical networks.[166]

Finally, and perhaps most extraordinarily, is the case of the indiscriminate compromise of tens of thousands Microsoft Exchange servers by the Chinese APT referred to as Hafnium.[167] Following a March 2, 2021, report from Microsoft describing the compromise,[168] Microsoft released out-of-cycle mitigations and the FBI and CISA launched an aggressive information campaign to motivate victims to remove the exploit and patch the zero-day vulnerabilities it targeted. Additionally, spurred by concerns that many victims were either unaware of or unable to remove the exploit that "could have been used to maintain and escalate persistent, unauthorized access to U.S. networks," the FBI proactively removed the exploit from "hundreds" of on-premise Microsoft Exchange servers owned by US private organizations.[169] This is a promising example of enhanced operational tempo and public-private alignment that can better meet the demands of security in the cyber strategic environment.

Engaged Citizenry

The engagement of the public has traditionally begun with the basic need for cyber awareness. For decades, most US citizens remained unaware of the threats to them at the individual level. Many Western States' national approaches focused on education and awareness of basic cyber hygiene[170] under a model of shared responsibility.[171] This framed the individual security problem in the context of private interest—government would take care of large-scale attacks, while citizens could protect their identity, personal assets, and privacy through better passwords and awareness of phishing links.[172] A shift toward a more active, anticipatory, engaged citizenry is what cyber persistence theory would expect. In the case of the United States, such a shift is emerging. In 2020, the National Cybersecurity Alliance's cyber education framing, while not fully one of civic duty, couched education efforts in the context of interconnectedness and the need for cumulative action, linking the individual level to national security calling on all to "do your part."[173]

There are also examples of non-governmental initiatives aimed at providing security as a public good.[174] In 2020, the Carnegie Endowment for International Peace published its report, *International Strategy to Better Protect the Financial System Against Cyber Threats*, offering a vision and recommendations for the international community to protect the global financial system from cyber threats.[175] The HoneyNet Project is a nonprofit security research organization with a mission to "to learn the tools, tactics and motives involved in computer and network attacks, and share the lessons learned."[176] The nonprofit StopBadware's mission is to protect the public from malware-distributing or malware-infected websites.[177] The aforementioned nonprofit Cyber Threat Alliance "is working to improve the cybersecurity of our global digital ecosystem by enabling near real-time, high-quality cyber threat information sharing among companies and organizations in the cybersecurity field."[178]

In 2016, the CEOs of eight banks—Bank of America, BNY Mellon, Citigroup, Goldman Sachs, JPMorgan Chase, Morgan Stanley, State Street, and Wells Fargo—came together in the Financial Systemic Analysis and Resilience Center (FSARC) to proactively identify ways to enhance the resilience of the critical infrastructure underpinning much of the US financial system. The FSARC's mission was to proactively identify, analyze, assess, and coordinate activities to mitigate systemic risk to the US financial system from current and emerging cyber security threats through focused operations and enhanced collaboration between participating firms, industry partners, and the US government.[179] In January 2021, the nonprofit Institute for Security and Technology formed the Ransomware Task Force, comprising thirty-two formal members charged with developing a comprehensive framework of actionable solutions to significantly mitigate the ransomware threat.[180]

Adding the Plus (+)

A whole-of-nation framework is a significant undertaking for any State. If one succeeded alone in an interconnected world, a lack of effective cyber security in other countries would keep the potential for exploitation open on one's flank. Thus, the undertaking of expanding WON to allies, while substantial, is required to secure the cyber strategic environment. Aligning the intergovernmental, private sector, and citizens of allies and similarly concerned States must inform national cyber policies and strategies.

There is evidence of the United States seizing the initiative by working by, with, and through allies and partners—a WON+ approach. The US DOS encourages allies and partners to limit their deployments of Huawei 5G networking platforms. DHS's CISA is seeking to extend its reach internationally by establishing information-sharing relationships with foreign critical infrastructure owners.[181] The US government is accepting invitations from allies and partners such as Montenegro, North Macedonia, Ukraine, and Estonia for USCYBERCOM to deploy "hunt forward" teams to operate side by side with these partners to secure certain designated networks in those countries.[182]

In sum, while not driven by US national cyber strategic guidance to persist in seizing and maintaining the initiative to set and maintain the conditions of security in and through cyberspace using a whole-of-government, whole-of-nation, or WON+ framework, since 2018 (and in some cases before) a number of US government policies and private sector and/or nonprofit initiatives, some independent of and some collaborative or coordinated with the US government, are consistent with this core strategic prescription of cyber persistence theory.

US Choices

From 2010 to 2021, US national cyber strategy centered on a strategy of deterrence, even as policymakers and cyber practitioners repeatedly expressed frustration with the results of that strategy. In 2018, a new strategy aligning with cyber persistence theory's structural imperative and strategic prescriptions (a new paradigm) emerged from DoD and USCYBERCOM: defend forward/persistent engagement. Although White House and congressional actions tacitly supported an acceptance of this new paradigm, its core strategic principle has not yet been deliberately and evenly adopted across the whole of government.

This is consistent with Kuhn's observations that a shift is more akin to a conversion experience—a product of persuasion rather than proof.[183] It gradually spreads across a community, beginning with a few scientists who sense that their new paradigm is on the right track.[184] It is these initial supporters "who will develop it to the point where hardheaded arguments can be produced and multiplied," resulting in not a single group conversion, but rather in "an increasing shift in the distribution of professional allegiances."[185]

Absent a US government-wide conversion, it should be expected that the legacy of deterrence as the central US security strategy will continue to cast a long shadow, limiting opportunities to better protect and advance US national interests in and through cyberspace and to promote cyber stability. Cyber persistence theory would expect that such misalignment with the logic of the cyber strategic environment will have negative consequences for the security of any State that does not adjust to the initiative persistence required to ensure security.

To avoid such an outcome, the prescriptions of cyber persistence theory suggest that the United States needs to anchor its national cyber strategy on the core strategic principle of persistence in seizing the initiative to set and maintain the conditions of security in and through cyberspace. Key departments and agencies should be tasked with drafting and coordinating cyber strategies grounded in the same core strategic principle under a WON+ framework.[186] Broadly institutionalizing the principle through coordinated strategic guidance will lay the groundwork for a more effective strategic approach better aligned to the security demands of the cyber strategic environment.

Conclusion

We began this book with the observation that misapplication of theory cannot only lead to misunderstanding of behavior and dynamics, but can also produce ineffective policy prescription. In international security relations, a misalignment between strategic approach and strategic environment does not simply position States to sub-optimize their outcomes—it creates the potential for catastrophic failure. Deterrence is essential for stability in the nuclear strategic environment. It and other coercive strategies can advance interests in the conventional strategic environment. The extension of deterrence as a principal anchor into a strategic realm not of the same logic, however, will negatively impact States that adopt such postures. The cyber strategic environment requires its own theory to capture its own distinctive logic of exploitation. Whereas security requires States to triumph in war in the conventional environment and avoid war in the nuclear environment, States in the cyber strategic environment may have a true alternative to war through which to achieve strategically relevant outcomes. Understanding how States will leverage that alternative is the central question of early twenty-first-century international security. It is our hope that this exposition of cyber persistence theory will not only advance more rigorous academic examination of the complexities of the cyber strategic environment, but lead to better management by those entrusted to secure interests and values in and through cyberspace.

NOTES

Chapter 1

1. Carl von Clausewitz, *On War*, ed. Michael Howard and Petter Paret (Princeton, NJ: Princeton University Press, 1984), 578.
2. Political scientists Jon Lindsay and Erik Gartzke have discussed cross-domain coercion via deterrence, which in the context of military planning should be understood again as using different means, including cyber, within the logic of coercion that exists in either the conventional or nuclear strategic environments. It does not hold, however, that we can apply coercive strategies to a non-coercive environment and expect policy success. Jon Lindsay and Erik Gartzke, eds., *Cross-Domain Deterrence: Strategy in an Era of Complexity* (Oxford: Oxford University Press, 2019).
3. Thomas S. Kuhn, *The Structure of Scientific Revolutions*, 4th ed. (Chicago: University of Chicago Press, 2012).
4. Kuhn actually uses the term "change" more often than "shift" in his writings. For our purposes in applying Kuhn's *structure* of reasoning through a paradigmatic thinking model, we posit in Chapter 2 that the anomalies introduced first by nuclear weapons and then by networked computing to security each require new ways of "puzzle-solving" the challenge of national security, including changes in the way States think, organize, and act from the traditional realm of conventional warfare. Our nuance here is twofold. First, seeking security in the conventional, nuclear, and cyber strategic environments requires distinct paradigms—theories, concepts, logic, and prescriptions—that coexist. Since States must secure in each environment, they must change and align thinking to the particular security puzzle they face. Second, applying the logic of one strategic environment to another is a misapplication that can lead to problems and even the disaster of losing one's security. See Ian Hacking, "Introductory Essay," in *The Structure of Scientific Revolutions*, by Thomas S. Kuhn, 4th ed. (Chicago: University of Chicago Press, 2012), x–xxiii.
5. Michael Warner, "A Brief History of Cyber Conflict," in *Ten Years In: Implementing Strategic Approaches to Cyberspace*, ed. Jacquelyn G. Schneider, Emily O. Goldman, and Michael Warner (Newport, RI: Naval War College Press, 2020), 31–46. https://digital-commons.usnwc.edu/cgi/viewcontent.cgi?article=1044&context=usnwc-newport-papers.
6. The White House, *National Security Decision Directive Number 145*, September 17, 1984, https://fas.org/irp/offdocs/nsdd145.htm.
7. Yoon Kyu-sik, "North Korea's Cyber Warfare Capability and Threat," *Military Review* [Gunsanondan] 68 (Winter 2011): 64–95, cited in Daniel A. Pinkston, "North Korean Cyber Threats," in *Confronting an "Axis of Cyber"?: China, Iran, North Korea, Russia in Cyberspace*, ed. Fabio Rugge (Milan: Ledizioni-LediPublishing, 2018), 96.
8. John Arquilla and David Ronfeldt, "Cyberwar Is Coming!," *Comparative Strategy* 12 (April–June 1993): 141–165, https://doi.org/10.1080/01495939308402915.

9. In the case of the United States when the Department of Defense (DoD) issued its *Joint Doctrine for Information Operations* (JP 3-13) in 1996, information warfare was recast as information operations (IO), and computer network attack subsumed under IO. US Department of Defense, *Joint Publication 3-13, Information Operations* (Washington, DC: Department of Defense, 2012).
10. William J. Lynn III, "Defending a New Domain: The Pentagon's Cyberstrategy," *Foreign Affairs* (September/October 2010), https://www.foreignaffairs.com/articles/united-states/2010-09-01/defending-new-domain.
11. "NATO Recognises Cyberspace as a 'Domain of Operations' at Warsaw Summit," https://ccdcoe.org/incyder-articles/nato-recognises-cyberspace-as-a-domain-of-operations-at-warsaw-summit/.
12. The following literature review borrows from and builds on a complementary discussion found in Richard J. Harknett and Max Smeets, "Cyber Campaigns and Strategic Outcomes: The Other Means," *Journal of Strategic Studies* (Spring 2020): 1–34, https://doi.org/10.1080/01402390.2020.1732354. On the discourse among policymakers about the nature of the cyber threat, see Myriam Dunn Cavelty, *Cyber-Security and Threat Politics: US Efforts to Secure in the Information Age* (Abingdon: Routledge, 2007), https://www.researchgate.net/publication/277714726_Cyber-Security_and_Threat_Politics_US_Efforts_to_Secure_the_Information_Age.
13. For a comprehensive overview up to 2012, see Jason Healey, ed., *A Fierce Domain: Conflict in Cyberspace, 1986 to 2012* (Vienna, VA: Cyber Conflict Studies Association, 2013).
14. Arquilla and Ronfeldt, "Cyberwar Is Coming!" Also see John Arquilla and David Ronfeldt, *In Athena's Camp: Preparing for Conflict in the Information Age* (Santa Monica, CA: Rand Corporation, 1997), https://www.rand.org/pubs/monograph_reports/MR880.html; and John Arquilla and David Ronfeldt, *Networks and Netwars: The Future of Terror, Crime, and Militancy* (Santa Monica, CA: Rand Corporation, 2001), https://www.rand.org/pubs/monograph_reports/MR1382.html.
15. In 1997 the [US] President's Commission on Critical Infrastructure Protection released a report noting that, although it did not "discover . . . an immediate threat sufficient to warrant a fear of imminent national crisis, it did find reasons to implement new measures, especially in the area of cyber security." A consequence of the report was the drafting of Presidential Decision Directive 63 (PDD-63) in May 1998, which called for a range of actions intended to improve the nation's ability to protect "critical infrastructure" from physical and cyber attacks. See President's Commission on Critical Infrastructure Protection, *Critical Foundations: Protecting America's infrastructures* (Washington, DC: The Commission, 1997), https://sgp.fas.org/library/pccip.pdf. Additionally, David Betz and Tim Stevens noted that "[p]opular discourse on cyberwar tends to focus on the vulnerability of the 'physical layer' of cyberspace to cyber-attack and the ways in which this may permit even strong powers to be brought to their knees by weaker ones, perhaps bloodlessly." David J. Betz and Tim Stevens, *Cyberspace and the State: Towards a Strategy for Cyber-Power* (London: Routledge, 2011). Finally, securitization literature offers an alternative view that concerns over critical infrastructure are overstated. See, for example, Cavelty, *Cyber-Security and Threat Politics*; Ralf Bendrath, "The Cyberwar Debate: Perception and Politics in US Critical Infrastructure Protection," *Information & Security: An International Journal* 47 (2001): 80–103, https://infosec-journal.com/article/cyberwar-debate-perception-and-politics-us-critical-infrastructure-protection; and Rachel Yould, "Beyond the American Fortress: Understanding Homeland Security in the Information Age," 74–97, and Ralf Bendrath, "The American Cyber-Angst and the Real World," 49–73, both in *Bombs and Bandwidth: The Emerging Relationship between Information Technology and Security*, ed. Robert Latham (New York: The New Press, 2003).
16. On Estonia, see Eneken Tikk, Kadri Kaska, and Liis Vihul, *International Cyber Incidents: Legal Considerations* (Tallinn: NATO CCDCOE, 2010), https://ccdcoe.org/uploads/2018/10/legalconsiderations_0.pdf; and Rain Ottis, "Analysis of the 2007 Cyber Attacks against Estonia from the Information Warfare Perspective," in *Proceedings of the 7th European Conference on Information Warfare and Security, Plymouth 2008* (Reading: Academic Publishing Limited, 2008), 163–168, https://ccdcoe.org/library/publications/analysis-of-the-2007-cyber-attacks-against-estonia-from-the-information-warfare-perspective/. On

Georgia, see Tikk, Kaska, and Vihul, *International Cyber Incidents*; Paulo Shakarian, "The 2008 Russian Cyber Campaign Against Georgia," *Military Review* (November–December 2011): 63–68; Ronald J. Deibert, Rafal Rohizinski, and Masashi Crete-Nishihata, "Cyclone in Cyberspace: Information Shaping and Denial in the 2008 Russia-Georgia War," *Security Dialogue* 43 (February 15, 2012): 3–24, https://journals.sagepub.com/doi/full/10.1177/0967010611431079; and David M. Hollis, "Cyberwar Case Study: Georgia 2008," *Small Wars Journal* (January 2011): 1–9, https://smallwarsjournal.com/blog/journal/docs-temp/639-hollis.pdf. This is not to suggest that discussions have not expanded in scope. For example, scholars have been exploring in more detail the destructive potential of cyberattacks, the costs of conducting cyberattacks, the offense-defense balance, and the diffusion of capabilities.

17. Mike McConnell, "Cyberwar Is the New Atomic Age," *New Perspectives Quarterly* 26, no. 3 (Summer 2009): 72–77, https://onlinelibrary.wiley.com/doi/pdf/10.1111/j.1540-5842.2009.01103.x.
18. Richard A. Clarke and Robert K. Knake, *Cyber War* (New York: Ecco, 2010). A year earlier, William J. Lynn III talked about the Pentagon building defenses to prepare for the next cyber war. Lynn, "Defending a New Domain."
19. Gary McGraw, "Cyber War Is Inevitable (Unless We Build Security In)," *Journal of Strategic Studies* 36, no. 1 (2013): 109–119, http://doi.org/cp6f. For a similar argument, John Arquilla, "Cyberwar Is Already Upon Us," *Foreign Policy*, February 27, 2012, https://foreignpolicy.com/2012/02/27/cyberwar-is-already-upon-us/.
20. See Thomas Rid, "Cyber War Will Not Take Place," *Journal of Strategic Studies* 35, no. 1 (2012): 5–32, https://www.tandfonline.com/doi/full/10.1080/01402390.2011.608939; Thomas Rid, *Cyber War Will Not Take Place* (Oxford: Oxford University Press, 2013); Thomas Rid, "Think Again: Cyberwar," *Foreign Policy*, February 27, 2012, https://foreignpolicy.com/2012/02/27/think-again-cyberwar/; and Thomas Rid, "Is Cyberwar Real?," in response to Jarno Limnéll, Foreign Affairs (March/April 2014), https://www.foreignaffairs.com/articles/commons/2014-02-12/cyberwar-real.
21. As Rid noted, most cyber operations do not even meet one of these criteria. For a similar analysis, also drawing upon these three elements, see Thomas G. Mahnken, "Cyber War and Cyber Warfare," in *America's Cyber Future: Security and Prosperity in the Information Age*, Vol. 2, ed. Kristin Lord and Travis Sharp (Washington DC: Center for New American Security, 2011), 53–62.
22. Rid, "Cyber War Will Not Take Place." For a theory-driven response to Rid, see John Stone, "Cyber War Will Take Place!," *Journal of Strategic Studies* 36, no. 1 (2013): 101–108, https://www.tandfonline.com/doi/full/10.1080/01402390.2012.730485.
23. "Even the most successful forms of cyberwar (such as cyber espionage) do not presage much of a transformation." Erik Gartzke, "The Myth of Cyberwar: Bringing War in Cyberspace Back Down to Earth," *International Security* 38, no. 2 (2013): 41–73, https://www.mitpressjournals.org/doi/pdf/10.1162/ISEC_a_00136.
24. Ibid.
25. Ibid.
26. Adam P. Liff, "Cyberwar: A New 'Absolute Weapon'?: The Proliferation of Cyberwarfare Capabilities and Interstate War," *Journal of Strategic Studies* 35, no. 3 (2012): 401–428, https://www.tandfonline.com/doi/full/10.1080/01402390.2012.663252. Also see Timothy J. Junio, "How Probable Is Cyber War? Bringing IR Theory Back In to the Cyber Conflict Debate," *Journal of Strategic Studies* 36, no. 1 (2013): 125–133, https://www.tandfonline.com/doi/full/10.1080/01402390.2012.739561; and Adam P. Liff, "The Proliferation of Cyberwarfare Capabilities and Interstate War, Redux: Liff Responds to Junio," *Journal of Strategic Studies* 36, no. 1 (2013): 134–138, https://www.tandfonline.com/doi/full/10.1080/01402390.2012.733312.
27. Martin Libicki, *Cyberdeterrence and Cyberwar* (Santa Monica, CA: Rand Corporation, 2009), https://www.rand.org/content/dam/rand/pubs/monographs/2009/RAND_MG877.pdf. As Libicki noted: "Can strategic cyberwar induce political compliance the way, say, strategic airpower would? Airpower tends to succeed when societies are convinced that matters will only get worse. With cyberattacks, the opposite is more likely. As systems are attacked, vulnerabilities are revealed and repaired or routed around. As systems become more

hardened, societies become less vulnerable and are likely to become more, rather than less, resistant to further coercion." Also see Martin C. Libicki, "Cyberspace Is Not a Warfighting Domain," *I/S: A Journal of Law and Policy for the Information Society* 8, no. 2 (2012): 325–340, https://moritzlaw.osu.edu/ostlj/?file=2012%2F02%2F4.Libicki.pdf.

28. Robert J. Art and Kelly M. Greenhill, "Coercion: An Analytical Overview," in *Coercion: The Power to Hurt in International Politics*, ed. Kelly M. Greenhill and Peter Krause (New York: Oxford University Press, 2018), 5.
29. Seminal works include Thomas C. Schelling, *The Strategy of Conflict* (Cambridge, MA: Harvard University Press, 1960) and *Arms and Influence* (New Haven, CT: Yale University Press, 1966); and Alexander L. George, David K. Hall, and William R. Simons, *Limits of Coercive Diplomacy* (Boston: Little, Brown & Company, 1971).
30. John Lewis Gaddis, *The Long Peace: Inquiries into the History of the Cold War* (Oxford: Oxford University Press, 1989).
31. Harknett and Smeets, "Cyber Campaigns and Strategic Outcomes."
32. A cyber operations tracker created by the Council on Foreign Relations (CFR) Digital and Cyberspace Policy program identifies 426 publicly known State-sponsored incidents that have occurred since 2005. None of the events rises to the level of an armed-attack equivalent or a "significant cyber incident." https://www.cfr.org/cyber-operations/#CyberOperations.
33. For examples of research since 2015 that have a coercion-theory focus, see Jon R. Lindsay, "Tipping the Scales: The Attribution Problem and the Feasibility of Deterrence against Cyberattack," *Journal of Cybersecurity* 1, no. 11 (2015): 53–67, https://academic.oup.com/cybersecurity/article/1/1/53/2354517; Uri Tor, "'Cumulative Deterrence' as a New Paradigm for Cyber Deterrence," *Journal of Strategic Studies* 40, nos. 1–2 (2017): 92–117, https://www.tandfonline.com/doi/full/10.1080/01402390.2015.1115975; Jon R. Lindsay and Erick Gartzke, "Coercion through Cyberspace: The Stability-Instability Paradox Revisited," in *Coercion*, ed. Greenhill and Krause (New York: Oxford University Press, 2018), 176–204; Erica D. Borghard and Shawn W. Lonergan, "The Logic of Coercion in Cyberspace," *Security Studies* 26, no. 3 (2017): 452–481, https://www.tandfonline.com/doi/full/10.1080/09636412.2017.1306396; Erik Gartzke and John R. Lindsay, "Weaving Tangled Webs: Offense, Defense and Deception in Cyberspace," *Security Studies* 24, no. 2 (2015): 316–348, https://doi.org/10.1080/09636412.2015.1038188; Joseph S. Nye Jr., "Deterrence and Persuasion in Cyberspace," *International Security* 41, no. 3 (Winter 2016/17): 44–71, https://www.mitpressjournals.org/doi/pdf/10.1162/ISEC_a_00266; and Lucas Kello, *The Virtual Weapon and International Order* (New Haven, CT: Yale University Press, 2017).
34. There are some scholars who agree with us that war was not the right frame, but do not accept the construct of strategic competition through exploitation. Rather, they claim that State behaviors in and through cyberspace are better understood as an intelligence contest rather than as strategic competition. See, for example, the alternative views on this topic by Joshua Rovner, Michael Warner, Jon R. Lindsay, Michael P. Fischerkeller, Richard J. Harknett, and Nina Kollars in "Policy Roundtable: Cyber Conflict as an Intelligence Contest," *Texas National Security Review—Special Issue: Cyber Competition*, September 17, 2020, https://tnsr.org/roundtable/policy-roundtable-cyber-conflict-as-an-intelligence-contest/.

Chapter 2

1. Kuhn, *The Structure of Scientific Revolutions*, 4th ed., 85.
2. Structure is the stage-setting arrangement of units in a system, not an interaction-level process variable. Structure is not an "event" to which units react, but the defined stage upon which they must play out their existence and upon which they interact with each other. For a Realist example of process-level description of structure, see Glenn H. Snyder, "Process Variables in Neorealist Theories," *Security Studies* 5, no. 3 (1996): 167–192, https://doi.org/10.1080/09636419608429279. On the utility of structuralism, see Richard J. Harknett and Hasan Yalcin, "The Struggle for Autonomy: A Realist Structural Theory of International Relations," *International Studies Review* 14, no. 4 (December 2012): 499–521, https://www.jstor.org/sta

ble/41804152. The default to parsimony and fundamentals, of course, is also the main critique of structuralism.

3. The blurred lines between strategic environments will be addressed later. Much is to be gained by thinking through the distinctiveness of these environments first.
4. Anthony Giddens, *Central Problems in Social Theory* (London: Macmillan, 1979); Robert Keohane, ed., *Neorealism and Its Critics* (New York: Columbia University Press, 1986).
5. See F. H. Hinsley, *Sovereignty*, 2nd ed. (Cambridge: Cambridge University Press). For fuller discussion see Richard J. Harknett, "The Nuclear Condition and the Soft Shell of Territoriality," in *Globalisation: Theory and Practice*, 3rd ed., ed. Eleonore Kaufman and Gillian Youngs (London: Continuum, 2008), 277–288.
6. Classic examinations of the difference between absolute and relative gains can be found in Robert Powell, "Absolute and Relative Gains in International Relations Theory," *American Political Science Review* 85 (December 1991): 1303–1329, https://doi.org/10.2307/1963947; and Joseph Greico, "The Relative-Gains Problem for International Cooperation Comment," *American Political Science Review* 87 (September 1993): 729, https://www.jstor.org/stable/2938747.
7. Kenneth Waltz, *Theory of International Politics* (Reading, MA: Addison-Wesley, 1979).
8. Robert Jervis, "Cooperation under the Security Dilemma," *World Politics* 30, no. 2 (1978): 167–214, https://doi.org/10.2307/2009958.
9. John Mearsheimer, *The Tragedy of Great Power Politics* (New York: W. W. Norton, 2001).
10. W. L. Shirer, *The Collapse of the Third Republic: An Inquiry into the Fall of France in 1940* (New York: Simon & Schuster, 1969); Charles River, ed., *The Maginot Line: The History of the Fortifications That Failed to Protect France from Nazi Germany during World War II* (N.p.: Create Space Publishing, 2017). The historical material on André Maginot and the Maginot Line comes from several sources. This publication is a good general discussion.
11. John Mearsheimer, *Conventional Deterrence* (Ithaca, NY: Cornell University Press, 1983).
12. Annika Mombauer, "A Reluctant Military Leader? Helmuth von Moltke and the July Crisis of 1914," *War in History* 6, no. 4 (1999): 417–446, https://journals.sagepub.com/doi/pdf/10.1177/096834459900600403; Jack Snyder, *The Ideology of the Offensive: Military Decision Making and the Disasters of 1914* (Ithaca, NY: Cornell University Press, 1984).
13. Bernard Brodie, ed., *The Absolute Weapon: Atomic Power and World Order* (New York: Harcourt, Brace, 1946), 76.
14. While the National Security Act of 1947, the emergence of nuclear deterrence theory, and the conduct of crisis management for over seventy years point to a general if not specific adoption of the notion that security must be conceptualized in a distinct way due to the presence of nuclear weapons, full adherence to the notion of a nuclear revolution is open to debate. See Brendan Green, *The Revolution That Failed: Nuclear Competition, Arms Control, and the Cold War* (Cambridge: Cambridge University Press, 2020). For more on the coexistence of conventional and nuclear paradigms and how they have interacted, see T.V. Paul, Richard J. Harknett, and James J. Wirtz, eds., *The Absolute Weapon Revisited: Nuclear Arms and the Emerging International Order* (Ann Arbor: University of Michigan Press, 1998), Chapters 1 and 3.
15. Aside from variants of deterrence based on the technology being leveraged (conventional deterrence, nuclear deterrence, cyber deterrence) and territorial-based variants (strategic and extended deterrence), there exist a multitude of variations in forms of applications and conditions (mutual assured destruction, war-fighting, cumulative).
16. Patrick Morgan, *Deterrence* (Beverly Hills, CA: Sage Publications, 1977), 32; Lawrence Freedman, *Deterrence* (Cambridge: Polity Press, 2008).
17. This definition borrows and builds from the extensive literature but most specifically from the work of Patrick Morgan, who himself builds on Richard Brody in defining deterrence as the attempt by decision makers in one nation or group of nations to restructure the set of alternatives available to decision makers in another nation or group of nations by posing a threat to their key values. The restructuring is an attempt to exclude armed aggression (resorting to war) from consideration (Morgan, *Deterrence*, 32). Morgan also distinguishes between immediate and general deterrence in his definition. The case studies being analyzed here relate to instances of immediate deterrence defined by two conditions: a potential

challenger actively considering an attack, and a deterrer that is aware of this threat. This is different from general deterrence in which a deterrer deploys forces based on the contingency of a possible attack even though there is no indication of an imminent assault.

18. The term "opponent" is used here interchangeably with adversary, decision maker or policymaker. In each instance we do not assume a unitary actor. Although referred to in the singular, opponent may be defined as "a collection of those individuals, groups, and organizations with significant input into the decision-making process. The composition of the decision-maker is likely to change not only as the structure of the process changes over (long or extremely short) periods of time but also as the definition of the decision problem changes." Edward Rhodes, *Power and MADness* (New York: Columbia University Press, 1989), 74. The terms "challenger," "initiator," and "deterrer" also come under this definition. Paul Bracken forwards the idea that in war, institutional perspectives are much more important than unitary utility assessment perspectives for predicting and understanding decision outcomes. We agree with Bracken and therefore include organizations and their procedures into the definition of "decision maker." This idea was discussed in Paul Bracken, "National Security Organization and Conflict Termination," presented at Haverford College, April 27, 1990.
19. The conveyance of deterrent threats by words alone does not necessarily provide for a deterrent situation. See Robert Jervis, *Perception and Misperception in International Politics* (Princeton, NJ: Princeton University Press, 1976).
20. Thomas C. Schelling, *The Strategy of Conflict* (Cambridge, MA: Harvard University Press, 1960), 5–13.
21. Morgan, *Deterrence*, 30.
22. The "value" accorded a weapon is determined by both the particular characteristics of the weapon and the manner (skill, will, and strategy) in which it is employed. See Richard J. Harknett, "The Logic of Conventional Deterrence and the End of the Cold War," *Security Studies* 4, no. 1 (Autumn 1994): 86–114, https://doi.org/10.1080/09636419409347576.
23. See, for example, Glenn H. Snyder, *Deterrence and Defense* (Princeton, NJ: Princeton University Press, 1961), 3–6.
24. Thomas C. Schelling, *Arms and Influence* (New Haven, CT: Yale University Press, 1966), 2. Schelling continues the distinction considering deterrence the "power to hurt" and defense the "power to oppose." 26.
25. Deterrence by denial assumes that an attacker can anticipate attrition of their forces and, fearing that outcome, would avoid attacking. See Mearsheimer, *Conventional Deterrence*. As we will discuss later, attrition is not a feature of the cyber strategic environment and, thus, deterrence by denial is untenable as a primary strategy within that environment. Importantly, the term "attrition" has been used in a different manner by some cyber practitioners. For example, Sally Walker, former cyber director for the United Kingdom's (UK) Government Communications Headquarters (GCHQ), has said that "cyber warfare is attritional." Similarly, Ciaran Martin, former chief executive officer of the UK's National Cyber Security Centre, has spoken of "attritional competition." This use of the term, however, is in regard to the erosive effects cumulative cyber campaigns have on sources of national power (a perspective we hold), not in regard to attriting another's cyber capabilities or the threat to do so as part of deterrence by denial. See Deborah Haynes, "Into the Grey Zone: The 'Offensive Cyber' Used to Confuse Islamic State Militants and Prevent Drone Attacks," *SkyNews Podcast*, February 8, 2021, Sally Walker at 13:20,https://news.sky.com/story/into-the-grey-zone-the-offensive-cyber-used-to-confuse-islamic-state-militants-and-prevent-drone-attacks-12211740?utm_campaign=wp_the_cybersecurity_202&utm_medium=email&utm_source=newsletter&wpisrc=nl_cybersecurity202; and National Security Institute, "Silverado Debate: Cyber Offense versus Cyber Defense," January 21, 2021, Ciaran Martin at 29:08, https://nationalsecurity.gmu.edu/nsi-silverado-debate-cyber-offense-vs-cyber-defense/.
26. For a full discussion of the concept of contestability and deterrence as contestable versus incontestable costs, see Harknett, "The Logic of Conventional Deterrence" and Richard J. Harknett, "State Preferences, Systemic Constraints, and the Absolute Weapon," in *The Absolute Weapon Revisited*, ed. Paul, Harknett, and Wirtz, 65–100.
27. Martin van Creveld, *Technology and War* (New York: Free Press, 1989), 224–225.

28. William Kincade, "New Military Capabilities: Propellants and Implications," in *The Uncertain Course: New Weapons, Strategies and Mind-sets*, ed. Carl G. Jacobsen (New York: Oxford University Press, 1987), 69–70.
29. van Creveld, *Technology and War*, 228.
30. von Clausewitz, *On War*, Book 1, Chapter 7, 119–121.
31. The actual date of the writing of *The Art of War* is debated, but the general view understands it to be some 2,000 years ago. See Mark Cartwright, *The Art of War*, https://www.ancient.eu/The_Art_of_War/, for a brief synopsis of this debate.
32. John Mueller has presented the counterargument that it is conventional firepower that has deterred great power war along with economic incentives to void Industrial Age conflict. See Mueller, "The Escalating Irrelevance of Nuclear Weapons," *The Absolute Weapon Revisited: Nuclear Arms and the Emerging International Order* (Ann Arbor: University of Michigan Press, 1997), 73–98.
33. Rid, "Cyber War Will Not Take Place."
34. For other works drawing upon this cyber war notion, see Arquilla, "Rebuttal Cyberwar Is Already Upon Us"; McGraw, "Cyber War Is Inevitable (Unless We Build Security In)"; and Isabelle Duyvesteyn, "Between Doomsday and Dismissal: Cyber War, the Parameters of War, and Collective Defense," *Atlantisch Perspectief* 38, no. 7 (2014): 20–24, https://www.jstor.org/stable/e48504447.
35. This argument is made most directly by Harknett and Smeets, "Cyber Campaigns and Strategic Outcomes."
36. See US Department of Defense, *DoD Cyber Strategy* (April 2015), https://archive.defense.gov/home/features/2015/0415_cyber-strategy/final_2015_dod_cyber_strategy_for_web.pdf; Libicki, *Cyberdeterrence and Cyberwar*; and Libicki, "Cyberspace Is Not a Warfighting Domain."
37. See Liff, "Cyberwar: A New 'Absolute Weapon'?"; Junio, "How Probable Is Cyber War?"; and Liff, "The Proliferation of Cyberwarfare Capabilities and Interstate War, Redux."
38. In cyber lexicon, a vulnerability is a weakness in an information system, system security procedures, internal controls, implementation, data, or users that could be exploited or triggered by a threat source. See National Institute of Standards and Technology, Computer Security Resource Center, https://csrc.nist.gov/glossary/term/vulnerability.
39. See Richard J. Harknett, John Callaghan, and Rudi Kaufmann, "Leaving Deterrence Behind: Warfighting and National Cybersecurity," *Journal of Homeland Security and Emergency Management* 7, no. 1 (Spring 2010): 1–24, https://doi.org/10.2202/1547-7355.1636; and Emily Goldman and Richard Harknett, "The Search for Cyber Fundamentals," *Journal of Information Warfare* 15, no. 2 (Spring 2016): 81–88, https://www.jstor.org/stable/26487534.
40. This section is revised and adapted from Richard J. Harknett, "Integrated Security: A Strategic Response to Anonymity and the Problem of the Few," *Contemporary Security Policy* 24, no. 1 (April 2003): 13–45, https://doi.org/10.1080/13523260312331271809.
41. James Gleck, *Chaos: Making a New Science* (New York: Penguin Books, 1987), 103.
42. David Gelernter, *Mirror Worlds: Or the Day Software Puts the Universe in a Shoebox . . . How It Will Happen and What It Will Mean* (New York: Oxford University Press, 1991), 54.
43. Gleck, *Chaos*, 98–103.
44. Tom Simonite, "Moore's Law Is Dead. Now What?," *MIT Technology Review* (May 2016), https://www.technologyreview.com/2016/05/13/245938/moores-law-is-dead-now-what/.
45. The article quotes Intel's head of silicon engineering as still expecting to produce a 100-fold increase in computing power by 2030. David Rotman, "We're Not Prepared for the End of Moore's Law," *MIT Technology Review* (February 2020), https://www.technologyreview.com/2020/02/24/905789/were-not-prepared-for-the-end-of-moores-law/.
46. Further alternatives include specializing chip production to link with algorithms necessary for particular functions, although this raises concerns of the general purpose nature of the IC. See discussion in Rotman, "We're Not Prepared for the End of Moore's Law," and more in-depth policy analysis in Hasan N. Kahn, David Hounshell, and Erica Fuchs, "Science and

Research Policy at the End of Moore's Law," *Nature Electronics* 1 (January 2018): 14–21, https://doi.org/10.1038/s41928-017-0005-9.

47. The original US Defense Department plan created what was known as the ARPANET. Its basic goal was to create a communication system that was both redundant and independent. For background see Katie Hafner and John Markoff, *CyberPunk: Outlaws and Hackers on the Computer Frontier* (New York: Simon & Schuster, 1991), 263–282. The link between innovation, warfare, and societal organization again seems apparent.
48. The idea that information warfare is fundamentally new is contested by many who point to the emphasis that has always been placed on knowing where your enemy was, what his plans were, and the capabilities that supported his plans as well as the importance of denying the same information to your enemy. Authors have pointed to the writings of Sun Tzu as an example. John Arquilla points to the Mongol approach to war as another example. See Arquilla and Ronfeldt, "Cyberwar Is Coming."
49. When the ARPANET was first developed, modem speed was about one-tenth the speed of the slowest modems in the late 1980s; Hafner and Markoff, *CyberPunk*, 265. By the 2020s, connection speed was keeping pace with innovation of content delivery, whether it is streaming video or high-end graphic interfaces for gaming. While bandwidth remains an issue, it is an obstacle that State actors do not concern themselves with as a fundamental obstacle to most cyber activity.
50. This is an author observation based on interaction with hundreds of high school students over several years in cyber awareness camps and presentations.
51. We are grateful to Graham Fairclough for this observation.
52. Statistics on Internet use and growth rates are relatively consistent across multiple data collectors. A reasonable source that aggregates from multiple sources is the site "Internet World Stats," https://www.internetworldstats.com/emarketing.htm. Another source is https://www.statista.com/topics/1145/internet-usage-worldwide/.
53. Numbers used at the World Economic Forum: https://www.weforum.org/agenda/2019/09/chart-of-the-day-how-many-websites-are-there/.
54. An exemplar is the US White House, *International Strategy for Cyberspace: Prosperity, Security, and Openness in a Networked World*, May 2011, https://obamawhitehouse.archives.gov/sites/default/files/rss_viewer/international_strategy_for_cyberspace.pdf.
55. An early recognition of this reality appears in Willis H. Ware, *Security Controls for Computer Systems: Report of Defense Science Board Task Force on Computer Security* (Santa Monica, CA: Rand Corporation, 1970), https://www.rand.org/pubs/reports/R609-1.html.
56. This was recognized in a 1991 National Research Council report that noted: "We are at risk. Increasingly, America depends on computers. They control power delivery, communications, aviation, and financial services. They are used to store vital information, from medical records to business plans to criminal records. Although we trust them, they are vulnerable—to the effects of poor design and insufficient quality control, to accident, and perhaps most alarmingly, to deliberate attack. The modern thief can steal more with a computer than with a gun. Tomorrow's terrorist may be able to do more damage with a keyboard than with a bomb." The report also presciently noted: "But, as far as we can tell, there has been no successful systematic attempt to subvert any of our critical computing systems. Unfortunately, there is reason to believe that our luck will soon run out. Thus far we have relied on the absence of malicious people who are both capable and motivated. We can no longer do so." National Research Council, *Computers at Risk: Safe Computing in the Information Age* (Washington, DC: The National Academies Press, 1991), 7, https://doi.org/10.17226/1581.
57. In January 2021 a group of Reddit users banded together and through synchronized herd activity drove the stock price of GameStop to the point where some major Wall Street firms shorting the stock lost billions of dollars in assets in a matter of several days. The implications for such organized herd activity on financial markets revealed that the movement of finance was not the province primarily of large powerful financial firms, which now could be hurt by the few. Noah Kulwin, "Reddit, Elites and the Dream of GameStop to the Moon," *New York Times*, January 29, 2021, https://www.nytimes.com/2021/01/28/opinion/reddit-games top-robinhood-hedge-fund.html?searchResultPosition=4.

58. Jeremy Hsu, "The Strava Heat Map and the End of Secrecy," *Wired*, January 28, 2018, https://www.wired.com/story/strava-heat-map-military-bases-fitness-trackers-privacy/.
59. European Union, "Complete Guide to GDPR Compliance," https://gdpr.eu/.
60. Although the logical flow of cyber persistence theory is like structural realist international relations theory (but not in content), cyber persistence theory intends to explain the conditions associated with the cyber strategic environment and not general international relations. As an example of similar logic, John Mearsheimer, in particular, focused on the organizing principle of anarchy and the condition of self-help it produced, combined with the uncertainty that revolves around relative power, to argue that States had to be fearful of falling behind and thus had to seek as much power as possible. In cyber persistence theory, there is no uncertainty about the fact that devices, systems, and networks are continually reconfigured and that such reconfiguration can impact national security. See Mearsheimer, *The Tragedy of Great Power Politics*.
61. Bernard Brodie, "The Continuing Relevance of *On War*," in Carl Von Clausewitz, *On War*, ed. and trans. Michael Howard and Peter Paret (Princeton, NJ: Princeton University Press, 1984), 52.
62. Clausewitz, *On War*, 87.
63. Brodie, "The Continuing Relevance of *On War*," 55.
64. For full development of the counter argument to the "cyber as war thesis," see Harknett and Smeets, "Cyber Campaigns and Strategic Outcomes."

Chapter 3

1. The phrase "in and through" refers, respectively, to cyber operations or campaigns originating in cyberspace with the objective of generating effects in cyberspace, and cyber operations or campaigns originating in cyberspace with the objective of generating effects in one of the physical domains (air, land, maritime, and space), including, for example, cyber-enabled operations in the information environment and cyber-enabled armed-attack equivalent effects.
2. Waltz, *Theory of International Politics*.
3. Thomas C. Schelling, *Arms and Influence* (New Haven, CT: Yale University Press, 2008), https://yalebooks.yale.edu/book/9780300143379/arms-and-influence. This argument is made in Michael P. Fischerkeller, *What Is the Purpose of the Cyber Mission Force?* (Alexandria, VA: Institute for Defense Analyses, 2019).
4. TrendMicro, https://www.trendmicro.com/vinfo/us/security/definition/exploit.
5. Herbert S. Lin, "Offensive Cyber Operations and the Use of Force," *Journal of National Security Law and Policy* 4, no. 63 (2010), https://jnslp.com/wp-content/uploads/2010/08/06_Lin.pdf.
6. US Department of Defense, *Joint Publication 3-12, Cyberspace Operations* (Washington, DC: Department of Defense, June 8, 2018), https://www.jcs.mil/Portals/36/Documents/Doctrine/pubs/jp3_12.pdf.
7. Preclusion is not the same as deterrence by denial. The latter is premised on the notion that an adversary perceives a strategic opportunity to act and chooses to not act because the benefits do not outweigh the costs. Preclusion, alternatively, reduces the number of potential strategic opportunities.
8. "Tactical" refers to individual engagements. It "is the employment, ordered arrangement, and directed actions of forces in relation to each other." The operational, or campaign, level of analysis "links the tactical employment of forces to national strategic objectives." See US Department of Defense, *Joint Publication 3-0, Operations* (Washington, DC: Department of Defense, 2017). Efforts at supply chain exploitation through hardcoded backdoors in the firmware of systems expected to be strategically distributed or through compromised software supply chains, for example, represent an operational, or campaign, level of action with strategic intent, rather than the grand strategy level in our classification. Whether or not a supply chain exploitation could result in strategic effects is a function of scope and/or scale. For an example of scale, see Catalin Cimpanu, "Backdoor Accounts Discovered in 29 FTTH Devices from Chinese Vendor C-Data," *ZDNet*, July 10, 2020, https://www.zdnet.com/google-amp/article/backdoor-accounts-discovered-in-29-ftth-devices-from-chinese-vendor-c-data/. For examples of software supply chain exploitation, see *Microsoft Security Intelligence Report* 24,

January–December 2018, https://info.microsoft.com/SIRv24Report.html; Craig Timberg and Ellen Nakashima, "Federal Investigators Find Evidence of Previously Unknown Tactics Used to Penetrate Government Networks," *Washington Post*, December 17, 2020, https://www.washingtonpost.com/business/technology/government-warns-new-hacking-tactics-russia/2020/12/17/bba43fd8-408c-11eb-a402-fba110db3b42_story.html; and Josephine Wolff, "The SolarWinds Hack Is Unlike Anything We Have Ever Seen Before," *Slate*, December 18, 2020, https://slate.com/technology/2020/12/solarwinds-hack-malware-active-breach.html.

9. Michael P. Fischerkeller and Richard J. Harknett, "Persistent Engagement, Agreed Competition, Cyberspace Interaction Dynamics and Escalation," *Cyber Defense Review: Special Edition* (2019), https://cyberdefensereview.army.mil/Portals/6/CDR-SE_S5-P3-Fischerkeller.pdf.
10. See David Sacks, "China's Huawei Is Winning the 5G Race. Here's What the United States Should Do to Respond," *Council on Foreign Relations*, March 29, 2021, https://www.cfr.org/blog/china-huawei-5g; The White House, "Message to the Congress on Securing the Information and Communications Technology and Services Supply Chain," May 15, 2019, https://www.whitehouse.gov/briefings-statements/message-congress-securing-information-communications-technology-services-supply-chain/; and Lucy Fischer, "Downing Street Plans New 5G Club of Democracies," *The Times*, May 29, 2020, https://www.thetimes.co.uk/article/downing-street-plans-new-5g-club-of-democracies-bfnd5wj57.
11. Michael R. Pompeo, "Announcing the Expansion of the Clean Network to Safeguard America's Assets," August 5, 2020, https://www.state.gov/announcing-the-expansion-of-the-clean-network-to-safeguard-americas-assets/.
12. Recall that a vulnerability is defined as a weakness in an information system, system security procedures, internal controls, implementation, data, or users that could be exploited or triggered by a threat source. These three layers of cyberspace, when coupled with associated processes, are often understood collectively as "cyber terrain." US Department of Defense, *Joint Publication 3-12*.
13. In Kaspersky's 2020 annual incident report, external attack vectors targeted vulnerabilities in the physical and logical layers in 42 percent of all incidents reported and the cyber persona layer in 43 percent of incidents. See "Kaspersky Incident Response Analyst Report—2020," https://media.kasperskycontenthub.com/wp-content/uploads/sites/43/2020/08/06094905/Kaspersky_Incident-Response-Analyst_2020.pdf. These findings are similar to those in other security vendor annual incident reports. See, for example, "Verizon Business 2020 Data Breach Investigations Report," https://enterprise.verizon.com/resources/reports/dbir/; and "M-Trends 2020, FireEye Mandiant Services Special Report," https://content.fireeye.com/m-trends/rpt-m-trends-2020.
14. Dan Altman, "Advancing without Attacking: The Strategic Game around the Use of Force," *Security Studies* 27, no. 1 (2018): 58–88, https://doi.org/10.1080/09636412.2017.1360074.
15. James D. Fearon, "Threats to Use Force: Costly Signals and Bargaining in International Crises" (PhD diss., University of California, Berkeley, 1992).
16. James D. Fearon, "Signaling Foreign Policy Interests: Tying Hands versus Sinking Costs," *Journal of Conflict Resolution* 41, no. 1 (February 1997): 68–90, https://doi.org/10.1177%2F0022002797041001004; Paul K. Huth, "Deterrence and International Conflict: Empirical Findings and Theoretical Debates," *Annual Review of Political Science* 2, no. 1 (June 1999): 25–48, https://doi.org/10.1146/annurev.polisci.2.1.25; James D. Morrow, "The Strategic Setting of Choices: Signaling, Commitment, and Negotiation in International Politics," in *International Relations: A Strategic Choice Approach*, ed. David Lake and Robert Powell (Princeton, NJ: Princeton University Press, 1999), 86–91; and Branislav L. Slantchev, *Military Threats: The Costs of Coercion and the Price of Peace* (Cambridge: Cambridge University Press, 2011).
17. Emily O. Goldman, "The Cyber Paradigm Shift," in *Ten Years In: Implementing Strategic Approaches to Cyberspace*, ed. Jacquelyn G. Schneider, Emily O. Goldman, and Michael Warner (Newport, RI: Naval War College Press, 2020), 31–46, https://digital-commons.usnwc.edu/cgi/viewcontent.cgi?article=1044&context=usnwc-newport-papers.

18. Dan Altman, "By *Fait Accompli*, Not Coercion: How States Wrest Territory from Their Adversaries," *International Studies Quarterly* 61, no. 4 (2017): 881–891, https://doi.org/10.1093/isq/sqx049.
19. Alexander L. George, "Strategies for Crisis Management," in *Avoiding War: Problems of Crisis Management*, ed. Alexander L. George (Boulder, CO: Westview Press, 1991), 377–394.
20. Dan Altman, "Red Lines and *Faits Accomplis* in Interstate Coercion and Crisis" (PhD diss., Massachusetts Institute of Technology, June 2015) , https://dspace.mit.edu/bitstream/handle/1721.1/99775/927329080-MIT.pdf?sequence=1&isAllowed=y.
21. Ibid.
22. Ibid.
23. Altman, "By Fait Accompli, Not Coercion."
24. This definition represents a slight adjustment from our first iteration of the concept. See Michael P. Fischerkeller, "The Fait Accompli and Persistent Engagement in Cyberspace," *War on the Rocks*, June 24, 2020, https://warontherocks.com/2020/06/the-fait-accompli-and-persistent-engagement-in-cyberspace/.
25. There is no open source reporting regarding the method of exploitation (i.e., whether it was a direct hack or one of several other methods of exploitation, including, for example, a successful phishing campaign).
26. US Senate Committee on Armed Services Hearing, "Review Testimony on United States Special Operations Command and United States Cyber Command in Review on the Defense Authorization Request for Fiscal Year 2020 and the Future Years Defense Program," February 14, 2019, https://www.armed-services.senate.gov/imo/media/doc/19-13_02-14-19.pdf.
27. Richard J. Harknett, "SolarWinds: The Need for Persistent Engagement," *Lawfare*, December 23, 2020, https://www.lawfareblog.com/solarwinds-need-persistent-engagement.
28. Thus, an initial gain could be considered a "technical gain," which may be the objective itself, or it may support follow-on objectives seeking additional technical, political, economic, or military gains.
29. William Turton and Kartikay Mehrotra, "FireEye Discovered SolarWinds Breach While Probing Own Hack," *Bloomberg*, December 14, 2020 (updated December 15, 2020), https://www.bloomberg.com/news/articles/2020-12-15/fireeye-stumbled-across-solarwinds-breach-while-probing-own-hack.
30. US Senate Committee on Armed Services Hearing, "Review Testimony on United States Special Operations Command and United States Cyber Command."
31. Brendan I. Koerner, "Inside the Cyberattack that Shocked the U.S. Government," *Wired*, October 23, 2016, https://www.wired.com/2016/10/inside-cyberattack-shocked-us-government/.
32. *United States of America v. Zhu Hua and Zhang Shilong*, https://www.justice.gov/opa/page/file/1122671/download.
33. For analyses of the campaign, see Jack Stubbs, Joseph Menn, and Christopher Bing, "Inside the West's Failed Fight against China's 'Cloud Hopper' Hackers," *Reuters*, June 26, 2019, https://www.reuters.com/investigates/special-report/china-cyber-cloudhopper/; and Rob Barry and Dustin Volz, "Ghosts in the Clouds: Inside China's Major Corporate Hack," *Wall Street Journal*, December 30, 2019, https://www.wsj.com/articles/ghosts-in-the-clouds-inside-chinas-major-corporate-hack-11577729061.
34. "Verizon 2019 Data Breach Investigation Report," https://enterprise.verizon.com/resources/reports/dbir/2019/introduction/.
35. "Breakout time" is the time it takes for an intruder to begin moving laterally, beyond the initial beachhead, to other systems in the network. See Adam Myers, "First-Ever Adversary Ranking in 2019 Global Threat Report Highlights the Importance of Speed," *CrowdStrike Blog*, February 19, 2019, https://www.crowdstrike.com/blog/first-ever-adversary-ranking-in-2019-global-threat-report-highlights-the-importance-of-speed/; and Michael Busselen, "CrowdStrike CTO Explains 'Breakout Time'—A Critical Metric in Stopping Breaches," *CrowdStrike Blog*, June 6, 2018, https://www.crowdstrike.com/blog/crowdstrike-cto-explains-breakout-time-a-critical-metric-in-stopping-breaches/.

36. Mandiant, "Mandiant Exposes APT1—One of China's Cyber Espionage Units & Releases 3,000 Indicators," February 9, 2013, https://www.fireeye.com/blog/threat-research/2013/02/mandiant-exposes-apt1-chinas-cyber-espionage-units.html.
37. For this reason, we also exclude from our definition Altman's verb "imposes" because it harkens to a key phrase—*cost imposition*—that is tightly coupled with coercion theory and strategic bargaining. For the same reason, we exclude both Altman's and George's presumption that escalation would characterize a defender's potential response.
38. Michael P. Fischerkeller and Richard J. Harknett, "Persistent Engagement and Cost Imposition: Distinguishing between Cause and Effect," *Lawfare*, February 6, 2020, https://www.lawfareblog.com/persistent-engagement-and-cost-imposition-distinguishing-between-cause-and-effect.
39. That the source of vulnerability is not an ambiguous coercive demand has another important implication—it eliminates Altman's rationale for coupling the cyber *fait accompli* with coercive bargaining in cyberspace.
40. Fischerkeller and Harknett, "Persistent Engagement, Agreed Competition, Cyberspace Interaction Dynamics and Escalation."
41. Gartzke and Lindsay, "Weaving Tangled Webs."
42. Consider that over 22,000 vulnerabilities were reported in 2019 for the physical and logical layers of cyberspace, of which 61 percent are considered to present medium risk and 33 percent are considered to present high to critical risk. Additionally, almost half of those vulnerabilities (47%) have a public exploit available to malicious cyber actors, and more than a third (40.2%) do not have a publicly available solution, such as a software upgrade, workaround, or software patch. When these numbers are considered against the number of systems in cyberspace that may have these vulnerabilities, the magnitude of the abundance becomes clear. If one also takes into account the vulnerabilities introduced at the cyber persona layer of cyberspace through poor user cyber hygiene (e.g., weak passwords, carelessness in responding to phishing campaigns, and visiting websites that may download malware), the magnitude is further amplified. See Eduard Kovacs, "Over 22,000 Vulnerabilities Disclosed in 2019: Report," *SecurityWeek*, February 18, 2020, https://www.securityweek.com/over-22000-vulnerabilities-disclosed-2019-report; and Dina Bekerman and Sharit Yarushalmi, "The State of Vulnerabilities in 2019," January 23, 2020, https://www.imperva.com/blog/the-state-of-vulnerabilities-in-2019/. For a description of how risk is calculated, see https://www.first.org/cvss/specification-document. Note that these vulnerabilities are also present in open source software, as the 2022 discovery of the weakness in Log4j made clear. Eric Geller, "Lesson from Log4j: Open-source Software Improvements Need Help from Feds," *Politico*, January 6, 2022, https://www.politico.com/news/2022/01/06/open-source-software-help-526676..
43. For a case study on strategic significance, see Michael P. Fischerkeller, "Influencing China's Rise through Defend Forward / Persistent Engagement in Cyberspace," *Asia Policy Journal* 15, no. 4 (October 2020): 65–89, https://www.nbr.org/publication/opportunity-seldom-knocks-twice-influencing-chinas-trajectory-via-defend-forward-and-persistent-engagement-in-cyberspace/.
44. For a discussion of how a strategic approach based on unilateral action differs from "traditional bargaining," see Charles E. Osgood, *An Alternative to War or Surrender* (Urbana: University of Illinois Press, 1962).
45. United Nations, "Report of the Panel of Experts Established Pursuant to Resolution 1874 (2009) S/2019/691," https://www.securitycouncilreport.org/atf/cf/%7B65BFCF9B-6D27-4E9C-8CD3-CF6E4FF96FF9%7D/S_2019_691.pdf.
46. Tim Maurer and Arthur Nelson, "COVID-19's Other Virus: Targeting the Financial System," *Strategic Europe*, April 21, 2020, https://carnegieeurope.eu/strategiceurope/81599.
47. Michael P. Fischerkeller and Richard J. Harknett, "Cyber Persistence Theory, Intelligence Contests, and Strategic Competition," *Texas National Security Review: Special Issue—Cyber Competition*, September 17, 2020, https://tnsr.org/roundtable/policy-roundtable-cyber-conflict-as-an-intelligence-contest/.
48. See, respectively, Erik Gartzke, "The Myth of Cyberwar: Bringing War in Cyberspace Back Down to Earth," *International Security* 38, no. 2 (Fall 2013): 41–73, https://www.mitpressj

ournals.org/doi/pdf/10.1162/ISEC_a_00136; and Emily O. Goldman and Michael Warner, "Why a Digital Pearl Harbor Makes Sense . . . and Is Possible," *Carnegie Endowment for International Peace*, October 16, 2017, https://carnegieendowment.org/2017/10/16/why-digital-pearl-harbor-makes-sense-.-.-.-and-is-possible-pub-73405.

49. For two different examples and forms of contestation over control of a botnet network's command and control infrastructure, see Ellen Nakashima, "Cyber Command Has Sought to Disrupt the World's Largest Botnet, Hoping to Reduce Its Potential Impact on the Election," *Washington Post*, October 9, 2020, https://www.washingtonpost.com/national-security/cyber-command-trickbot-disrupt/2020/10/09/19587aae-0a32-11eb-a166-dc429b380d10_story.html; National Cyber Security Centre and National Security Agency, "Advisory: Turla Group Exploits Iranian APT to Expand Coverage of Victims," October 21, 2019, https://www.ncsc.gov.uk/news/turla-group-exploits-iran-apt-to-expand-coverage-of-victims; and Jai Vijayan, "Russian Hackers Using Iranian APT's Infrastructure in Widespread Attacks," *DARKReading*, October 21, 2019, https://www.darkreading.com/attacks-breaches/russian-hackers-using-iranian-apts-infrastructure-in-widespread-attacks/d/d-id/1336134. On key cyber terrain, see David Raymond, Gregory Conti, Tom Cross, and Michael Nowatkowski, "Key Terrain in Cyberspace: Seeking the High Ground," in *2014 6th International Conference on Cyber Conflict*, ed. P. Brangetto, M. Maybaum, and J. Stinissen (Tallinn: NATO CCDCOE Publications, 2014), 287–300; and US Department of Defense, *Joint Publication 3-12*. There are reported instances of accidental or inadvertent engagements resulting from States pursuing cyber *faits accomplis*—for example, among advanced persistent threat groups using the same compromised website for use in water-holing attacks. See Juan Andrés Guerrero-Saade and Costin Raiu, "Walking in Your Enemy's Shadow: When Fourth Party Collection Becomes Attribution Hell," *virusBULLETIN*, October 20, 2017, https://www.virusbulletin.com/blog/2017/10/vb2017-paper-walking-your-enemys-shadow-when-fourth-party-collection-becomes-attribution-hell/. An apt metaphor for such occurrences is Longfellow's "two ships passing in the night," used to describe two or more people who encounter one another in a transitory, incidental manner. Henry Wadsworth Longfellow, "Part Third, The Theologian's Tale, Elizabeth," in *Tales of a Wayside Inn* (1863).
50. See, for example, Michael Markey, Jonathon Pearl, and Benjamin Bahney, "How Satellites Can Save Arms Control," *Foreign Affairs*, August 5, 2020. https://www.foreignaffairs.com/articles/asia/2020-08-05/how-satellites-can-save-arms-control. States could also use non-cyber means of interaction to communicate the same message.
51. These opportunities will likely increase by at least one order of magnitude with the full deployment of 5G networks and the connection thereto of so-called Internet-of-Things devices.
52. This example speaks to a defender being "unable" to respond. For an example of non-governmental infrastructure, consider the Society for Worldwide Interbank Financial Telecommunications (SWIFT), a global member-owned cooperative and the world's leading provider of secure financial messaging services that has been exploited by North Korean cyber operators. See https://www.swift.com/about-us.
53. This example speaks to a defender being unwilling to respond.
54. On "noise" in cyberspace, see Edward Skoudis, "Evolutionary Trends in Cyberspace," in *Cyberpower and National Security*, ed. Franklin D. Kramer, Stuart H. Starr, and Larry K. Wentz (Washington, DC: National Defense University Press, 2009), 147–170, https://ndupress.ndu.edu/Portals/68/Documents/Books/CTBSP-Exports/Cyberpower/Cyberpower-I-Preface.pdf?ver=2017-06-16-115055-553. We review the literature on "noise" and tacit bargaining in Chapter 5. Failure to respond could stem from a target State's inability to fully comprehend the cyber manifestations of their national interests that are being engaged (i.e., key cyber terrain) and/or failing to detect the behavior once engaged. Given the intrusion data cited previously, defenders should strive for a direct cyber engagement "interaction cycle" response time of less than 19 minutes, which equals 2019's best "breakout" time of the most sophisticated cyber actor. Anything longer risks losing both the attention of the intruder and control of the key cyber terrain being contested. This window is a high bar, of course, and may only be achievable with support from machine learning or other AI capabilities. Until and if such capabilities are widely adopted, direct cyber engagement interactions among

peers will likely be characterized by cycle times of hours if not days. For examples, see the exemplifications of direct cyber engagements presented in Chapter 4.

55. This example also speaks to a defender being unable to respond. In 2019, Verizon reported average "time to discovery" as being "months"; see "Verizon 2019 Data Breach Report." This is consistent with other reports such as, for example, "Kaspersky Incident Response Analyst Report—2020," and "M-Trends 2020, FireEye Mandiant Services Special Report."
56. See, for example, Herbert S. Lin, "Escalation Dynamics and Conflict Termination in Cyberspace," *Strategic Studies Quarterly* 6, no. 3 (Fall 2012): 46–70, https://www.jstor.org/stable/26267261; Martin C. Libicki, *Crisis and Escalation in Cyberspace* (Santa Monica, CA: RAND Corporation, 2012); Lawrence J. Cavaiola, David C. Gompert, and Martin Libicki, "Cyber House Rules: On War, Retaliation and Escalation," *Survival* 57, no. 1 (2015): 81–104, https://doi.org/10.1080/00396338.2015.1008300; David C. Gompert and Martin Libicki, "Cyber Warfare and Sino-American Crisis Instability," *Survival* 56, no. 4 (2014): 7–22, https://doi.org/10.1080/00396338.2014.941543; and Jason Healy, "Triggering the New Forever War in Cyberspace," *The Cipher Brief*, April 1, 2018, https://www.thecipherbrief.com/triggering-new-forever-war-cyberspace.
57. Herman Kahn (with a new introduction by Thomas C. Schelling), *On Escalation: Metaphors and Scenarios* (London: Routledge, 2017), 3.
58. Ibid., 4–6.
59. A notable exception is Austin Carson, *Secret Wars: Covert Conflict in International Politics* (Princeton, NJ: Princeton University Press, 2018).
60. Kahn, *On Escalation*, 4, fn. 3. Kahn attributed this term to Max Singer.
61. Charles Lipson, "Why Are Some International Agreements Informal?," *International Organization* 45, no. 4 (October 1991): 495–538, https://www.jstor.org/stable/2706946. Archives show US and Soviet pilots directly engaged in *covert* air-to-air and anti-air operations in Korea and Vietnam. See Carson, *Secret Wars: Covert Conflict in International Politics*</IBT>.
62. Ibid, 6.
63. Schelling described focal points as a determining differentia, a range of behavior, or a combination of both around which actors converge through interest. Schelling, *The Strategy of Conflict*.
64. Thomas C. Schelling, "Reciprocal Measures for Arms Stabilization," *Daedalus* 89, no. 4 (Fall 1960): 892–914, https://doi.org/10.1080/00396336108440237.
65. Ibid.
66. Osgood, *An Alternative to War or Surrender*.
67. Schelling, "Reciprocal Measures for Arms Stabilization."
68. Michael P. Fischerkeller and Richard J. Harknett, "A Response on Persistent Engagement and Agreed Competition," *Lawfare*, June 27, 2019, https://www.lawfareblog.com/response-persistent-engagement-and-agreed-competition.
69. The arguments in this section represent a progression in thinking over our earlier publications describing agreed competition and competitive interaction. See, for example, Michael P. Fischerkeller and Richard J. Harknett, "What Is Agreed Competition in Cyberspace?," *Lawfare*, February 19, 2019, https://www.lawfareblog.com/what-agreed-competition-cyberspace; and Fischerkeller and Harknett, "Persistent Engagement, Agreed Competition, and Cyberspace Interaction Dynamics and Escalation."
70. Although "the cyber agreed competition phenomenon" is a lengthy phrase, we have opted not to adopt a shorthand version in this book. This choice was informed by Kenneth Waltz's observation of the confusion and criticisms that emerged, in international relations scholarship and policy circles, over balance-of-power theory. For Waltz, the balance of power concept is a structurally derived phenomenon. And yet, he observed, it also came to be understood by scholars and policymakers as, for example, a foreign policy recommendation—something to be pursued rather than something that was structurally derived. And so we repeat "the cyber agreed competition phenomenon" phrase to reinforce its structural derivation and to help ensure it is not misunderstood as a policy recommendation (or in any other way). See Waltz, *Theory of International Politics*, 118–123.

71. In 2015, a United Nations Group of Government Experts representing twenty States endorsed a set of "voluntary and non-binding . . . limiting norms" and "good practices and positive duties." The former include the following norms: States should not knowingly allow their territory to be used for internationally wrongful acts using Information Communication Technology (ICT), States should not conduct or knowingly support ICT activity that intentionally damages critical infrastructure, States should take steps to ensure supply chain security and should seek to prevent the proliferation of malicious ICT and the use of harmful hidden functions, States should not conduct or knowingly support activity to harm the information systems of another State's emergency response teams (CERT/CSIRTS) and should not use their own teams for malicious international activity, and States should respect the UN resolutions that are linked to human rights on the Internet and to the right to privacy in the digital age. The latter include the following practices and duties: States should cooperate to increase stability and security in the use of ICTs and to prevent harmful practices; States should consider all relevant information in case of ICT incidents; States should consider how best to cooperate to exchange information, to assist each other, and to prosecute terrorist and criminal use of ICTs; States should take appropriate measures to protect their critical infrastructure; States should respond to appropriate requests for assistance by other States whose critical infrastructure is subject to malicious ICT acts; and States should encourage responsible reporting of ICT vulnerabilities and should share remedies to these. See United Nations, "Group of Governmental Experts on Developments in the Field of Information and Telecommunications in the Context of International Security: A/70/174," July 22, 2015, https://www.un.org/ga/search/view_doc.asp?symbol=A/70/174.
72. This learning curve was recognized in the context of the coercive environment of the nuclear strategic environment of the early Cold War. Schelling refers to such a possibility when noting that the Americans and Soviets "may yet develop tacit understandings about zones and traffic rules for submarines, and may (or may not) develop a tradition for leaving each other's reconnaissance satellites alone." Schelling, "Reciprocal Measures for Arms Stabilization."
73. Kahn, *On Escalation.*
74. US-Soviet maritime interactions in the 1960s serve as an example that eventually led, in 1972, to the Agreement Between the Government of The United States of America and the Government of The Union of Soviet Socialist Republics on the Prevention of Incidents On and Over the High Seas. See https://2009-2017.state.gov/t/isn/4791.htm.
75. We approach theory development similarly to the structurally derived deductive reasoning of Ken Waltz. It is important to note, however, that the constructs of strategic environments are subsets to Waltz's overall anarchic system. Waltz understood balance of power to be the systemic strategic phenomena of international relations. We understand strategic environments to be the conditions under which the pursuit of security is practiced with attention to the overall distribution of power. In cyber persistence theory, cyber agreed competition emerges as a phenomenon tied to the distinct exploitative nature of the cyber strategic environment. As such, cyber agreed competition describes the specific behaviors tied to cyberspace that can impact the overall systemic distribution of power (and since they are salient to that distribution, they are actions that should be understood as strategic in intent and potential outcome). On structurally derived systemic understandings of balance of power, see Waltz, *Theory of International Politics.*
76. For a similar argument, see James A. Lewis, "Rethinking Cyber Security: Strategy, Mass Effects, and States," *Center for Strategic and International Studies,* January 2018, https://www.csis.org/analysis/rethinking-cybersecurity.
77. Lindsay and Gartzke also argue that States are incentivized to not encourage military retaliation to ensure they maintain connectivity, given the political and social benefits of connectivity. See Lindsay and Gartzke, "Coercion through Cyberspace: The Stability-Instability Paradox Revisited." Borghard and Lonergan also argue that offensive cyber operations are poor tools for supporting cyber escalation dynamics based on technical foundations of cyber capabilities, including, but not limited to, the role of access, non-universal lethality of offensive cyber capabilities, and the temporal nature of access and capability development and maintenance. See Erica D. Borghard and Shawn W. Lonergan, "Cyber Operations as Imperfect Tools

of Escalation," *Strategic Studies Quarterly* 13, no. 3 (Fall 2019): 122–145, https://www.jstor.org/stable/26760131.

78. See, for example, "Cyber Warfare in the 21st Century: Threats, Challenges, and Opportunities." Committee on Armed Services, US House of Representatives, March 1 2017, https://www.gpo.gov/fdsys/pkg/CHRG-115hhrg24680/pdf/CHRG-115hhrg24680.pdf; Cavaiola, Gompert, and Libicki, "Cyber House Rules: On War, Retaliation and Escalation"; Gompert and Libicki, "Cyber Warfare and Sino-American Crisis Instability"; and Healy, "Triggering the New Forever War in Cyberspace."
79. Kahn, *On Escalation.* Many others have offered definitions (see previous endnote) but those tend to include the mechanisms of escalation. Our interest lies in the dynamic itself.
80. A 2015 cyberattack on the French magazine *L'Express* may be an example of an accidental action by undisciplined members of Russia's Fancy Bear APT group. See Feike Hacquebord, "Two Years of Pawn Storm: Examining an Increasingly Relevant Threat," *TrendMicro*, https://documents.trendmicro.com/assets/wp/wp-two-years-of-pawn-storm.pdf. An accidental outcome is an operational action resulting in direct effects that are unintended by those who ordered them—for example, malware may spread unintentionally beyond the designated target. The global spread of NotPetya malware is often cited as example. See Andy Greenburg, "The Untold Story of NotPetya, the Most Devastating Cyberattack in History," *Wired*, September 22, 2018, https://www.wired.com/story/notpetya-cyberattack-ukraine-russia-code-crashed-the-world/.
81. Fischerkeller and Harknett argued, for example, that cyber capabilities offer unique opportunities for reversing limited damage that kinetic capabilities do not. Thus, States could breach the ceiling of agreed competition in a controllable and temporary manner. See Fischerkeller and Harknett, "Persistent Engagement, Agreed Competition, Cyberspace Interaction Dynamics and Escalation."
82. For an example of scholars exploring such conditions, see Jason Healey and Robert Jervis, "Escalation Inversion and Other Oddities of Situational Cyber Stability," *Texas National Security Review* 3, no. 4 (Fall 2020), https://tnsr.org/2020/09/the-escalation-inversion-and-other-oddities-of-situational-cyber-stability/.
83. Pedro Palandrani and Andrew Little, "A Decade of Change: How Tech Evolved in the 2010s and What's in Store for the 2020s," *Mirae Asset*, February 10, 2020, https://www.globalxetfs.com/a-decade-of-change-how-tech-evolved-in-the-2010s-and-whats-in-store-for-the-2020s/.
84. For an overview of expert opinion on AI, see Vincent C. Müller and Nick Bostrom, "Future Progress in Artificial Intelligence: A Survey of Expert Opinion," in *Fundamental Issues of Artificial Intelligence*, ed. Vincent C. Müller (Berlin: Springer, 2016), 555–572. Scholarship focusing on the nexus of AI and security studies includes Joe Burton and Simona R. Soare, "Understanding the Strategic Implications of the Weaponization of Artificial Intelligence," *IEEE Xplore*, July 11, 2019, https://ieeexplore.ieee.org/document/8756866; Kareem Ayoub and Kenneth Payne, "Strategy in the Age of Artificial Intelligence," *Journal of Strategic Studies* 39, no. 5–6 (2015), https://doi.org/10.1080/01402390.2015.1088838; Heather Roff, *Advancing Human Security through Artificial Intelligence* (London: Chatham House, 2017); Michael C. Horowitz, "Artificial Intelligence, International Competition, and the Balance of Power," *Texas National Security Review* 1, no. 3 (2018), https://tnsr.org/2018/05/artificial-intelligence-international-competition-and-the-balance-of-power/; Kenneth Payne, *Strategy, Evolution, and War: From Apes to Artificial Intelligence* (Washington, DC: Georgetown University Press, 2018); Heather Roff, "COMPASS: A New AI-Driven Situational Awareness Tool for the Pentagon?," *Bulletin of the Atomic Scientists*, May 10, 2018, https://thebulletin.org/2018/05/compass-a-new-ai-driven-situational-awareness-tool-for-the-pentagon/; Kenneth Payne, "Artificial Intelligence: A Revolution in Strategic Affairs?," *Survival* 60, no. 5 (2018), https://doi.org/10.1080/00396338.2018.1518374; and Michael C. Horowitz, Gregory C. Allen, Elsa B. Kania, and Paul Scharre, "Strategic Competition in an Era of Artificial Intelligence," *Center for a New American Security*, 2018, https://www.cnas.org/publications/reports/strategic-competition-in-an-era-of-artificial-intelligence.
85. On fuzzing, see Gary J. Saavedra, Kathryn N. Rodhouse, Daniel M. Dunlavy, and Philip W. Kegelmeyer, "A Review of Machine Learning Applications in Fuzzing," https://deepai.

org/publication/a-review-of-machine-learning-applications-in-fuzzing; and James Fell, "A Review of Fuzzing Tools and Methods" (originally published in *PenTest*, March 2017), https://dl.packetstormsecurity.net/papers/general/a-review-of-fuzzing-tools-and-methods.pdf. For a novel approach to using fuzzing tools, see Timothy Nosco, Jared Ziegler, Zechariah Clark, Davy Marrero, Todd Finkler, Andrew Barbarello, and W. Michael Petullo, "The Industrial Age of Hacking," presented at 29th USENIX Security Symposium, August 12–14, 2020, https://www.usenix.org/conference/usenixsecurity20/presentation/nosco. On symbolic execution, see Thanassis Avgerinos, Alexandre Rebert, Sang Kil Cha, and David Brumley, "Enhancing Symbolic Execution with Veritesting," https://users.ece.cmu.edu/~aavgerin/papers/veritesting-icse-2014.pdf.

86. See Christopher Whyte, "Problems of Poison: New Paradigms and 'Agreed' Competition in the Era of AI-Enabled Cyber Operations," in *2020 12th International Conference on Cyber Conflict—20/20 Vision: The Next Decade*, ed. T. Jančárková, L. Lindström, M. Signoretti, I. Tolga, and G. Visky (Tallinn: NATO CCDCOE, 2020), 215–232. For more details on this case, see Lior Keshet, "An Aggressive Launch: TrickBot Trojan Rises with Redirection Attacks in the UK," *Security Intelligence* (2016); and Darrel Rendell, "Understanding the Evolution of Malware," *Computer Fraud & Security* 1 (2019): 17–19, https://doi.org/10.1016/S1361-3723(19)30010-7.
87. For example, Dave Baggett argues that phishing and spear phishing as attack vectors may be obsolete by 2024. See *Implications of Artificial Intelligence for Cybersecurity: Proceedings of a Workshop* (Washington, DC: National Academies Press, 2019), http://nap.edu/25488. Others argue, however, that AI will make spear phishing an even more effective approach for compromising systems or networks. See Kaveh Waddell, "The Twitter Bot That Sounds Just Like Me," *The Atlantic*, August 18, 2016, https://www.theatlantic.com/technology/archive/2016/08/the-twitter-bot-that-sounds-just-like-me/496340/.
88. This capability is often referred to as security information and event management (SIEM). For a discussion of intrusion detection systems, see Karen Scarfone and Peter Mell, *National Institute of Standards and Technology: Guide to Intrusion Detection and Prevention Systems—Special Publication 800-94* (Washington, DC: US Department of Commerce, February 2007), https://nvlpubs.nist.gov/nistpubs/Legacy/SP/nistspecialpublication800-94.pdf.
89. Ibid.
90. See, for example, Will Hunt, *The Flight to Safety-Critical AI: Lessons in AI Safety from the Aviation Industry* (Berkeley: UC Berkeley, August 2020), https://cltc.berkeley.edu/wp-content/uploads/2020/08/Flight-to-Safety-Critical-AI.pdf.
91. Tyler Moore, quoted in *Implications of Artificial Intelligence for Cybersecurity*.
92. Consider, for example, that in 2019, the sophisticated Russia State-sponsored group APT28 compromised Norwegian parliament systems using a common technique called "brute forcing," which bombards accounts with passwords until one works. See Sean Lyngaas, "Norwegian Police Implicate Fancy Bear in Parliament Hack, Describe 'Brute Forcing' of Email Accounts," *CyberScoop*, December 8, 2020, https://www.cyberscoop.com/norwegian-police-implicate-fancy-bear-in-parliament-hack-describe-brute-forcing-of-email-accounts/.
93. Bruce Schneier, "Machine Learning to Detect Software Vulnerabilities," *Schneier on Security*, January 8, 2019, https://www.schneier.com/blog/archives/2019/01/machine_learnin.html.
94. Carson, *Secret Wars*. Carson argued that Americans and Soviets sought to minimize two escalation risks in limited wars—hawkish domestic pressures and poor communication—by relying on covert interventions. He referred to this as "tacit collusion" and argued that States do this "out of their own strategic interest," and not out of a shared expectation of reciprocity. Lipson argued that it is important not to exaggerate the scale of tacit cooperation (and we would add "collusion") by always referring to the existence of mutual interests, understandings, and expectations as tacit cooperation. Importantly, Lipson also argued that stable expectations can arise from either tacit cooperation or unilateral, independent actions around mutual interests (tacit coordination) based on stable Nash equilibria.
95. This also applies to tacit coordination.
96. See Thomas C. Schelling, "Bargaining, Communication, and Limited War," *Conflict Resolution* 1, no. 1 (March 1957): 19–36, https://www.jstor.org/stable/172548; George W. Downs and

David M. Rocke, "Tacit Bargaining and Arms Control," *World Politics: A Quarterly Journal of International Relations* 39, no. 3 (April 1987): 297–325, https://doi.org/10.2307/2010222; and George W. Downs and David M. Rocke, *Tacit Bargaining, Arms Races, and Arms Control* (Ann Arbor: University of Michigan Press, 1990), 3. Importantly, Downs and Rocke noted that tacit bargaining may include "negative feedback" but that does not *ipso facto* make the feedback coercive.

Chapter 4

1. The escalation dynamics hypotheses speak to the likelihood that the escalation scenarios offered in Chapter 3 will come to pass.
2. For an argument that private sector security vendor data can serve as credible evidence of State cyber behaviors, see J. D. Work, "Evaluating Commercial Cyber Intelligence Activity," *International Journal of Intelligence and CounterIntelligence* 33, no. 2 (2020): 278–308, https://doi.org/10.1080/08850607.2019.1690877. For a detailed presentation based on unsealed US criminal indictments of several highly publicized cyber campaigns, see Ben Buchanan, *The Hacker and the State: Cyber Attacks and the New Normal of Geopolitics* (Cambridge, MA: Harvard University Press, 2020).
3. Richard Bejtlich, "What Is APT and What Does It Want," *TaoSecurity*, January 16, 2010, https://taosecurity.blogspot.com/2010/01/what-is-apt-and-what-does-it-want.html. Alternatively, CrowdStrike uses the term to describe cyberattacks themselves: "An advanced persistent threat (APT) is a sophisticated, sustained cyberattack in which an intruder establishes an undetected presence in a network in order to steal sensitive data over a prolonged period of time. An APT attack is carefully planned and designed to infiltrate a specific organization, evade existing security measures and fly under the radar." CrowdStrike, "What Is an Advanced Persistent Threat (APT)?," November 18, 2019, https://www.crowdstrike.com/epp-101/advanced-persistent-threat-apt/.
4. Mandiant, "APT1: Exposing One of China's Cyber Espionage Units," http://it-report-lb-1-312482071.us-east-1.elb.amazonaws.com/.
5. FireEye, "Advanced Persistent Threat Groups," https://content.fireeye.com/apt-41/website-apt-groups.
6. Cybersecurity and Infrastructure Security Agency, "Alert (TA18-074A)," March 15, 2018, https://us-cert.cisa.gov/ncas/alerts/TA18-074A.
7. Dragos, "Allanite, Since 2017," https://www.dragos.com/threat/allanite/.
8. For a comprehensive list, see ThaiCERT (A Member of the Electronics Transactions Development Agency), *Threat Group Cards: A Threat Actor Encyclopedia*, June 19, 2019, https://www.thaicert.or.th/downloads/files/A_Threat_Actor_Encyclopedia.pdf.
9. Andy Greenberg, "France Ties Russia's Sandworm to a Multiyear Hacking Spree," *Wired*, February 15, 2021, https://www.wired.com/story/sandworm-centreon-russia-hack.
10. Cybersecurity and Infrastructure Security Agency, https://us-cert.cisa.gov/government-users/compliance-and-reporting/incident-definition#:~:text=An%20incident%20is%20the%20act,NIST%20Special%20Publication%20800%2D61.
11. *U.S. Federal Information Modernization Act of 2014, Annual Report to Congress* (Executive Office of the President of the United States, 2020), https://www.whitehouse.gov/wp-content/uploads/2020/05/2019-FISMARMAs.pdf.
12. *U.S. Federal Information Modernization Act of 2014, Annual Report to Congress* (Executive Office of the President of the United States, 2018), https://www.whitehouse.gov/wp-content/uploads/2019/08/FISMA-2018-Report-FINAL-to-post.pdf; *U.S. Federal Information Modernization Act of 2014, Annual Report to Congress* (Executive Office of the President of the United States, 2017), https://www.whitehouse.gov/wp-content/uploads/2017/11/FY2017FISMAReportCongress.pdf; and *U.S. Federal Information Modernization Act of 2014, Annual Report to Congress* (Executive Office of the President of the United States, 2016), https://www.whitehouse.gov/sites/whitehouse.gov/files/briefing-room/presidential-actions/related-omb-material/fy_2016_fisma_report%20to_congress_official_release_march_10_2017.pdf.

13. Andrea Thomas, "Germany Points Finger at Russia over Parliament Attack," *Wall Street Journal*, May 13, 2016, https://www.wsj.com/articles/germany-points-finger-at-russia-over-parliament-hacking-attack-1463151250.
14. "Targeting U.S. Technologies: A Trend Analysis of Reporting from Defense Industry," Defense Security Service, 2013, https://www.hsdl.org/?abstract&did=757213.
15. Koerner, "Inside the Cyberattack That Shocked the U.S. Government." Interestingly, the discovery was not motivated by an alert. It was unexpected; a consequence of a policy put in five months prior tasking security engineers to decrypt the Secure Sockets Layer (SSL) traffic that flows across the agency's digital network to get a clearer view of the data transiting in and out of the agency's systems.
16. Greenberg, "France Ties Russia's Sandworm to a Multiyear Hacking Spree."
17. Sean Lyngaas, "German Intelligence Agencies Warn of Russian Hacking Threats to Critical Infrastructure," *CyberScoop*, May 26, 2020, https://www.cyberscoop.com/german-intelligence-memo-berserk-bear-critical-infrastructure/.
18. Timberg and Nakashima, "Federal Investigators Find Evidence of Previously Unknown Tactics Used to Penetrate Government Networks." APT29 is a Russian State-sponsored APT.
19. Catalin Zimpanu, "DarkHotel Hackers Use VPN Zero-day to Breach Chinese Government Agencies," *ZDNet*, April 6, 2020, https://www.zdnet.com/article/darkhotel-hackers-use-vpn-zero-day-to-compromise-chinese-government-agencies/.
20. A September 14, 2020, alert from the Cybersecurity and Infrastructure Security Agency (CISA) addressing China APT activity noted, "CISA analysts consistently observe targeting, scanning, and probing of significant vulnerabilities within days of their emergence and disclosure. This targeting, scanning, and probing frequently leads to compromises at the hands of sophisticated cyber threat actors." See "Alert (AA20-258A), Chinese Ministry of State Security-Affiliated Cyber Threat Actor Activity," https://us-cert.cisa.gov/ncas/alerts/aa20-258a.
21. Kathleen Metrick, Parnian Najafi, and Jared Semrau, "Zero-Day Exploitation Increasingly Demonstrates Access to Money, Rather than Skill—Intelligence for Vulnerability Management, Part One," *FireEye*, April 6, 2020, https://www.fireeye.com/blog/threat-research/2020/04/zero-day-exploitation-demonstrates-access-to-money-not-skill.html; Lily Hay Newman, "The NSA Warns That Russia Is Attacking Remote Work Platforms," *Wired*, December 7, 2020, https://www.wired.com/story/nsa-warns-russia-attacking-vmware-remote-work-platforms/?bxid=5f996bca2f9a1f2c6c541de2&cndid=62549538&esrc=register-page&mbid=mbid%3DCRMWIR012019%0A%0A&source=EDT_WIR_NEWSLETTER_0_DAILY_ZZ&utm_brand=wired&utm_campaign=aud-dev&utm_content=A&utm_mailing=WIR_Daily_120720&utm_medium=email&utm_source=nl&utm_term=list1_p4; and National Security Agency, "Russian State-Sponsored Malicious Cyber Actors Exploit Known Vulnerability in Virtual Workspaces," December 7, 2020, https://www.nsa.gov/News-Features/Feature-Stories/Article-View/Article/2434988/russian-state-sponsored-malicious-cyber-actors-exploit-known-vulnerability-in-v/.
22. See Dan Goodin, "Microsoft Issues Emergency Patches for 4 Exploited 0-days in Exchange," *arsTECHNICA*, March 2, 2021, https://arstechnica.com/information-technology/2021/03/microsoft-issues-emergency-patches-for-4-exploited-0days-in-exchange/; and Ryan Naraine, "Microsoft: Multiple Exchange Server Zero-Days under Attack by Chinese Hacking Group," *SecurityWeek*, March 2, 2021, https://www.securityweek.com/microsoft-4-exchange-server-zero-days-under-attack-chinese-apt-group.
23. Ryan Naraine, "Microsoft Office Zero-Day Hit in Targeted Attacks," *SecurityWeek*, September 7, 2021, https://www.securityweek.com/microsoft-office-zero-day-hit-targeted-attacks.
24. Metrick, Najafi, and Semrau, "Zero-Day Exploitation Increasingly Demonstrates Access to Money, Rather than Skill."
25. Kevin Townsend, "New Law Will Help Chinese Government Stockpile Zero-Days," *SecurityWeek*, July 14, 2021, https://www.securityweek.com/new-law-will-help-chinese-government-stockpile-zero-days.
26. FireEye, "What Is a Zero-Day Exploit?," https://www.fireeye.com/current-threats/what-is-a-zero-day-exploit.html.

27. Recall in Chapter 3 that, of the 22,000 vulnerabilities reported in 2019 for the physical and logical layers of cyberspace, more than a third (40.2%) do not have a publicly available solution, such as a software upgrade, workaround, or software patch.
28. Ashley Leonard, "The Problem with Patching: 7 Top Complaints," *DARKReading*, April 22, 2016, https://www.darkreading.com/endpoint/the-problem-with-patching-7-top-complaints-/a/d-id/1325232.
29. Tara Seals, "Companies Take an Average of 100–120 Days to Patch Vulnerabilities," *InfoSecurity Magazine*, October 1, 2015, https://www.infosecurity-magazine.com/news/companies-average-120-days-patch/.
30. Andy Greenberg, "This Map Shows the Global Spread of Zero-Day Hacking Techniques," *Wired*, June 4, 2020, https://www.wired.com/story/zero-day-hacking-map-countries/.
31. Richard Yew, "How Fast Can You Patch? How to Buy Time during the Next Zero-day Vulnerability," *Medium*, October 1, 2018, https://medium.com/@VZMediaPlatform/how-fast-can-you-patch-how-to-buy-time-during-the-next-zero-day-vulnerability-6c5772ee3ba8.
32. United States Government Accountability Office, "DATA PROTECTION: Actions Taken by Equifax and Federal Agencies in Response to the 2017 Breach," August 2018, https://www.warren.senate.gov/imo/media/doc/2018.09.06%20GAO%20Equifax%20report.pdf. A second example illustrating how rapidly State-sponsored APT groups can exploit a vulnerability that has been publicly announced and for which a patch has been provided is the 2020 exploitation of F5 networking devices by Iranian-backed APT 33 and APT 34 groups. These groups were reportedly exploiting the vulnerability within two days of its announcement. See Catalin Cimpanu, "FBI Says an Iranian Hacking Group Is Attacking F5 Networking Devices," *ZDNet*, August 10, 2020, https://www.zdnet.com/article/fbi-says-an-iranian-hacking-group-is-attacking-f5-networking-devices/; and Catalin Cimpanu, "Hackers Are Trying to Steal Admin Passwords from F5 BIG-IP Devices," *ZDNet*, July 4, 2020, https://www.zdnet.com/article/hackers-are-trying-to-steal-admin-passwords-from-f5-big-ip-devices/#ftag=CAD-00-10aag7e.
33. Alfred Ng, "How the Equifax Hack Happened, and What Still Needs to Be Done," *C|Net*, September 7, 2018, https://www.cnet.com/news/equifaxs-hack-one-year-later-a-look-back-at-how-it-happened-and-whats-changed/.
34. This is consistent with data in the US Federal Information Modernization Act of 2014 annual reports to Congress referenced previously (i.e., nearly 25%, or approximately 7,000, of cyber security incidents reported by US federal government agencies in 2019 indicated no knowledge of the attack vector).
35. Mandiant, "Deep Dive into Cyber Reality: Security Effectiveness Report 2020," https://content.fireeye.com/security-effectiveness/rpt-security-effectiveness-2020-deep-dive-into-cyber-reality.
36. Every member of the G-7, including the United States, as well as Thailand and Australia, has banned "hacking back." In 2018, more than fifty countries—but not the United States—signed an agreement that private firms based in their nations are not allowed to hack back. See Scott Shackelford, "How Far Should Organizations Be Able to Go to Defend Against Cyberattacks?," *GCN*, February 19, 2019, https://gcn.com/articles/2019/02/19/hacking-back.aspx.
37. Laura Sullivan, "As China Hacked, U.S. Businesses Turned a Blind Eye," *NPR*, April 12, 2019, https://www.npr.org/2019/04/12/711779130/as-china-hacked-u-s-businesses-turned-a-blind-eye.
38. Ibid.
39. Institute of Directors, "Cybersecurity: Underpinning the Digital Economy," March 2016, https://www.iod.com/news/news/articles/Cyber-security-underpinning-the-digital-economy.
40. Companies in the UK are legally obligated under the GDPR to inform the Information Commissioner's Office if they suffer a breach involving personal information of customers or employees. Similar obligations exist under the Health Insurance Portability and Accountability Act of 1996 (HIPAA) in the United States or the Personal Information Protection and Electronic Documents Act (PIPEDA) in Canada.

41. Dan Swinhoe, "Why Businesses Don't Report Cybercrimes to Law Enforcement," *CSO*, May 30, 2019, https://www.csoonline.com/article/3398700/why-businesses-don-t-report-cybercrimes-to-law-enforcement.html. There also exists a stock-price-based incentive to not publicize a significant exploitation (via an 8-K report). Through a bluntly worded 2018 guidance, the US Securities and Exchange Commission sought to end the practice of omission. See US Securities and Exchange Commission, "Release Nos. 33-10459, 34-82746, Commission Statement and Guidance on Public Company Cybersecurity Disclosures, Security Exchange Commission," February 26, 2018, https://www.sec.gov/rules/interp/2018/33-10459.pdf.
42. Sean Lyngaas, "Norwegian Police Implicate Fancy Bear in Parliament Hack, Describe 'Brute Forcing' of Email Accounts," *CyberScoop*, December 8, 2020, https://www.cyberscoop.com/norwegian-police-implicate-fancy-bear-in-parliament-hack-describe-brute-forcing-of-email-accounts/.
43. Josh Fruhlinger, "Ransomware Explained: How It Works and How to Remove It," *CSO*, June 19, 2020, https://www.csoonline.com/article/3236183/what-is-ransomware-how-it-works-and-how-to-remove-it.html.
44. "The State of Ransomware 2020," https://www.sophos.com/en-us/medialibrary/Gated-Assets/white-papers/sophos-the-state-of-ransomware-2020-wp.pdf.
45. Michael Novinson, "The 11 Biggest Ransomware Attacks of 2020 (So Far)," *CRN*, June 30, 2020, https://www.crn.com/slide-shows/security/the-11-biggest-ransomware-attacks-of-2020-so-far-. In the referenced report, "significance" is measured by total costs of responding to the attack, including costs incurred by investigating the attack, rebuilding networks, restoring backups, paying the ransom, and putting preventative measures in place to avoid future incidents.
46. Gartzke, "The Myth of Cyberwar."
47. Harry Eckstein describes illustrative case studies as being often quite brief and falling short of the degree of detail needed either to explain a case fully or to test a theoretical proposition. Their aim is to give the reader a "feel" for a theoretical argument by providing a concrete example of its application, or to demonstrate the empirical relevance of a theoretical proposition by identifying at least one relevant case. Harry Eckstein. "Case Study and Theory in Political Science," in *Handbook of Political Science, Vol. 7: Strategies of Inquiry*, ed. Fred I. Greenstein and Nelson W. Polsby (Reading, MA: Addison-Wesley, 1975), 109.
48. United Nations, "Security Council Strengthens Sanctions on Democratic Republic of Korea, Unanimously Adopting Resolution 2321 (2016)," https://www.un.org/press/en/2016/sc12603.doc.htm.
49. Ibid.
50. United Nations, "Report of the Panel of Experts Established Pursuant to Resolution 1874 (2009), S/2019/691," https://www.securitycouncilreport.org/atf/cf/%7B65BFCF9B-6D27-4E9C-8CD3-CF6E4FF96FF9%7D/S_2019_691.pdf.
51. Michelle Nichols, "North Korea Took $2 Billion in Cyberattacks to Fund Weapons Program: U.N. Report," *Reuters*, August 5, 2019, https://www.reuters.com/article/us-northkorea-cyber-un/north-korea-took-2-billion-in-cyberattacks-to-fund-weapons-program-u-n-report-idUSKCN1UV1ZX.
52. "SWIFT Attackers' Malware Linked to More Financial Attacks," *Symantec Security Response*, May 26, 2016, https://www.symantec.com/connect/blogs/swift-attackers-malware-linked-more-financial-attacks.
53. See https://www.swift.com/about-us.
54. United States District Court for the Central District of California, *United States of America v. PARK JIN HYOK, also known as ("aka") "Jin Hyok Park," aka "Pak Jin Hek,"* Case No. MJ18-1749, https://www.justice.gov/opa/press-release/file/1092091/download.
55. Kim Zetter, "That Insane, $81M Bangladesh Bank Heist? Here's What We Know," *Wired*, May 17, 2016, https://www.wired.com/2016/05/insane-81m-bangladesh-bank-heist-heres-know/.
56. The regime continues apace with this activity. For example, in September 2020 it illicitly acquired through cyber exploitation $281 million worth of assets from a cryptocurrency exchange. See Michelle Nichols and Raphael Satter, "U.N. Experts Point Finger at North Korea

for $281 Million Cyber Theft, KuCoin Likely Victim," *Reuters*, February 9, 2021, https://www.reuters.com/article/us-northkorea-sanctions-cyber-idCAKBN2AA00Q.

57. United Nations, "Report of the Panel of Experts Established Pursuant to Resolution 1874 (2009), S/2019/691."
58. "U.S. Sanctions North Korean Hackers for Swift Hack, Wannacry and Other Cyberattacks That Fund Its Weapons Programs," *Bloomberg*, September 14, 2019, https://www.japantimes.co.jp/news/2019/09/14/asia-pacific/u-s-sanctions-north-korean-hackers-swift-hack-wannacry-cyberattacks-fund-weapons-programs/#.XkAVcjFKjIU.
59. Ed Caesar, "The Incredible Rise of North Korea's Hacking Army," *New Yorker*, April 19, 2021, https://www.newyorker.com/magazine/2021/04/26/the-incredible-rise-of-north-koreas-hacking-army.
60. See Matthew Ha and David Maxwell, "Kim Jong Un's All-Purpose Sword," *Foundation for the Defense of Democracies Report* (October 2018), https://www.fdd.org/analysis/2018/10/03/kim-jong-uns-all-purpose-sword/ and Kong Ji Young, Kim Kyoung Gon, and Lim Jong In, "The All-Purpose Sword: North Korea's Operations and Strategies," *2019 11th International Conference on Cyber Conflict*, https://ccdcoe.org/uploads/2019/06/Art_08_The-All-Purpose-Sword.pdf.
61. Stephanie Kleine-Ahlbrandt, "North Korea's Illicit Cyber Operations: What Can Be Done?," *38 North*, February 28, 2020, https://www.38north.org/2020/02/skleineahlbrandt022820/.
62. Ibid.
63. Ibid.
64. See, for example, National Counterintelligence and Security Center, *Foreign Economic Espionage in Cyberspace*, 2018, https://www.dni.gov/files/NCSC/documents/news/20180724-economic-espionage-pub.pdf.
65. Lisa Ferdinando, "DoD Officials: Chinese Actions Threaten U.S. Technological, Industrial Base," *DOD News*, June 21, 2018, https://www.defense.gov/Explore/News/Article/Article/1557188/.
66. For a report estimating annual damage in dollars, see Office of the United States Trade Representative: Executive Office of President, *Findings of the Investigation into China's Acts, Policies, and Practices Related to Technology Transfer, Intellectual Property, and Innovation under Section 301 of the Trade Act of 1974*, March 22, 2018, https://ustr.gov/sites/default/files/Section%20301%20FINAL.PDF. For a report on impact on US GDP (between 1% and 3%), see Executive Office of the President of the United States, *Annual Intellectual Property Report to Congress*, February 2019, https://www.whitehouse.gov/wp-content/uploads/2019/02/IPEC-2018-Annual-Intellectual-Property-Report-to-Congress.pdf.
67. That said, some push the economic angle even further by arguing massive IP theft will inevitably enable China to alter the current rules-based global economic system in ways that favor their interests over those of the United States.
68. See Andrea Gilli and Mauro Gilli, "Why China Has Not Caught Up Yet: Military-Technological Superiority and the Limits of Imitation, Reverse Engineering, and Cyber Espionage," *International Security* 43, no. 3 (Winter 2018/19): 141–189, https://www.mitpressjournals.org/doi/full/10.1162/isec_a_00337.
69. See Indermit S. Gill and Homi Kharas, "The Middle Income Trap Turns 10," *World Bank Group*, August 2015, http://documents.worldbank.org/curated/en/291521468179640202/pdf/WPS7403.pdf; and Jan Rudengren, Lars Rylander, and Claudia Rives Casanova, "It's Democracy, Stupid: Reappraising the Middle-Income Trap," *Institute for Security and Development Policy*, 2014, https://www.files.ethz.ch/isn/184240/2014-rudengren-rylander-casanova-reappraising-the-middle-income-trap.pdf.
70. *Findings of the Investigation into China's Acts.*
71. For a more in-depth and comprehensive presentation of this case, see Michael P. Fischerkeller, "Influencing China's Trajectory via Defend Forward / Persistent Engagement in Cyberspace.".
72. Hu Shuu, Zhu Changzheng, and Yang Zheyu, "Liu He on China's New Transformation Trail," *Caixin Online*, October 28, 2010, http://english.caing.com/2010-11-08/100196829.html. Japan, South Korea, and Singapore spent twenty-one, twenty-three, and twenty-nine years, respectively, as middle-income economies before moving up to the upper-income level.
73. Shuu, Changzheng, and Zheyu, "Liu He on China's New Transformation Trail."

74. See, respectively, Indermit Gill, "Future Development Reads: Xi Jingping, China's People's Party, and the Middle-Income Trap," *Brookings*, October 20, 2017, https://www.brookings.edu/blog/future-"development/2017/10/20/future-development-reads-xi-jinping-chinas-peoples-party-and-the-middle-income-trap/; and "China May Be Running Out of Time to Escape the Middle-Income Trap," *Asia Society*, October 2017, https://asiasociety.org/new-york/china-may-be-running-out-time-escape-middle-income-trap.
75. The volume is entitled *Trap or Wall: Real Challenges and Strategic Choice in China's Economy*, http://en.drc.gov.cn/2014-06/26/content_17617382.htm. The World Bank and Development Research Center of the State Council, the People's Republic of China, 2013, *China 2030: Building a Modern, Harmonious and Creative Society*, https://www.worldbank.org/en/news/feature/2012/02/27/china-2030-executive-summary.
76. Ambassador Chas W. Freeman, Jr. (USFS, Ret.), "China as a Great Power: Remarks to China Renaissance Capital Investors," *Middle East Policy Council*, https://mepc.org/speeches/china-great-power.
77. David Wertime, "Unpacking Xi Jinping's Pet Phrase for U.S.-China Ties," *Foreign Policy*, September 23, 2015, https://foreignpolicy.com/2015/09/23/unpacking-xi-jinpings-pet-phrase-new-model-of-great-power-relations-us-china-explainer/.
78. Chris Buckley and Keith Bradsher, "Xi Jinping's Marathon Speech: Five Takeaways," *New York Times*, October 18, 2017, https://www.nytimes.com/2017/10/18/world/asia/china-xi-jinping-party-congress.html.
79. Charles Clover, "Xi Jinping Signals Departure from Low-Profile Policy," *Financial Times*, October 20, 2017, https://www.ft.com/content/05cd86a6-b552-11e7-a398-73d59db9e399.
80. Ibid.
81. In his initial phone conversation with US president Joe Biden, President Xi stated that the countries must treat each other as equals. Shannon Tiezzi, "First Biden-Xi Phone Call Shows Not Much Has Changed in US-China Relations," *The Diplomat*, February 12, 2020, https://thediplomat.com/2021/02/first-biden-xi-phone-call-shows-not-much-has-changed-in-us-china-relations/.
82. See Damien Ma, "Can China Avoid the Middle Income Trap?," *Foreign Policy*, March 12, 2016, https://foreignpolicy.com/2016/03/12/can-china-avoid-the-middle-income-trap-five-year-plan-economy-two-sessions/; and Edward Wong, "China Aims for 6.5% Economic Growth over Next 5 Years, Xi Says," *New York Times*, November 3, 2015, https://www.nytimes.com/2015/11/04/world/asia/china-economic-growth-xi.html.
83. Erik Roth, Jeongmin Seong, and Jonathan Woetzel, "Gauging the Strength of Chinese Innovation," *McKinsey Quarterly* 4 (2015): 66–73, https://www.mckinsey.com/~/media/McKinsey/McKinsey%20Quarterly/Digital%20Newsstand/2015%20Issues%20McKinsey%20Quarterly/Agility.ashx; and McKinsey Global Institute, *The China Effect on Global Innovation*, October 2015, https://www.mckinsey.com/~/media/McKinsey/Featured%20Insights/Innovation/Gauging%20the%20strength%20of%20Chinese%20innovation/MGI%20China%20Effect_Full%20report_October_2015.ashx.
84. See "State Council Decision on Accelerating the Development of Strategic Emerging Industries," State Council, Guo Fa, 2010, no. 32 (October 10, 2010), https://chinaenergyportal.org/wp-content/uploads/2017/01/Development-Strategic-Emerging-Industries.pdf; and "China's Top Political Stresses Indigenous Innovation," *Xinhua English*, April 19, 2011, http://english.sina.com/china/p/2011/0419/369456.html.
85. See Office of the United States Trade Representative: Executive Office of President, *Section 301 Investigation and Hearing: China's Acts, Policies, and Practices Related to Technology Transfer, Intellectual Property, and Innovation*, October 10, 2017, https://ustr.gov/sites/default/files/enforcement/301Investigations/China%20Technology%20Transfer%20Hearing%20Transcript.pdf; Sean O'Connor, "How Chinese Companies Facilitate Technology Transfer from the United States," *U.S.-China Economic and Security Review Commission*, May 6, 2019, https://www.uscc.gov/sites/default/files/Research/How%20Chinese%20Companies%20Facilitate%20Tech%20Transfer%20from%20the%20US.pdf; and National Counterintelligence and Security Center, *Foreign Economic Espionage in Cyberspace*, 2018, https://www.dni.gov/files/NCSC/documents/news/20180724-economic-espionage-pub.pdf.

86. Fischerkeller, "Opportunity Seldom Knocks Twice." For a similar argument placed within a longer Chinese historical context, see Julian Baird Gewirtz, "China's Long March to Technological Supremacy: The Roots of Xi Jinping's Ambition to 'Catch Up and Surpass','" *Foreign Affairs*, August 27, 2019, https://www.foreignaffairs.com/articles/china/2019-08-27/chinas-long-march-technological-supremacy.
87. Fischerkeller, "Opportunity Seldom Knocks Twice."
88. See, respectively, Mandiant, "APT1: Exposing One of China's Cyber Espionage Units," http://it-report-lb-1-312482071.us-east-1.elb.amazonaws.com/; and "Targeting U.S. Technologies: A Trend Analysis of Reporting from Defense Industry," Defense Security Service, 2011, https://premium.globalsecurity.org/intell/library/reports/2011/2011-dss-targeting-us-tech.pdf. All 2010–2013 Defense Security Service reports were reviewed.
89. Mandiant, "APT1."
90. U.S.-China Economic and Security Review Commission, *Hearing on Chinese Intelligence Services and Espionage Operations, written testimony of Peter Mattis*, June 9, 2016, https://www.uscc.gov/sites/default/files/Peter%20Mattis_Written%20Testimony060916.pdf.
91. Mandiant, "APT1."
92. Ibid.
93. Michael Brown and Pavneet Singh, "China's Technology Transfer Strategy: How Chinese Investments in Emerging Technology Enable a Strategic Competitor to Access the Crown Jewels of U.S. Innovation," Defense Innovation Unit Experimental (DIUx), January 2018, https://admin.govexec.com/media/diux_chinatechnologytransferstudy_jan_2018_(1).pdf.
94. Ibid.
95. *Hearing before the U.S.-China Economic and Security Review Commission, Commercial Cyber Espionage and Barriers to Digital Trade in China, written Testimony of Jen Weedon, Manager, Threat Intelligence and Strategic Analysis, FireEye and Mandiant, Inc.*, June 15, 2015, https://www.uscc.gov/sites/default/files/Weedon%20Testimony.pdf.
96. Mandiant, "APT1."
97. Defense Security Service, "Targeting U.S. Technologies: A Trend Analysis of Reporting from Defense Industry," 2011, https://premium.globalsecurity.org/intell/library/reports/2011/2011-dss-targeting-us-tech.pdf.
98. DCSA reports provide regional, rather than country-specific data, but by almost any resource measure China overwhelms other members in the East Asian and Pacific region of which it is a member. When coupled with China's expressed economic ambitions, it is reasonable to conclude activity reporting for this group primarily represents Chinese behavior.
99. Defense Security Service, "Targeting U.S. Technologies: A Trend Analysis of Reporting from Defense Industry," 2012, https://www.dcsa.mil/Portals/69/documents/about/err/2012_Trend_Analysis_Report.pdf; Defense Security Service, "Targeting U.S. Technologies: A Trend Analysis of Reporting from Defense Industry," 2013.
100. Defense Security Service, "Targeting U.S. Technologies: A Trend Analysis of Reporting from Defense Industry," 2012, https://www.dcsa.mil/Portals/69/documents/about/err/2012_Trend_Analysis_Report.pdf; Defense Security Service, "Targeting U.S. Technologies: A Trend Analysis of Reporting from Defense Industry," 2013.
101. Defense Security Service, "Targeting U.S. Technologies: A Trend Analysis of Reporting from Defense Industry," 2014, https://www.dcsa.mil/Portals/69/documents/about/err/2014_Trend_Analysis_Report.pdf. Interestingly, this report notes that "The prominence of the SNA MO endured despite significant press coverage during FY13 detailing the results of Western research on and analysis of recent East Asia and the Pacific cyber activities. Reporting cataloged much of the infrastructure; command and control protocols; and tactics, techniques, and procedures (TTPs) East Asia and the Pacific cyber actors used. Industry submissions during the period immediately after the revelations decreased precipitately compared to the same period in 2012." The report's authors are referring, of course, to the release of the Mandiant Report.
102. Gilli and Gilli, "Why China Has Not Caught up Yet."
103. Ibid. This evidence does not discredit Gilli and Gilli's arguments; rather, it suggests conclusions from their important, but highly specific, analysis of state-of-the-art weapons

systems should not be generalized uncritically to other technologies and technological processes.

104. United States District Court, Western District of Pennsylvania, *United States of America v. Wang Dong, Sun Kailiang, Wen Xinyu, Huang Zhenyu, Gu Chunhui*, Criminal No. 14-118, https://www.justice.gov/iso/opa/resources/5122014519132358461949.pdf.
105. United States District Court, Southern District of California, *United States of America v. Zhang Zhang-Gui, Zha Rong, Chai Meng, Liu Chunliang, Gao Hong Kun, Zhuang Xiaowei, Ma Zhiqi, Li Xiao, Gu Gen, Tian Xi*, Case No. 13CR3132-H, https://www.justice.gov/opa/press-release/file/1106491/download.
106. It generally takes a few years, from intrusion to indictment, before cases of cyber-enabled IP theft become public knowledge through published indictments. Thus, given recent FBI comments regarding the number of IP theft cases currently under investigation, we should expect to see more indictments from this same period. See Mark Hosenball and David Brunnstrom, "To Counter Huawei, U.S. Could Take 'Controlling Stake' in Ericsson, Nokia: Attorney General," *Reuters*, February 6, 2020, https://www.reuters.com/article/us-usa-china-espionage/top-u-s-officials-to-spotlight-chinese-spy-operations-pursuit-of-american-secrets-idUSKBN2001DL?utm_campaign=wp_the_cybersecurity_202&utm_medium=email&utm_source=newsletter&wpisrc=nl_cybersecurity202.
107. United States District Court, Western District of Pennsylvania, *United States of America v. Wang Dong*.
108. Office of the United States Trade Representative: Executive Office of President, *Section 301 Investigation and Hearing*.
109. Ibid.
110. See John W. Miller, "Steelmaker Alleges Chinese Government Hackers Stole Plans for Developing New Steel Technology," *Wall Street Journal*, April 28, 2016, https://www.wsj.com/amp/articles/u-s-steel-accuses-china-of-hacking-1461859201; and David Lawder and Ruby Lian, "U.S. Panel Launches Trade Secret Theft Probe into China Steel," *Business News*, May 26, 2016, https://www.reuters.com/article/us-usa-china-steel-idUSKCN0YH2KX.
111. John W. Miller, "Steelmaker Alleges Chinese Government Hackers Stole Plans for Developing New Steel Technology."
112. Kenneth Rapoza, "Westinghouse Electric's Chinese 'Trojan Horse'," May 17, 2016, *Forbes*, https://www.forbes.com/sites/kenrapoza/2016/05/17/westinghouse-electrics-chinese-trojan-horse/amp/.
113. United States District Court, Western District of Pennsylvania, *United States of America v. Wang Dong*.
114. China General Nuclear Power Group, "China General Nuclear Power Group's Fangchenggang-3 Begins Construction with First Concrete Pour," January 8, 2016, https://electricenergyonline.com/article/organization/29681/559216/China-General-Nuclear-Power-Group-s-Fangchenggang-3-begins-construction-with-first-concrete-pour.htm.
115. Ibid.
116. Gopal Ratnam, "Underground Hackers and Spies Helped China Steal Jet Secrets: Crowdstrike Researchers Reveal Beijing's Efforts to Boost Its Own Domestic Aircraft Industry," *Rollcall*, October 15, 2019, https://www.rollcall.com/news/policy/hackers-spies-helped-china-steal-jet-secrets-report-says. Note that from 2010 to 2015, MSS cyber operators allegedly targeted several other aerospace firms, including Ametek and Honeywell, which manufacture aircraft parts.
117. Frank Fang, "Cybersecurity Firm Details How China Hacked Western Firms to Steal Aviation Tech," *Epoch Times*, October 16, 2019, https://www.theepochtimes.com/cybersecurity-firm-details-how-china-hacked-western-firms-to-steal-aviation-tech_3118899.html.
118. Keith Crane, Jill E. Luoto, Scott Warren Harold, David Yang, Samuel K. Berkowitz, and Xiao Wang, *The Effectiveness of China's Industrial Policies in Commercial Aviation Manufacturing* (Santa Monica, CA: Rand Corporation, 2014), https://www.rand.org/content/dam/rand/pubs/research_reports/RR200/RR245/RAND_RR245.pdf.
119. United States District Court, Southern District of California, *United States of America v. Zhang Zhang-Gui*.
120. Ibid.

121. Fang, "Cybersecurity Firm Details How China Hacked Western Firms." Stephen Trimble, "China Completes Assembly of First High-Bypass Turbofan Engine," *Flight Global*, December 29, 2017, https://www.flightglobal.com/systems-and-interiors/china-completes-assembly-of-first-high-bypass-turbofan-engine/126587.article.
122. Fang, "Cybersecurity Firm Details How China Hacked Western Firms."
123. International Monetary Fund, "Word Economic Outlook Database," https://www.imf.org/en/Countries/CHN#data. This same average is found in data provided by "Trading Economics: National Bureau of Statistics of China," https://tradingeconomics.com/china/gdp-growth-annual. World Bank data for 2017 to 2019 shows an average of 6.5 percent growth. See World Bank, https://data.worldbank.org/country/china.
124. Cornell University, INSEAD, and WIPO, *Global Innovation Index 2019* (Ithaca, Fontainebleau, and Geneva: WIPO, 2019), https://www.globalinnovationindex.org/userfiles/file/reportpdf/GII2019-keyfinding-E-Web3.pdf.
125. "GDP Revisions Put China on Target to Double Economy, but Data Doubts Remain," *Reuters*, November 21, 2019, https://www.reuters.com/article/us-china-economy-gdp/gdp-revisions-put-china-on-target-to-double-economy-but-data-doubts-remain-idUSKBN1XW04C. A far more aggressive claim that China "falsifies economic statistics" has been made by the US-China Economic and Security Review Commission. See "2019 Report to Congress of the US-China Economic and Security Review Commission," November 2019, https://www.uscc.gov/sites/default/files/2019-11/2019%20Annual%20Report%20to%20Congress.pdf.
126. Frank Tang, "China Set to Break Key Economic Barrier despite Trade War, but Can It Avoid the Middle Income Trap?," *South China Morning Post*, January 1, 2020, https://www.scmp.com/economy/china-economy/article/3044124/china-set-break-key-economic-barrier-despite-trade-war-can-it.
127. The emphasis on *consistently* is important. That China uses a consistent methodology year-over-year does not challenge the argument that growth was sustained post-2015. Rather, it argues for a caveat that growth was sustained but at a rate lower than NBS announced while being *consistent* with lower, unpublished rates from prior years. For a discussion of NBS's methodology, see Wi Chen, Xilu Chen, Chang-Tai Hsieh, and Zheng (Michael) Song, "A Forensic Examination of China's National Accounts," *Brookings Papers on Economic Activity*, https://www.brookings.edu/wp-content/uploads/2019/03/bpea-2019-forensic-analysis-china.pdf.
128. Recorded Future, Insikt Group, "How North Korea Revolutionized the Internet as a Tool for Rogue Regimes," February 2020, https://www.recordedfuture.com/north-korea-internet-tool/.
129. Buckley and Bradsher, "Xi Jinping's Marathon Speech."
130. Eckstein, "Case Study and Theory in Political Science."
131. See, for example, Paul Diehl and Gary Goertz, *War and Peace in International Rivalry* (Ann Arbor: University of Michigan Press, 2000); and William Thompson, "Identifying Rivals and Rivalries in World Politics," *International Studies Quarterly* 45, no. 4 (December 2001): 557–586, https://www.jstor.org/stable/3096060.
132. Diehl and Goertz, *War and Peace in International Rivalry*.
133. Ryan C. Maness, Brandon Valeriano, and Benjamin Jensen, *Code Book for the Dyadic Cyber Incident and Dispute Dataset* (Version 1.1), http://www.brandonvaleriano.com/uploads/8/1/7/3/81735138/dcid_1.1_codebook.pdf. The dataset itself is available at https://drryanmaness.wixsite.com/cyberconflict/cyber-conflict-dataset.
134. See, respectively, James P. Klein, Gary Goertz, and Paul F, Diehl, "The New Rivalry Dataset: Procedures and Patterns," *Journal of Peace Research* 43, no. 3 (May 2006): 331–348, https://www.jstor.org/stable/27640320; and Thompson, "Identifying Rivals and Rivalries in World Politics."
135. Maness, Valeriano, and Jensen, *Code Book*.
136. Ibid.
137. Ibid.
138. Dmitri Alperavitch, "Revealed: Operation Shady RAT" (McAfee White Paper, 2011), http://www.a51.nl/sites/default/files/pdf/wp_operation_shady_rat.pdf. It has been

argued that ShadyRAT is a China State-sponsored group. Bruce Sterling, "Operation Shady RAT," *Wired,* August 3, 2011, https://www.wired.com/2011/08/operation-shady-rat/.

139. Martin C. Libicki, "Correlations between Cyberspace Attacks and Kinetic Attacks," in *2020 12th International Conference on Cyber Conflict—20/20 Vision: The Next Decade,* ed. T. Jančárková, L. Lindström, M. Signoretti, I. Tolga, and G. Visky (Tallinn: NATO CCDCOE, 2020), 199–213, https://ccdcoe.org/uploads/2020/05/CyCon_2020_11_Libicki.pdf.
140. Ibid. The emphasis is Libicki's.
141. Ibid. For an earlier explanation of why such retaliation is unlikely to be seen as a credible deterrent, see Richard J. Harknett, John Callaghan, and Rudi Kaufmann, "Leaving Deterrence Behind: War-fighting and National Cybersecurity," *Journal of Homeland Security and Emergency Management* 7, no. 1 (Spring 2010): 1–24, https://doi.org/10.2202/1547-7355.1636.
142. See, for example, Pierluigi Pagolini, "IDF Hit Hamas, It Is the First Time a State Launched an Immediate Physical Attack in Response to a Cyber Attack," *SecurityAffairs,* May 6, 2019, https://securityaffairs.co/wordpress/85022/cyber-warfare-2/idf-hit-hamas.html.
143. Robert Chesney, "Crossing a Cyber Rubicon? Overreactions to the IDF's Strike on the Hamas Cyber Facility," *Lawfare,* May 6, 2019, https://www.lawfareblog.com/crossing-cyber-rubicon-overreactions-idfs-strike-hamas-cyber-facility.
144. Judah Ari Gross, "IDF Says It Thwarted a Hamas Cyber Attack during Weekend Battle," *The Times of Israel,* May 5, 2019, https://www.timesofisrael.com/idf-says-it-thwarted-a-hamas-cyber-attack-during-weekend-battle/. Interestingly, in terms of cyber *faits accomplis,* 2019–2020 cyber interactions between Israel and Iran, even those probing critical infrastructures, remained limited to attempts to unbalance the other side without nearing an explicit crossover to kinetic-equivalent effects. This may be the best example of an intense cyber interaction between rivals that one might expect escalation and yet none occurred, primarily, it seems, because they were not seeking to coerce each other but rather set and reset conditions of security in parallel operations. See J. D. Work and Richard J. Harknett, "Troubled Vision: Understanding Recent Israel-Iranian Offensive Cyber Exchanges," *Atlantic Council of the United States Issue Brief,* July 22, 2020, https://www.atlanticcouncil.org/in-depth-research-reports/issue-brief/troubled-vision-understanding-israeli-iranian-offensive-cyber-exchanges/.
145. Kevin Townsend, "92% of External Web Apps Have Exploitable Security Flaws or Weaknesses: Report," *SecurityWeek,* October 30, 2018, https://www.securityweek.com/92-external-web-apps-have-exploitable-security-flaws-or-weaknesses-report. Shadow IT applications are defined as IT systems, devices, software, applications, and services being used without explicit IT department approval.
146. For more on the challenge to security posed by shadow, legacy, and abandoned applications, see Application Security Services, "Murder of Cybersecurity by Legacy Applications," *ImmuniWeb,* September 21, 2018, https://www.immuniweb.com/blog/murder-of-cybersecurity-by-legacy-applications.html. For the impact that moving to a cloud environment has on the same, see "A Game Changer in IT Security," *MIT Technology Review Insights,* September 8, 2021, https://www.technologyreview.com/2021/09/08/1034262/a-game-changer-in-it-security/.
147. United States Senate, Committee on Homeland Security and Governmental Affairs, *Federal Cybersecurity: America's Data Still at Risk,* August 2021, https://www.hsgac.senate.gov/imo/media/doc/Federal%20Cybersecurity%20-%20America%27s%20Data%20Still%20at%20Risk%20%28FINAL%29.pdf, 25.
148. Ibid., iv.
149. Josh Margolin and Ivan Pereira, "Outdated Computer System Exploited in Florida Water Treatment Plant Hack," *ABC News,* https://abcnews.go.com/amp/US/outdated-computer-system-exploited-florida-water-treatment-plant/story?id=75805550.
150. Catalin Cimpanu, "Centreon Says Only 15 Entities Were Targeted in Recent Russian Hacking Spree," *ZDNet,* February 16, 2021, https://www.zdnet.com/google-amp/article/centreon-says-only-15-entitites-were-targeted-in-recent-russian-hacking-spree/.
151. François Delerue, Alix Desforges, and Aude Géry, "A Close Look at France's New Military Cyber Strategy," *War on the Rocks,* April 23, 2019, https://warontherocks.com/2019/04/

a-close-look-at-frances-new-military-cyber-strategy/; and Matthias Schulze, "German Military Cyber Operations Are in a Legal Gray Zone," *Lawfare*, April 8, 2020, https://www.lawfareblog.com/german-military-cyber-operations-are-legal-gray-zone.

152. Dan Sabbagh, "UK Unveils National Cyber Force of Hackers to Target Foes Digitally," *The Guardian*, November 19, 2020, https://www.theguardian.com/technology/2020/nov/19/uk-unveils-national-cyber-force-of-hackers-to-target-foes-digitally?utm_campaign=wp_the_cybersecurity_202&utm_medium=email&utm_source=newsletter&wpisrc=nl_cybersecurity202.
153. Arthur P. B. Laudrain, "France's New Offensive Cyber Doctrine," *Lawfare*, February 26, 2019, https://www.lawfareblog.com/frances-new-offensive-cyber-doctrine,
154. Schulze, "German Military Cyber Operations Are in a Legal Gray Zone."
155. The term "botnet" (short for "robot network") is a network of computers infected by malware that is under the control of a single managing entity.
156. Ellen Nakashima, "New Details Emerge about 2014 Russian Hack of the State Department: It Was 'Hand to Hand Combat'," *Washington Post*, April 3, 2017, https://www.washingtonpost.com/world/national-security/new-details-emerge-about-2014-russian-hack-of-the-state-department-it-was-hand-to-hand-combat/2017/04/03/d89168e0-124c-11e7-833c-503e1f6394c9_story.html.
157. Ibid.
158. David E. Sanger, Nicole Perlroth, and Julian E. Barnes, "As Understanding of Russian Hacking Grows, So Does Alarm," *New York Times*, January 2, 2020, https://www.nytimes.com/2021/01/02/us/politics/russian-hacking-government.html?referringSource=articleShare.
159. Nakashima, "New Details Emerge about 2014 Russian Hack of the State Department."
160. Matthew J. Schwartz, "Dridex Malware Campaign Disrupted," *BankInfoSecurity*, October 14, 2015, https://www.bankinfosecurity.com/dridex-malware-campaign-disrupted-a-8590.
161. US Department of Justice, "Bugat Botnet Administrator Arrested and Malware Disabled," October 13, 2015, https://www.justice.gov/opa/pr/bugat-botnet-administrator-arrested-and-malware-disabled.
162. Shane Schick, "Dridex Trojan Remains a Risk Even Following Takedown Operation and FBI Arrest," *SecurityIntelligence*, October 19, 2015, https://securityintelligence.com/news/dridex-trojan-remains-a-risk-even-following-takedown-operation-and-fbi-arrest/.
163. Eduard Kovacs, "Dridex Still Active after Takedown Attempt," *SecurityWeek*, October 19, 2015, https://www.securityweek.com/dridex-still-active-after-takedown-attempt.
164. US Department of Homeland Security officials fear that a ransomware attack on US state or local voter registration offices and related systems could disrupt preparations for the election or cause confusion or long lines on Election Day. Ellen Nakashima, "Cyber Command Has Sought to Disrupt the World's Largest Botnet, Hoping to Reduce Its Potential Impact on the Election," *Washington Post*, October 9, 2020, https://www.washingtonpost.com/national-security/cyber-command-trickbot-disrupt/2020/10/09/19587aae-0a32-11eb-a166-dc429b380d10_story.html.
165. Brian Krebs, "Attacks Aimed at Disrupting the Trickbot Botnet," *Krebs on Security*, October 2, 2020, https://krebsonsecurity.com/2020/10/attacks-aimed-at-disrupting-the-trickbot-botnet/.
166. Brian Krebs, "Report: U.S. Cyber Command behind Trickbot Tricks," October 10, 2020, *Krebs on Security*, https://krebsonsecurity.com/2020/10/report-u-s-cyber-command-behind-trickbot-tricks/.
167. Nakashima, "Cyber Command Has Sought to Disrupt the World's Largest Botnet."
168. Nakashima cited Mark Arena, chief executive officer of Intel 471, and Krebs cited Alex Holden, chief information security officer and president of Hold Security.
169. Liviu Arsene and Radu Tudorica, "Trickbot Is Dead: Long Live Trickbot!" *BitDefender*, November 23, 2020, https://labs.bitdefender.com/2020/11/trickbot-is-dead-long-live-trickbot/.
170. Quoted in Tim Starks, "It's Hard to Keep a Big Botnet Down: TrickBot Sputters Back toward Full Health," *CyberScoop*, November 30, 2020, https://www.cyberscoop.com/trickbot-status-microsoft-cyber-command-takedown/.

171. Quoted in Nakashima, "Cyber Command Has Sought to Disrupt the World's Largest Botnet, Hoping to Reduce Its Potential Impact on the Election."
172. Liviu Arsene and Radu Tudorica, "Trickbot Is Dead: Long Live Trickbot!"
173. "Report of the Group of Governmental Experts on Advancing Responsible State Behaviour in Cyberspace in the Context of International Security," May 28, 2021, https://front.un-arm.org/wp-content/uploads/2021/06/final-report-2019-2021-gge-1-advance-copy.pdf.
174. Although it is certainly the case that, within the context of armed conflict, cyber capabilities could be employed in manners aligned with the coercive logic of the nuclear and conventional strategic environment, it should not be assumed that they will be used in that manner in that context.
175. Chris Bing, "U.S. Cyberwarriors Are Getting Better at Fighting ISIS Online, Says Top General," *CyberScoop*, May 23, 2017, https://www.cyberscoop.com/paul-nakasone-isis-cyber-attacks-army-cyber-command/.
176. Joseph Cox, "How U.S. Military Hackers Prepared to Hack the Islamic State," *Vice*, August 1, 2018, https://www.vice.com/en_us/article/ne5d5g/how-us-military-cybercom-hackers-hacked-islamic-state-documents.
177. Dina Temple-Raston, "How the U.S. Hacked Isis," *NPR*, September 26, 2019, https://www.npr.org/2019/09/26/763545811/how-the-u-s-hacked-isis.
178. Ellen Nakashima, "U.S. Military Cyber Operation to Attack ISIS Last Year Sparked Heated Debate over Alerting Allies," *Washington Post*, May 9, 2017, https://www.washingtonpost.com/world/national-security/us-military-cyber-operation-to-attack-isis-last-year-sparked-heated-debate-over-alerting-allies/2017/05/08/93a120a2-30d5-11e7-9dec-764dc781686f_story.html. Evidence of these target sets is also present in highly redacted documents on the operations received by Motherboard through the Freedom of Information Act. See https://www.documentcloud.org/documents/4624362-Cybercom-Operation-Glowing-Symphony-Documents.html.
179. Nakashima, "U.S. Military Cyber Operation to Attack ISIS."
180. Lydia Khalil from the Lowy Institute, quoted in Stephanie Borys, "Licence to Hack: Using a Keyboard to Fight Islamic State," *ABC News Australia*, December 17, 2019, https://www.abc.net.au/news/2019-12-18/inside-the-islamic-state-hack-that-crippled-the-terror-group/11792958?nw=0. This is consistent with General Nakasone's assessment made three years after Operation Glowing Symphony commenced: "we were seeing a series of videos and posts and media products that were high-end. We haven't seen that recently." Quoted in Temple-Raston, "How the U.S. Hacked Isis."
181. Quoted in Temple-Raston, "How the U.S. Hacked ISIS."
182. Ibid.
183. Given that open source data detail significant volumes of State cyber behavior in and through cyberspace over the last decade or more, one counterfactual that is not considered is that States are exercising significant restraint in and through cyberspace.
184. Schelling, *Arms and Influence*, 21. This should not be confused with how the term "brute force attack" is understood in the cyber lexicon, where it describes a cryptographic hack that relies on guessing possible combinations of a targeted password until the correct password is discovered.
185. The oft-cited example of such activity between States not in militarized crises or at war is the Stuxnet campaign. But legal experts have concluded that, although the campaign could be considered a use of force, it did not constitute an armed attack. Kim Zetter, "Legal Experts: Stuxnet Attack on Iran Was Illegal 'Act of Force'," *Wired*, March 25, 2013, https://www.wired.com/2013/03/stuxnet-act-of-force/.

Chapter 5

1. Schelling captured these dynamics in his application of the game of chicken. Schelling, *The Strategy of Conflict*.
2. See Chapter 4 for a discussion of factors that may act as constraints.
3. This approach could include, for example, a "zero trust" defense-in-depth security architecture. Zero trust is described as providing a collection of concepts and ideas designed

to minimize uncertainty in enforcing accurate, least privilege per-request access decisions in information systems and services in the face of a network viewed as compromised. See Scott Rose, Oliver Borchert, Stu Mitchell, and Sean Connelly, *Zero Trust Architecture* (Gaithersburg, MD: National Institute of Standards and Technology, 2020), https://nvlpubs.nist.gov/nistpubs/SpecialPublications/NIST.SP.800-207.pdf; and NSA Cybersecurity Information Sheet, "Embracing a Zero Trust Security Model," February 25, 2021, https://www.nsa.gov/News-Features/Feature-Stories/Article-View/Article/2515176/nsa-issues-guidance-on-zero-trust-security-model/.

4. For concerns regarding winning too much in competition, see Admiral Charles A. Richard, "Forging 21st-Century Strategic Deterrence," *Proceedings*, February 21, 2021, https://www.usni.org/magazines/proceedings/2021/february/forging-21st-century-strategic-deterrence.
5. The Soviet empire collapsed without major great power war, so there are modern conditions that would suggest such a dramatic shift in the distribution of power is possible without armed conflict being the reaction to arrest the extreme of State dissolution.
6. On the important role uncertainty plays in stability, see James D. Fearon, "Rationalist Explanations for War," *International Organization* 49, no. 3 (Summer 1995): 379–414, https://web.stanford.edu/group/fearon-research/cgi-bin/wordpress/wp-content/uploads/2013/10/Rationalist-Explanations-for-War.pdf; James D. Morrow, "Capabilities, Uncertainty, and Resolve: A Limited Information Model of Crisis Bargaining," *American Journal of Political Science* 33, no. 4 (November 1989): 941–972, https://www.jstor.org/stable/2111116; Robert Powell, *In the Shadow of Power: States and Strategies in International Politics* (Princeton, NJ: Princeton University Press, 1999); Charles L. Glaser, *Rational Theory of International Politics: The Logic of Competition and Cooperation* (Princeton, NJ: Princeton University Press, 2010); and Downs and Rocke, "Tacit Bargaining and Arms Control."
7. See Paul Keal, *Unspoken Rules and Superpower Dominance* (London: Macmillan, 1983); Lipson, "Why Are Some International Agreements Informal?"; Schelling, *The Strategy of Conflict*; and Osgood, *An Alternative to War or Surrender*. Some may argue that addressing uncertainty could instead be accomplished through explicit bargaining (i.e., diplomacy) vice tacit, cyber operational persistence. It may well be the case that explicit bargaining would be more effective under certain conditions. This, and the potential complementarity of explicit and tacit approaches, is explored in the next section on dynamics.
8. Not Petya is a case of collateral damage from a militarized cyber conflict. There is no indication that Russia intended, for example, to significantly disrupt the Maersk shipping company's operations via their malware attack on Ukraine, but the software update upload was not contained to the Ukrainian conflict. Whether this was sloppiness or a lack of concern, the fact that there were no notable consequences for the collateral damage suggests that States and companies treated the incident as accidental or inadvertent. A global disruption did not lead to an escalation dynamic.
9. For a list detailing thirty-nine commercial airline incidents both in wartime and peacetime conditions, see https://en.wikipedia.org/wiki/List_of_airliner_shootdown_incidents.
10. Ben Buchanan, *The Cyber Security Dilemma: Hacking, Trust, and Fear Between Nations* (London: Oxford University Press, 2016).
11. For example, several States have explicitly stated that many existing conventions of international law apply to the cyber context. See "Report of the Group of Governmental Experts on Developments in the Field of Information and Telecommunications in the Context of International Security," June 24, 2013, https://www.un.org/ga/search/view_doc.asp?symbol=A/68/98. Schelling noted that existing conventions were one of many potential sources of focal points that could be the basis for tacit coordination. Schelling, *A Strategy of Conflict*. This is an important observation as it argues that some preexisting (to cyberspace itself) explicit and formal prohibitive and permissive conventions intended to moderate State behavior; for example, UN Charter articles 2(4) and 51 speaking to use of force and armed attack, respectively, appear to serve as focal points for tacit coordination (through cyber *faits accomplis*) and tacit bargaining (through direct cyber engagement).
12. We say "additional" tacit coordination efforts because tacit coordination can be supported by observations of different types of behaviors, not merely State's cyber behaviors.

13. See, for example, Lipson, "Why Are Some International Agreements Informal?"; and Alexander L. George, "US-Soviet Global Rivalry: Norms of Competition," *Journal of Peace Research* 23, no. 3 (September 1986): 247–262, https://www.jstor.org/stable/423823.
14. Lipson, "Why Are Some International Agreements Informal?"
15. Ibid. Lipson argued that although formal agreements often contain clauses permitting renegotiation, the process is often slow, cumbersome and impractical.
16. Downs and Rocke, "Tacit Bargaining and Arms Control."
17. It is also important to consider the role that private industry could play in this area. As most of the physical and logical infrastructure comprising the cyberspace strategic environment—the "pubic core of the Internet"—is owned and operated by the private sector, they are well-positioned to contribute. As an example, see "A Digital Geneva Convention to Protect Cyberspace," *Microsoft Policy Papers,* February 14, 2017, https://query.prod.cms.rt.microsoft.com/cms/api/am/binary/RW67QH. For an assessment of Microsoft's proposal, see Valentin Jeutner, "The Digital Geneva Convention: A Critical Appraisal of Microsoft's Proposal," *Journal of International Humanitarian Legal Studies* 10, no. 1 (June 2019): 158–170, https://doi.org/10.1163/18781527-01001009. For a discussion on States' responsibilities to secure the "public core" of the Internet, see Dennis Broeders, "Aligning the International Protection of 'The Public Core of the Internet' with State Sovereignty and National Security," *Journal of Cyber Policy* 2, no. 3 (November 2017): 366–376, https://doi.org/10.1080/23738871.2017.1403640.
18. For a literature review, see Martha Finnemore and Kathryn Sikkink, "International Norm Dynamics and Political Change," *International Organization* 52, no. 4 (Autumn 1998): 887–917, https://www.jstor.org/stable/2601361.
19. In international law, *opinio juris* is the subjective element used to judge whether, in the context of determining the existence of a customary international law rule, a State believes that it is legally obliged to do or refrain from doing a particular act. See *United Nations Report of the International Law Commission, Sixty-eighth session (May 2–June 10 and July 4–August 12, 2016), A/71/10,* https://documents-dds-ny.un.org/doc/UNDOC/GEN/G16/184/25/PDF/G1618425.pdf?OpenElement.
20. *Opinio juris* is the first element. Cyber *faits accomplis* and direct cyber engagement would constitute "State practice," the second element. The third key element, addressed later in this chapter, is "general practice." See *United Nations Report of the International Law Commission.*
21. For a brief overview, see Michael N. Schmitt, "Taming the Lawless Void: Tracking the Evolution of International Law Rules for Cyberspace," *Texas National Security Review* 3, no. 3 (Autumn 2020), https://tnsr.org/2020/07/taming-the-lawless-void-tracking-the-evolution-of-international-law-rules-for-cyberspace/.
22. Liisi Adamson, "International Law and Cyber Norms: A Continuum?," in *Governing Cyberspace: Behavior, Power and Diplomacy,* ed. Dennis Broeders and Bibi van den Berg (Lanham, MD: Rowman & Littlefield, 2020), https://rowman.com/WebDocs/Open_Access_Governing_Cyberspace_Broeders_and_van_den_Berg.pdf.
23. "Report of the Group of Governmental Experts on Developments in the Field of Information and Telecommunications in the Context of International Security," June 24, 2013.
24. Anthea Roberts, *Is International Law International?* (Oxford: Oxford University Press, 2017).
25. Adamson, "International Law and Cyber Norms."
26. "Report of the Group of Governmental Experts on Developments in the Field of Information and Telecommunications in the Context of International Security," July 22, 2015, https://www.un.org/ga/search/view_doc.asp?symbol=A/70/174. The report also confirmed that the fundamental protections of international humanitarian law—necessity, proportionality, humanity, and distinction—applied in cyberspace.
27. Michael Schmitt and Liis Vihul, "International Cyber Law Politicized: The UN GGE's Failure to Advance Cyber Norms," *JustSecurity,* June 30, 2017, https://www.justsecurity.org/42768/international-cyber-law-politicized-gges-failure-advance-cyber-norms/.
28. See Adam Segal, "The Development of Cyber Norms at the United Nations Ends in Deadlock, Now What?," June 19, 2017, https://www.cfr.org/blog/development-cyber-norms-united-nations-ends-deadlock-now-what; Elaine Korzak, "UN GGE on Cybersecurity: The End of an Era?," *The Diplomat,* July 31, 2017, https://thediplomat.com/2017/07/un-gge-on-cybers

ecurity-have-china-and-russia-just-made-cyberspace-less-safe/; and Schmitt, "Taming the Lawless Void."

29. Schmitt and Vihul, "International Cyber Law Politicized."
30. Schmitt, "Taming the Lawless Void."
31. Alex Grigsby, "The United Nations Doubles Its Workload on Cyber Norms, and Not Everyone Is Pleased," *Council of Foreign Relations*, November 15, 2018, https://www.cfr.org/blog/united-nations-doubles-its-workload-cyber-norms-and-not-everyone-pleased.
32. Russia secured support in the United Nations for a draft instrument on cybercrime, despite the existence of the Budapest Convention on Cybercrime, which was opened for signature in 2001 and currently has seventy State signatories. See UN General Assembly, "Countering the Use of Information and Communications Technologies for Criminal Purposes, Resolution 74/247," December 27, 2019, https://undocs.org/en/A/RES/74/247; Draft United Nations Convention on Cooperation in Combating Cybercrime, Russia, 2017, Annexed to UN General Assembly, "Letter dated 11 October 2017 from the Permanent Representative of the Russian Federation to the United Nations Addressed to the Secretary-General, UN Doc. A/C.3/72/12*," October 16, 2017, https://undocs.org/A/C.3/72/12. The existing instrument is Council of Europe, "Convention on Cybercrime," *European Treaty Series No. 185*, November 23, 2001, https://www.europarl.europa.eu/meetdocs/2014_2019/documents/libe/dv/7_conv_budapest_/7_conv_budapest_en.pdf.
33. Schmitt, "Taming the Lawless Void." Related, Tallinn Manual 2.0 notes that there are very few treaties that directly deal with cyber operations and those that have been adopted are of limited scope. Michael N. Schmitt and Liis Vihul, eds. *Tallinn Manual 2.0 on the International Law Applicable to Cyber Operations*, 2nd ed. (Cambridge: Cambridge University Press, 2017).
34. United Nations General Assembly, "Open-ended Working Group on Developments in the Field of Information and Telecommunications in the Context of International Security: A/AC.290/2021/CRP.2," March 10, 2021, https://front.un-arm.org/wp-content/uploads/2021/03/Final-report-A-AC.290-2021-CRP.2.pdf.
35. See, respectively, Pavlina Ittelson and Vladimir Radunovic, "What's New with Cybersecurity Negotiations? UN Cyber OEWG Final Report Analysis," *Diplo*, March 19, 2021, https://www.diplomacy.edu/blog/whats-new-cybersecurity-negotiations-un-cyber-oewg-final-report-analysis; and Arindrajit Basu, Irene Poetranto, and Justin Lau, "The UN Struggles to Make Progress on Securing Cyberspace," *Carnegie Endowment for International Peace*, May 19, 2021, https://carnegieendowment.org/2021/05/19/un-struggles-to-make-progress-on-securing-cyberspace-pub-84491.
36. Michael N. Schmitt, "The Sixth United Nations GGE and International Law in Cyberspace," *JustSecurity*, June 10, 2021, https://www.justsecurity.org/76864/the-sixth-united-nations-gge-and-international-law-in-cyberspace/. Emphasis in the original.
37. "Report of the Group of Governmental Experts on Advancing Responsible State Behaviour in Cyberspace in the Context of International Security," May 28, 2021, https://front.un-arm.org/wp-content/uploads/2021/06/final-report-2019-2021-gge-1-advance-copy.pdf.
38. See https://www.whitehouse.gov/briefings-statements/remarks-homeland-security-advisor-thomas-p-bossert-cyber-week-2017/.
39. Martha Finnemore and Duncan B. Hollis, "Constructing Norms for Global Cybersecurity," *American Journal of International Law* 110, no. 3 (July 2016): 425–479, https://doi.org/10.1017/S0002930000016894.
40. The *2011 White House International Strategy for Cyberspace* noted the United States would "work with like-minded states to establish an environment of expectations, or norms of behavior."
41. See http://www.g20.utoronto.ca/2015/151116-communique.pdf.
42. See https://www.mofa.go.jp/files/000246367.pdf.
43. It is reasonable to argue that, when the 2015 G20 Communiqué was released, China and Russia (members of the G20) were members of a like-minded group. Their subsequent regression from 2015 GGE consensus, however, has placed them outside of that group.
44. Eckstein, "Case Study and Theory in Political Science."
45. See Gary King, Robert O. Keohane, and Sidney Verba, *Designing Social Inquiry: Scientific Inference Qualitative Research* (Princeton, NJ: Princeton University Press, 1994); Barbara

Geddes, "How the Cases You Choose Affect the Answers You Get: Selection Bias in Comparative Politics," in *Political Analysis, vol. 2*, ed. James A. Stimson (Ann Arbor: University of Michigan Press, 1990), 131–150; Christopher H. Achen and Duncan Snidal, "Rational Deterrence Theory and Comparative Case Studies," *World Politics: A Quarterly Journal of International Relations* 41, no. 2 (January 1989): 143–169, https://doi.org/10.2307/2010405; Christopher H. Achen, *The Statistical Analysis of Quasi-Experiments* (Berkeley: University of California Press, 1986); and Gary King, *Unifying Political Methodology: The Likelihood Theory of Statistical Inference* (Cambridge: Cambridge University Press, 1989).

46. The same can be said for the International Code of Conduct for Information Security (the "Code"), which was submitted for consideration to the UN General Assembly in January 2015 by the founding member States of the Shanghai Cooperation Organization (SCO): China, Russia, Kazakhstan, Kyrgyzstan, Tajikistan, and Uzbekistan. The "Code," according to its sponsors, is intended "to push forward the international debate on international norms on information security, and help forge an early consensus on this issue." See UN General Assembly, "Letter dated 9 January 2015 from the Permanent Representatives of China, Kazakhstan, Kyrgyzstan, the Russian Federation, Tajikistan and Uzbekistan to the United Nations addressed to the Secretary-General, U.N. Doc. A/69/273," 2015, http://www.un.org/Docs/journal/asp/ws.asp?m=A/69/723. For an analysis of the "Code," see Sarah McKune, "An Analysis of the International Code of Conduct for Information Security," *The Citizen Lab*, September 28, 2015, https://citizenlab.ca/2015/09/international-code-of-conduct/.
47. Importantly, it is not sufficient to merely raise costs. The costs must be *prohibitive* for an agreement to result in its intended effect.
48. This is recognized by private sector technology leaders who, consequently, have called for a legally binding framework. For example, Brad Smith, president of Microsoft, notes that "First, the G7 declaration is focused on voluntary, non-binding state behavior during peacetime. The challenge with voluntary norms is that they are just that: 'voluntary.' I believe we need to push ourselves further and set our sights higher to pursue a legally binding framework that would codify rules for governments and thus help prevent extraordinary damage." See Brad Smith, "Growing Consensus on the Need for an International Treaty on Nation State Attacks," *Microsoft*, April 13, 2017, https://blogs.microsoft.com/on-the-issues/2017/04/13/growing-consensus-need-international-treaty-nation-state-attacks/.
49. Michael N. Schmitt, "Grey Zones in the International Law of Cyberspace," *Yale Journal of International Law Online*, https://cpb-us-w2.wpmucdn.com/campuspress.yale.edu/dist/8/1581/files/2017/08/Schmitt_Grey-Areas-in-the-International-Law-of-Cyberspace-1cab8kj.pdf.
50. See https://www.mofa.go.jp/files/000246367.pdf. Emphasis added.
51. The 2021 GGE report approaches this notion cautiously by noting, "In accordance with the Group's mandate, an official compendium of voluntary national contributions of participating governmental experts on the subject of how international law applies to the use of ICTs by States will be made available on the website of the United Nations Office of Disarmament affairs in the original language of submission without translation." See "Report of the Group of Governmental Experts on Advancing responsible State Behaviour in Cyberspace in the Context of International Security," May 28, 2021.
52. Schmitt, "Grey Zones in the International Law of Cyberspace."
53. Robert E. Barnsby and Shane R. Reeves, "Give Them an Inch, They'll Take a Terabyte: How States May Interpret Tallinn Manual 2.0's International Human Rights Law Chapter," *Texas Law Review* 95, no. 1515 (2017): https://texaslawreview.org/wp-content/uploads/2017/11/Barnsby.Reeves.pdf. Schmitt draws the same conclusion. Schmitt, "Taming the Lawless Void."
54. Kenneth L. Adelman, "Arms Control: Arms Control with and without Agreements," *Foreign Affairs* 63, no. 2 (Winter 1984/85): 240–263, https://www.foreignaffairs.com/articles/1984-12-01/arms-control-arms-control-and-without-agreements.
55. Schelling, "Reciprocal Measures for Arms Stabilization."
56. This has been referred to as "catalytic" escalation in cyberspace. See Lin, "Escalation Dynamics and Conflict Termination in Cyberspace."
57. Schelling, "Reciprocal Measures for Arms Stabilization."

58. Ibid.
59. Schelling, *The Strategy of Conflict*.
60. Adamson, "International Law and Cyber Norms: A Continuum?," 28.
61. Martti Koskenniemi, "International Cyber Law: Does It Exist and Do We Need It?," *European Cyber Diplomacy Dialogue, EU Cyber Direct, Opening Lecture,* April 9, 2019, https://eucyberdirect.eu/content_events/professor-martti-koskenniemi-international-cyber-law-does-it-exist-and-do-we-need-it-european-cyber-diplomacy-dialogue-2019-2/.
62. *United Nations Report of the International Law Commission, Sixty-eighth session.*
63. Ibid. (for additional unilateral, independent expressions of *pinion juris* that are also accepted as evidence of the same).
64. Schelling, "Reciprocal Measures for Arms Stabilization."
65. See https://ccdcoe.org/.
66. Schmitt and Vihul, eds., *Tallinn Manual 2.0.*
67. Schmitt, "Grey Zones in the International Law of Cyberspace." Importantly, Schmitt admits that these two areas are by no means exhaustive, as "states wishing to operate in the grey zones [of international law] will seize opportunity wherever it presents itself."
68. Schmitt and Vihul, eds., *Tallinn Manual 2.0,* 11, 17.
69. Eric Talbot Jensen, "Tallinn Manual 2.0: Highlights and Insights," *Georgetown Journal of International Law* 48 (2017): 735–778, https://papers.ssrn.com/sol3/papers.cfm?abstract_id=2932110#.
70. Schmitt, "Grey Zones in the International Law of Cyberspace."
71. Colonel (Retired) Gary Corn, "Tallinn Manual 2.0—Advancing the Conversation," *JustSecurity,* February 15, 2017, https://www.justsecurity.org/37812/tallinn-manual-2-0-advancing-conversation/#more37812.
72. Jensen, "Tallinn Manual 2.0: Highlights and Insights." From the perspective of cyber persistence theory, interconnectedness challenges the most rigid application of sovereignty as a rule, since, as a rule, it enshrines a segmentation of territory globally that does not capture the core structural feature of cyberspace.
73. For an argument that current international law should be adapted to account for both exploitation and coercion, see Michael P. Fischerkeller, "Current International Law Is Not an Adequate Regime for Cyberspace," *Lawfare,* April 22, 2021, https://www.lawfareblog.com/current-international-law-not-adequate-regime-cyberspace.
74. Although the UN Charter does not explicitly mention the principle of non-intervention, Articles 2(1), (3), and (4) are nevertheless often looked to as a basis for the rule of international law regarding non-intervention. Note that the Tallinn process international group of experts did not significantly rely on the UN General Assembly's Declaration on the Inadmissibility of Intervention and Interference, as they agreed that the declaration did not represent customary international law in its entirety. See Edward McWhinney, "General Assembly Resolution 2131 (XX) OF 21 December 1965 Declaration on the Inadmissibility of Intervention in the Domestic Affairs of States and the Protection of Their Independence and Sovereignty," https://legal.un.org/avl/pdf/ha/ga_2131-xx/ga_2131-xx_e.pdf. Also, see Friendly Relations Declaration, UN General Assembly, 1970.
75. United Nations General Assembly, "Resolution 2625 (XXV), annex, Declaration on Principles of International Law Concerning Friendly Relations and Co-operation among States in Accordance with the Charter of the United Nations," October 24, 1970, https://www.un.org/ruleoflaw/files/3dda1f104.pdf.
76. The point at which international human rights law, such as the rights to freedom of expression or privacy, takes domestic regulation beyond the confines of the *domaine réservé* remains unsettled. This sets the stage for divergent views on acceptable and unacceptable behaviors between, for example, the United States and China and Russia. The 2018 US National Cyber Strategy states, "We will also work to prevent authoritarian states that view the open Internet as a political threat from transforming the free and open Internet into an authoritarian web under their control, under the guise of security or countering terrorism." China and Russia, on the other hand, cite their views on sovereignty when assertively and at times aggressively managing the Internet content available to their populations. As an example of aggressive management, consider the report that China initiated an intense, six-day DDoS attack against

two GitHub pages: one hosting the anti-censorship page GreatFire.org and the second a mirror site of the Chinese edition of the *New York Times*. See https://arstechnica.com/information-technology/2015/03/massive-denial-of-service-attack-on-github-tied-to-chinese-government/.

77. Schmitt, "Grey Zones in the International Law of Cyberspace."
78. "Letter of 5 July 2019 from the Minister of Foreign Affairs to the President of the House of Representatives on the international legal order in cyberspace, Appendix: International Law in Cyberspace," https://www.government.nl/documents/parliamentary-documents/2019/09/26/letter-to-the-parliament-on-the-international-legal-order-in-cyberspace.
79. Schmitt and Vihul, eds., *Tallinn Manual 2.0*.
80. Harriet Moynihan, "The Application of International Law to State Cyberattacks," *Chatham House International Law Program*, December 2019, https://www.chathamhouse.org/sites/default/files/publications/research/2019-11-29-Intl-Law-Cyberattacks.pdf.
81. Oxford Institute for Ethics, Law and Armed Conflict, "Applying International Law in Cyberspace: Protections and Prevention," May 18–19, 2020, https://www.elac.ox.ac.uk/files/elacvirtualworkshopapplyingiltocyberspacereport18-19may2020pdf. Schmitt argued, however, that this "interpretive creativity" does little, if anything, to provide greater clarity, as it merely shifts the challenge of applying the non-intervention rule from ascertaining when "no reasonable choice" is present to determining when "making the choice difficult" is present. Schmitt, "Taming the Lawless Void."
82. Colonel (Retired) Gary Corn and Eric Jensen, "The Technicolor Zone of Cyberspace—Part I: Analyzing the Major U.K. Speech on International Law of Cyber," *JustSecurity*, May 30, 2018, https://www.justsecurity.org/57217/technicolor-zone-cyberspace-part/. We would be remiss to not highlight the parallels between the debates surrounding the concept of coercion in the international law and international relations and security communities. Scholars and policymakers in both communities are struggling to adapt a concept (and its associated terms) that has been central, over the last seventy years, to the formation of international law and national security strategies intended to prohibit a use of force (or threat thereof) to achieve State objectives. This book, of course, argues that the cyberspace strategic environment demands a paradigmatic change in international relations and security thinking (i.e., that coercion should not be a central concept around which cyber strategies should be built and that efforts to redefine coercion are not serving the community well). Cyber persistence theory, in this context, would suggest that scholars and policymakers in international law community should also adopt that mindset.
83. Fischerkeller, "Current International Law Is Not an Adequate Regime for Cyberspace."
84. The notion of "without more" was also taken up by the participating experts and they could not reach consensus. For example, there was majority, but not total agreement that the exfiltration of data violated no rule of international law. However, some experts believe that at some point the exfiltration might be so severe as to make it unlawful. See Jensen, "The Tallinn Manual 2.0: Highlights and Insights."
85. Schmitt and Vihul, eds., *Tallinn Manual 2.0*.
86. Fischerkeller, "Current International Law Is Not an Adequate Regime for Cyberspace."
87. See United Nations General Assembly, "Resolution 56/83 Annex, United Nations Document A/CN.4/L. 778, Responsibility of States for Internationally Wrongful Acts, Chapter II," May 30, 2011, https://documents-dds-ny.un.org/doc/UNDOC/LTD/G11/614/25/PDF/G1161425.pdf?OpenElement.
88. International Law Commission, "Report of the International Law Commission on the Work of Its Fifty-Third Session, Draft Articles on Responsibility of States for Internationally Wrongful Acts, with Commentaries, art. 22, UN Doc. A/56/10," https://legal.un.org/ilc/texts/instruments/english/commentaries/9_6_2001.pdf.
89. Whereas countermeasures are acts that might otherwise be considered wrongful, acts of retorsion under any circumstance do not interfere with the target State's rights under international law. An example is a diplomatic démarche. See https://opil.ouplaw.com/view/10.1093/law:epil/9780199231690/law-9780199231690-e983.
90. In support of this position, the experts cited Articles on State Responsibility, Article 15.

91. We argue in Chapter 7 that State instruments such as unsealed federal indictments and public attributions could also serve as evidence if those instruments referenced violations of international obligations. Most have yet to do so, however. Finnemore and Hollis adopt a less stringent perspective, arguing that no specific references to violations are necessary. Martha Finnemore and Duncan B. Hollis, "Beyond Naming and Shaming: Accusations and International Law in Cybersecurity," *European Journal of International Law* 31, no. 3 (August 2020): 969–1003. https://doi.org/10.1093/ejil/chaa056.
92. Jeremy Wright, Attorney General of the UK, "Cyber and International Law in the 21st Century," *Chatham House*, May 23, 2018, https://www.chathamhouse.org/event/cyber-and-international-law-21st-century. This view was then codified in Foreign, Commonwealth and Development Office, *Application of International Law to States' Conduct in Cyberspace: UK Statement*, June 3, 2021, https://www.gov.uk/government/publications/application-of-international-law-to-states-conduct-in-cyberspace-uk-statement/application-of-international-law-to-states-conduct-in-cyberspace-uk-statement.
93. Republic of France, Ministry of the Armies, "International Law Applied to Operations in Cyberspace (*Droit international appliqué aux operations dans le cyberspace*)," 2019, https://www.defense.gouv.fr/content/download/567648/9770527/file/international+law+applied+to+operations+in+cyberspace.pdf.
94. For a summary, see Przemysław Roguski, "France's Declaration on International Law in Cyberspace: The Law of Peacetime Cyber Operations, Part I," *OpinioJuris*, September 24, 2019, http://opiniojuris.org/2019/09/24/frances-declaration-on-international-law-in-cyberspace-the-law-of-peacetime-cyber-operations-part-i/. For a dissenting view on whether the Ministry of Armies document is acceptable evidence of an expression of *opinio juris* and that the French view is one of "pure sovereignty," see Colonel (Retired) Gary Corn, "Punching on the Edges of the Grey Zone: Iranian Cyber Threats and State Cyber Responses. Chatham House Report on Cyber Sovereignty and Non-Intervention—Helpful or Off Target?," *JustSecurity*, February 11, 2020, https://www.justsecurity.org/68622/punching-on-the-edges-of-the-grey-zone-iranian-cyber-threats-and-state-cyber-responses/. Corn argued that the document "does not claim to be the official position of the French government. It was written and published by the French Ministère des Armées (MdA), in the same vein as the DoD Law of War Manual which does not necessarily reflect the views of the U.S. Government as a whole. . . . [and] at no point does it assert unequivocally that a violation of the principle of sovereignty constitutes a breach of an international obligation."
95. Republic of France, Ministry of the Armies, "International Law Applied to Operations in Cyberspace."
96. Moynihan, "The Application of International Law to State Cyberattacks."
97. Statements of US government legal advisors are considered as evidence given that "Published opinions of government legal advisers may likewise shed light on a State's legal position, though not if the State declined to follow the advice." "United Nations Report of the International Law Commission, Sixty-eighth session (May 2–June 10 and July 4–August 12, 2016), A/71/10."
98. Michael N. Schmitt, "The Defense Department's Measured Take on International Law in Cyberspace," *JustSecurity*, March 11, 2020, https://www.justsecurity.org/69119/the-defense-departments-measured-take-on-international-law-in-cyberspace/.
99. Paul C. Ney Jr., "DOD General Counsel Remarks at U.S. Cyber Command Legal Conference," March 2, 2020, https://www.defense.gov/Newsroom/Speeches/Speech/Article/2099378/dod-general-counsel-remarks-at-us-cyber-command-legal-conference/.
100. Ibid.
101. Ibid. The complexity of the issue is echoed by the Israeli Deputy Attorney General (International Law), who noted, "There are diverging views regarding whether sovereignty is merely a principle, from which legal rules are derived, or a binding rule of international law in itself, the violation of which could be considered an internationally wrongful act. This issue has many facets, and while I will not offer any definitive position for the time being." See Roy Schondorf, "Israel's Perspective on Key Legal and Practical Issues Concerning the Application of International Law to Cyber Operations," *EJIL:Talk!*, December 9, 2020,

https://www.ejiltalk.org/israels-perspective-on-key-legal-and-practical-issues-concerning-the-application-of-international-law-to-cyber-operations/.

102. New Zealand, *The Application of International Law to State Activity in Cyberspace*, https://www.mfat.govt.nz/assets/Peace-Rights-and-Security/International-security/International-Cyber-statement.pdf.
103. "Letter of 5 July 2019 from the Minister of Foreign Affairs to the President of the House of Representatives on the International Legal Order in Cyberspace, Appendix: International Law in Cyberspace."
104. Pre-Draft Report of the OEWG–ICT: Comments by Austria, March 13, 2020, https://front.un-arm.org/wp-content/uploads/2020/04/comments-by-austria.pdf.
105. Statement by Mr. Richard Kadlčák, Special Envoy for Cyberspace Director of Cybersecurity Department, February 11, 2020, https://www.nukib.cz/download/publications_en/CZ%20Statement%20-%20OEWG%20-%20International%20Law%2011.02.2020.pdf. There is a significant disconnect, however, between these States' expressions of *opinio juris* with regard to sovereignty and their associated practices. As Jack Goldsmith and Alex Loomis note, in over two decades of increasingly notable cyber operations and campaigns by State-sponsored APTs, "states are not shy about publicly denouncing cyberattacks or violations of sovereignty generally"; however, even where "attacks caused harm and disruption" and were "attributed to specific aggressors, either officially or by credible news outlets and experts," "many were strongly condemned—but not as violations of international law." Jack Goldsmith and Alex Loomis, "Defend Forward and Sovereignty," in *The United States' Defend Forward Cyber Strategy: A Comprehensive Legal Analysis*, ed. Jack Goldsmith (Oxford: Oxford University Press, 2022), 151–180.
106. Republic of France, Ministry of the Armies, "International Law Applied to Operations in Cyberspace."
107. New Zealand, *The Application of International Law to State Activity in Cyberspace.*
108. Wright, Attorney General of the UK, "Cyber and International Law in the 21st Century"; and Foreign, Commonwealth and Development Office, *Application of International Law to States' Conduct in Cyberspace: UK Statement.*
109. Ney, "DoD General Counsel Remarks at U.S. Cyber Command Legal Conference."
110. Schondorf, "Israel's Perspective on Key Legal and Practical Issues Concerning the Application of International Law to Cyber Operations."
111. *Australia's International Cyber Engagement Strategy: 2019 International Law Supplement*, https://www.dfat.gov.au/publications/international-relations/international-cyber-engagement-strategy/aices/chapters/2019_international_law_supplement.html.
112. Japan, *Cybersecurity Strategy* (Provisional Translation), July 27, 2018, https://nisc.go.jp/eng/pdf/cs-senryaku2018-en.pdf.
113. "Letter of July 5, 2019, from the Minister of Foreign Affairs to the President of the House of Representatives on the International Legal Order in Cyberspace, Appendix: International Law in Cyberspace."
114. Republic of France, Ministry of the Armies, "International Law Applied to Operations in Cyberspace."
115. Brian J. Egan, "International Law and Stability in Cyberspace," *Berkeley Journal of International Law* 35, no. 1 (2017): 169–180, https://www.law.berkeley.edu/wp-content/uploads/2016/12/BJIL-article-International-Law-and-Stability-in-Cyberspace.pdf.
116. Ney, "DoD General Counsel Remarks at U.S. Cyber Command Legal Conference."
117. *Australia's International Cyber Engagement Strategy.*
118. Wright, Attorney General of the UK, "Cyber and International Law in the 21st Century." and Foreign, Commonwealth and Development Office, *Application of International Law to States' Conduct in Cyberspace: UK Statement.*
119. Schondorf, "Israel's Perspective on Key Legal and Practical Issues Concerning the Application of International Law to Cyber Operations."
120. Foreign, Commonwealth and Development Office, *Application of International Law to States' Conduct in Cyberspace: UK Statement.*
121. Republic of France, Ministry of the Armies, "International Law Applied to Operations in Cyberspace."

122. Brussels Summit Communiqué, June 14, 2021, https://www.nato.int/cps/en/natohq/news_185000.htm.
123. Stefan Soesanto, "When Does a 'Cyber Attack' Demand Retaliation? NATO Broadens Its View," *DefenseOne*, June 30 2021, https://www.defenseone.com/ideas/2021/06/when-does-cyber-attack-demand-retaliation-nato-broadens-its-view/175028/.
124. Ibid.
125. Kersti Kaljulaid, President, Republic of Estonia, Speech at the Opening of the International Conference on Cyber Conflict (CyCon) 2019, May 29, 2019, https://www.president.ee/en/official-duties/speeches/15241-president-of-the-republic-at-the-opening-of-cycon-2019/index.html. Note that no other State has yet to endorse this position and one, France, has rejected it. That said, in 2019, before assuming the duties of US Deputy Assistance Secretary of Defense for cyber policy, Thomas Wingfield noted that ""I think that's a more effective way [collective countermeasures] to solve the problem, and I think that is the general [direction] of international law. . . . But I would also say we're not there yet and states are in the process of moving international law in that direction." Quoted in Shannon Vavra, "Pentagon's Next Cyber Policy Guru Predicts More Collective Responses in Cyberspace," *CyberScoop*, November 21, 2019, https://www.cyberscoop.com/pentagons-next-cyber-policy-guru-predicts-collective-responses-cyberspace/.
126. New Zealand, *The Application of International Law to State Activity in Cyberspace.*
127. The debate, centered on Nicaragua's assistance to armed groups in three other States, consisted of distinct and separate acts. See Jeff Kosseff, "Collective Countermeasures in Cyberspace," *Notre Dame Journal of Comparative & International Law* 10, no. 1 (January 2020): 18–34, https://scholarship.law.nd.edu/cgi/viewcontent.cgi?article=1115&context=ndjicl. Also see Michael N. Schmitt, "Three International Law Rules for Responding Effectively to Hostile Cyber Operations," *JustSecurity*, July 13, 2021, https://www.justsecurity.org/77402/three-international-law-rules-for-responding-effectively-to-hostile-cyber-operations/.
128. Aside from Estonia, of course, with regard to collective countermeasures. See Colonel (Retired) Gary Corn and Eric Jensen, "The Technicolor Zone of Cyberspace, Part 2," *JustSecurity*, June 8, 2018, https://www.justsecurity.org/57545/technicolor-zone-cyberspace-part-2/.
129. Recall that we argue the voluntary and non-binding character of the GGE and G7 agreements makes them equivalent in substance with tacit agreements.
130. Schmitt, "Taming the Lawless Void."
131. Jens David Ohlin, "Did Russian Cyber Interference in the 2016 Election Violate International Law?," *Texas Law Review* 95, no. 7 (2017), https://texaslawreview.org/russian-cyber-interference-2016-election-violate-international-law/.
132. Corn and Jensen, "The Technicolor Zone of Cyberspace—Part I."
133. Schmitt, "Taming the Lawless Void."
134. Moynihan, "The Application of International Law to State Cyberattacks."
135. Limits may flow from State's policy choices, as well, as they assess the impact of too little vice too much certainty on securing national interests and ensuring stability. Schmitt argued, "For example, uncertainty over the nature of cyber operations that violate sovereignty not only lessens the likelihood that one's own remote operations will be styled as unlawful, but also lessens the risk that a target State will take countermeasures pursuant to the law of State responsibility." Schmitt, "Grey Zones in the International Law of Cyberspace."
136. Note that if data is stored via a cloud service, it may well be stored outside the sovereign territory of the State and the vendor may periodically "move" it to other data centers in other States (for various operational reasons).
137. To view a deep dive into the US electoral system, see https://verifiedvoting.org/.
138. Global Commission on the Stability of Cyberspace, *Advancing Cyber Stability: Final Report*, November 2019, Appendix B.1., https://cyberstability.org/wp-content/uploads/2020/02/GCSC-Advancing-Cyberstability.pdf.
139. Ibid., Appendix B.1.
140. George, "US-Soviet Global Rivalry: Norms of Competition"; and Keal, *Unspoken Rules and Superpower Dominance.*

141. To not do so would be equivalent to, for example, a software/hardware company announcing a vulnerability before releasing a patch for the same.
142. There is some debate around this point. Gary Corn noted that if States declare that they are bound by international law not to do something, it is not coherent for them to turn around and say they did not breach international law when they do it because of lack of State practice. (Author communication with Gary Corn, December 15, 2020.)
143. *United Nations Report of the International Law Commission, Sixty-eighth session (May 2–June 10 and July 4–August 12, 2016), A/71/10.*
144. Regarding dynamism, a relatively short period in which a general practice is followed is not an obstacle to determining that a corresponding rule of customary international law exists. Although a long duration may result in more extensive relevant practice, a considerable or fixed duration of a general practice is not a condition for the existence of a customary rule. However, some amount of time must elapse for a general practice to emerge; there is no such thing as "instant custom." See ibid.
145. Thomas C. Schelling and Morton H. Halperin, *Strategy and Arms Control* (McLean, VA: Pergamon Press, 1985), 86–87.
146. Martha Finnemore, "Cybersecurity and the Concept of Norms," *Carnegie Endowment for International Peace*, November 30, 2017, https://carnegieendowment.org/2017/11/30/cybersecurity-and-concept-of-norms-pub-74870.
147. *United Nations Report of the International Law Commission, Sixty-eighth session May 2–June 10 and July 4–August 12, 2016), A/71/10.*
148. See Schmitt and Vihuls, eds., *Tallinn Manual 2.0,* and Schmitt, "Taming the Lawless Void: Tracking the Evolution of International Law Rules for Cyberspace."
149. Downs and Rocke, "Tacit Bargaining and Arms Control."
150. We do not discuss the current geopolitical environment because we have already argued that the poor state of relations among great powers should be encouraging States to explore tacit approaches to arriving at mutual understandings.
151. Downs and Rocke, "Tacit Bargaining and Arms Control."
152. Finnemore and Hollis, "Constructing Norms for Global Cybersecurity."
153. Ibid. The concept of "noise" in tacit bargaining was borrowed from the signal-to-noise ratio issue long faced by scholars pursuing a general theory of communication. See Claude E. Shannon, "A Mathematical Theory of Communication," *The Bell System Technical Journal* 27 (October 1948): 379–423, 623–656, https://ieeexplore.ieee.org/document/6773024.
154. Noise could also comprise random system technical errors and "insider" user error or "insider" attempts at exploitation.
155. Buchanan, *The Cyber Security Dilemma.*
156. Koch notes that even though OEEC discriminatory policies technically violated the General Agreement on Tariffs and Trade Most Favored Nation provision, "the US not merely acquiesced but encouraged this program as part of its policy to strengthen Europe," thus, "the OEEC policy was tacitly accepted without any waiver being asked for" and, subsequently, GATT posed no barrier to discriminatory policies of the sort. See Karin Koch, *International Trade Policy and the GATT, 1947–1967* (Stockholm: Almquist and Wiksell, 1969); and Lipson, "Why Are Some International Agreements Informal?" See OPEC's Declaration of Cooperation as a specific example. https://www.opec.org/opec_web/en/publications/4580.htm.
157. See https://mtcr.info/guidelines-for-sensitive-missile-relevant-transfers/.
158. Notably, explicit, formal agreements would also confront the attribution challenge.
159. George W. Downs and David M. Rocke, *Tacit Bargaining, Arms Races, and Arms Control* (Ann Arbor: University of Michigan Press, 1990).
160. Downs and Rocke, "Tacit Bargaining and Arms Control."
161. Thus, for example, should a compromise be attributed to an adversary's intelligence service, the cyber infrastructure and personnel managing and operating on and through it need not be the target of a countermeasure. Other State organs are also permitted targets. Michael N. Schmitt, "Peacetime Cyber Responses and Wartime Cyber Operations under International Law: An Analytical *Vade Mecum,*" *Harvard National Security Journal* 8 (2017): 239–282, https://harvardnsj.org/wp-content/uploads/sites/13/2017/02/Schmitt-NSJ-Vol-8.pdf.

162. Kaljulaid, President, Republic of Estonia, Speech at the Opening of the International Conference on Cyber Conflict (CyCon) 2019.
163. Gary Corn and Eric Talbot Jensen, "The Use of Force and Cyber Countermeasures," *Temple International & Comparative Law Journal* 127 (2018), https://papers.ssrn.com/sol3/papers.cfm?abstract_id=3190253&download=yes.
164. Consider, for example, when addressing Iranian cyber operations targeting the US 2020 presidential elections, USCYBERCOM Commander General Paul Nakasone asserted that the US National Security Agency had been watching the Iranians for a while and so "[w]e [CYBERCOM] had a very, very good bead on what a number of actors were trying to do" and "[w]e provided early warning and followed [them very closely]. We weren't surprised by their actions." Nakashima, "U.S. Undertook Cyber Operation against Iran as Part of Effort to Secure the 2020 Election."
165. Kaljulaid, President, Republic of Estonia, Speech at the Opening of the International Conference on Cyber Conflict (CyCon) 2019. Consider, for example, that establishing a pattern of opportunistic behavior by APTs is not much of an empirical challenge. There is significant evidence that many APTs successfully exploit vulnerabilities within days of their public acknowledgment and, as we noted previously, APTs also often seek out zero-day vulnerabilities for exploitation.
166. Egan, "Remarks on International Law and Stability in Cyberspace." For a description of a hierarchy of evidentiary standards in international law, see James A. Green, "Fluctuating Evidentiary Standards for Self-Defence in the International Court of Justice," *International & Comparative Law Quarterly* 58, no. 1 (January 2009): 163–180, https://www.jstor.org/stable/20488277.
167. Tom Wheeler and David Simpson, "Why 5G Requires New Approaches to Cybersecurity," *Brookings*, September 3, 2019, https://www.brookings.edu/research/why-5g-requires-new-approaches-to-cybersecurity/. Wheeler and Simpson also identify a new avenue of physical attack—the short range, small-cell antennas deployed throughout urban areas—which are excluded from this discussion because pursuing this avenue would result in armed-attack equivalent effects and represent an escalation out of the cyber agreed competition phenomenon.
168. See Alexei Bulazel, Sophia d'Antoine, Perri Adams, and Dave Aitel, "The Risks of Huawei Risk Mitigation," *Lawfare*, April 24, 2019, https://www.lawfareblog.com/risks-huawei-risk-mitigation. The authors argued, "Providing a third party with a foothold into a network introduces an entirely new array of risks [that are incalculable]. Any single risk has the potential to flow into other parts of the system in ways difficult to protect against, exponentially increasing exposure. Second, mitigation is impossible. With the rising complexity of technologies, validating security properties to any significant level in third-party systems has become untenable."
169. Jarred Whitley, "Washington Bureaucrats Are Handing China Keys to 5G Kingdom," *The Hill*, February 22, 2019, https://thehill.com/opinion/technology/430801-washington-bureaucrats-are-handing-china-keys-to-5g-kingdom.
170. Wheeler and Simpson, "Why 5G Requires New Approaches to Cybersecurity."
171. Gilad David Maayan, "The IoT Rundown for 2020: Stats, Risks, and Solutions," *SecurityToday*, January 13, 2020, https://securitytoday.com/articles/2020/01/13/the-iot-rundown-for-2020.aspx.
172. Consumer IoT includes items such as light fixtures, home appliances, home routers, home printers, security cameras, smart watches, and voice assistance. Commercial IoT includes building management systems, wireless business systems, and applications in the healthcare and transport industries, such as smart pacemakers, monitoring systems, and vehicle-to-vehicle communication. Industrial IoT includes digital control systems and smart agriculture. Infrastructure IoT enables the connectivity of smart cities through the use of infrastructure sensors, management systems, and user-friendly user apps. Finally, military IoT comprises all applications of IoT technologies in the military field, such as robots for surveillance and human-wearable biometrics for combat. See Charles McLelland, "Optimising the Smart Office: A Marriage of Technology and People," *ZDNet*, March 1, 2018, https://www.zdnet.

com/article/optimising-the-smart-office-a-marriage-of-technology-and-people/; and David Maayan, "The IoT Rundown for 2020: Stats, Risks, and Solutions."

173. As Tom Burt, Microsoft Corporate Vice President, Customer Security & Trust, noted in September 2020, "IoT threats are constantly expanding and evolving. The first half of 2020 saw an approximate 35% increase in total attack volume compared to the second half of 2019." See Tom Burt, "Microsoft Report Shows Increasing Sophistication of Cyber Threats," *Microsoft*, September 29, 2020, https://blogs.microsoft.com/on-the-issues/2020/09/29/microsoft-digital-defense-report-cyber-threats/. Also see National Intelligence Council, *Global Trends 2040: A More Contested World*, 71, https://www.dni.gov/files/ODNI/documents/assessments/GlobalTrends_2040.pdf; Lily Hay Newman, "100 Million More IoT Devices Are Exposed—and They Won't Be the Last," *Wired*, April 13, 2021, https://www.wired.com/story/namewreck-iot-vulnerabilities-tcpip-millions-devices/; and Tara Seals, "IoT Attacks Skyrocket, Doubling in 6 Months," *threatpost*, September 6, 2021, https://threatpost.com/iot-attacks-doubling/169224/.
174. Garrett M. Graff, "How a Dorm Room Minecraft Scam Brought Down the Internet," *Wired*, December 13, 2017, https://www.wired.com/story/mirai-botnet-minecraft-scam-brought-down-the-internet/.
175. Lily Hay Newman, "What We Know about Friday's Massive East Coast Internet Outage," *Wired*, October 21, 2016, https://www.wired.com/2016/10/internet-outage-ddos-dns-dyn/.
176. See William Largent, "New VPNFilter Malware Targets at Least 500K Networking Devices Worldwide," *Cisco Talos*, May 23, 2018, https://blog.talosintelligence.com/2018/05/VPNFilter.html; and William Largent, "VPNFilter Update—VPNFilter Exploits Endpoints, Targets New Devices," *Cisco Talos*, June 6, 2018, https://blog.talosintelligence.com/2018/06/vpnfilter-update.html.
177. "FBI Seizes Domain Responsible for Major Russian Botnet," *eyerys*, May 24, 2018, https://www.eyerys.com/articles/timeline/fbi-seizes-domain-responsible-major-russian-botnet?page=14#event-a-href-articles-timeline-sharewareshareware-a.
178. Microsoft Security Response Center, "Corporate IoT—A Path to Intrusion," August 5, 2019, https://msrc-blog.microsoft.com/2019/08/05/corporate-iot-a-path-to-intrusion/.
179. For example, seven of command-and-control servers supporting the Trickbot botnet (discussed in Chapter 4) were not "traditional" servers but rather were IoT devices exploited by Trickbot administrators as part of its server infrastructure. See Tom Burt, "An Update on Disruption of TrickBot," *Microsoft on the Issues*, October 20, 2020, https://blogs.microsoft.com/on-the-issues/2020/10/20/trickbot-ransomware-disruption-update/.
180. Microsoft Security Response Center, "Corporate IoT—A Path to Intrusion."
181. Additionally, the onus of securing personal IoT devices falls on the user, not on the CISO and, just as with legacy corporate software, some "older" connected IoT devices may no longer be supported (i.e., with security updates) by manufacturers. See Davey Winder, "Google Chrome Update Gets Serious: Homeland Security (CISA) Confirms Attacks Underway," *Forbes*, November 15, 2020, https://www.forbes.com/sites/daveywinder/2020/11/15/google-chrome-update-gets-serious-homeland-security-cisa-confirms-attacks-underway/amp/. Gartner Glossary, Bring Your Own Device (BYOD), https://www.gartner.com/en/information-technology/glossary/bring-your-own-device-byod.
182. Microsoft Security Response Center, "Corporate IoT—A Path to Intrusion."
183. This has often been discussed using the international security concept of offense-defense balance (or advantage). Although we do not subscribe to the use of that concept in cyberspace, to be true to the words of scholars holding competing perspectives on the potential impact of AI on State behavior, we summarize their arguments using those terms. See, for example, Rebecca Slayton, "What Is the Cyber Offense-Defense Balance? Conceptions, Causes, and Assessment," *International Security* 41, no. 3 (Winter 2016/17): 72–109, https://www.belfercenter.org/publication/what-cyber-offense-defense-balance-conceptions-causes-and-assessment.
184. Bruce Schneier, "Artificial Intelligence and the Attack/Defense Balance," *Schneier on Security*, March 15, 2018, https://www.schneier.com/blog/archives/2018/03/artificial_inte.html.

185. The implied suggestion that CISO's of public agencies and private firms will allow AI to automatically patch identified vulnerabilities is not consistent with patching concerns expressed by the same. Delays in patching are not merely a function of a technical inability to do so quickly, presumably the basis of AI's advantage. Delays can also stem from, for example, concerns of operational disruptions from patching. Additionally, if a vulnerability is discovered in a machine-learning AI defensive system, one cannot simply patch it, rather, the algorithm must be re-trained. See Wyatt Hoffmann, *AI and the Future of Cyber Competition* (Center for Security and Emerging Technology, Georgetown University, January 2021), https://cset.georgetown.edu/publication/ai-and-the-future-of-cyber-competition/.
186. Schneier, "Artificial Intelligence and the Attack/Defense Balance."
187. Miles Brundage, Shahar Avin, Jack Clark, Helen Toner, Peter Eckersley, Ben Garfinkel, Allan Dafoe, Paul Scharre, Thomas Zeitzoff, Bobby Filar, Hyrum Anderson, Heather Roff, Gregory C. Allen, Jacob Steinhardt, Carrick Flynn, Seán Ó hÉigeartaigh, Simon Beard, Haydn Belfield, Sebastian Farquhar, Clare Lyle, Rebecca Crootof, Owain Evans, Michael Page, Joanna Bryson, Roman Yampolskiy, and Dario Amodei, "The Malicious Use of Artificial Intelligence: Forecasting, Prevention, and Mitigation," February 2018, https://img1.wsimg.com/blobby/go/3d82daa4-97fe-4096-9c6b-376b92c619de/downloads/MaliciousUseofAI.pdf?ver=1553030594217. This 2018 report considered AI technologies that are currently available (at least as initial research and development demonstrations) or were plausible in the next five years, with a focus on technologies leveraging machine learning.
188. See Andrew J. Lohn, *Hacking AI: A Primer for Policymakers on Machine Learning Cybersecurity* (Washington, DC: Center for Security and Emerging Technology, Georgetown University, 2020), https://cset.georgetown.edu/wp-content/uploads/CSET-Hacking-AI.pdf; and Andrew J. Lohn, *Poison in the Well: Securing the Shared Resources of Machine Learning* (Washington, DC: Center for Security and Emerging Technology, Georgetown University, June 2021), https://cset.georgetown.edu/publication/poison-in-the-well/.
189. Ben Garfinkel and Allan Dafoe, "Artificial Intelligence, Foresight, and the Offense-Defense Balance," *War on the Rocks*, December 19, 2019, https://warontherocks.com/2019/12/artificial-intelligence-foresight-and-the-offense-defense-balance/.
190. Sven Krasser, chief scientist and vice president of CrowdStrike, observes that even with a detection system with a 99 percent success rate, an attacker can defeat it with over a 99 percent chance of success with 500 tries. See National Academies of Sciences, Engineering, and Medicine, *Implications of Artificial Intelligence for Cybersecurity: Proceedings of a Workshop* (Washington, DC: The National Academies Press, 2019), 43.
191. Hoffmann, *AI and the Future of Cyber Competition.*
192. Hoffmann considers cyberspace a "domain," whereas we consider it a strategic environment.
193. See Emily O. Goldman and Leslie C. Eliason, eds., *The Diffusion of Military Technology and Ideas* (Stanford, CA: Stanford University Press, 2003); and Michael C. Horowitz, "Artificial Intelligence, International Competition, and the Balance of Power," *Texas National Security Review* 1, no. 3 (May 2018), https://tnsr.org/2018/05/artificial-intelligence-international-competition-and-the-balance-of-power/. Horowitz argued that if commercially driven AI continues to fuel innovation, and the types of algorithms militaries might one day use are closely related to civilian applications, advances in AI are likely to diffuse more rapidly to militaries around the world. However, he noted that that military applications of AI based more exclusively in defense research will then generate larger first-mover advantages for early adopters.
194. This phenomenon is captured in the infamous scene in *Wargames* in which the computer WOPR plays out every possible scenario of nuclear exchange and comes to the conclusion, "a strange game, the only winning move is not to play" and then importantly, asks whether the professor wants to play an entirely different game, "How about a nice game of chess?" Being competitive did not end, but it was diverted to a form of game that did not lead to the catastrophe of war.

Chapter 6

1. US Department of Defense, *Joint Publication 3-12*.
2. In Waltzian IR theory, the organizing principle of anarchy creates the fundamental condition of self-help and the logical focus on relative power. Such logic would be different if living in a hierarchical system (most national domestic structures, for example), but holds as long as a world government does not appear. As long as cyberspace is defined by interconnectedness, constant contact is a condition that must be addressed. If that core feature were to change, for example, were cyberspace to become fully segmented at the global level, a different logic would prevail. It is important to note that interconnectedness is understood as a systemic construct as, obviously, within networks of networks technical segmentation of servers exists, but their cumulation comprising cyberspace meets the standard of interconnectedness.
3. To be more specific, if a State wants to be globally networked, it must assume continuous action, which is the logical extension of a condition of constant contact. Deterrence success is measured as the absence of proscribed action, a measure that cannot be achieved in an environment of continuous action. Defense in this environment is possible, but only in the moment of the particular known configuration of the network, which changes with each new update of hardware-software-process that links them. As noted above, in structural logic, if you change the organizing feature of a structure, so too does the core condition and dynamic change. If the global Internet were to become nationally balkanized, the structural condition of constant contact and the imperative to persist would be altered, thereby requiring the fundamental security question to be re-conceived again. On balkanization, see Chris C. Demcheck, "Uncivil and Post-Western Cyber Westphalia: Changing Interstate Power Relations of the Cybered Age," *Cyber Defense Review* (Spring 2016): 49–74, https://cyberdefensereview.army.mil/Portals/6/Documents/CDR%20Journal%20Articles/Uncivil%20and%20Post-Western%20Cyber%20Wesphalia_Demchak.pdf?ver=2018-07-31-093713-953.
4. For example, in 2016, US Presidential Policy Directive 41 defined a "significant cyber incident" as "A cyber incident that is (or group of related cyber incidents that together are) likely to result in demonstrable harm to the national security interests, foreign relations, or economy of the United States or to the public confidence, civil liberties, or public health and safety of the American people." The White House, Presidential Policy Directive 41—United States Cyber Incident Coordination. July 26, 2016, https://obamawhitehouse.archives.gov/the-press-office/2016/07/26/presidential-policy-directive-united-states-cyber-incident.
5. This prescription aligns with the analysis of Richard Harknett and Max Smeets, "Cyber Campaigns and Strategic Outcomes: The Other Means," *Journal of Strategic Studies* (Spring 2020): 1–34, DOI: 10.1080/01402390.2020.1732354.
6. Robert Chesney, "CYBERCOM's Out-of-Network Operations: What Has and Has Not Changed over the Past Year?," *Lawfare*, May 9, 2019, https://www.lawfareblog.com/cybercoms-out-network-operations-what-has-and-has-not-changed-over-past-year. The relevant section of the National Defense Authorization Act is available at https://casetext.com/statute/united-states-code/title-10-armed-forces/subtitle-a-general-military-law/part-i-organization-and-general-military-powers/chapter-19-cyber-matters/section-394-authorities-concerning-military-cyber-operations.
7. As noted in Chapter 3, preclusion is not the same as deterrence by denial. The latter is premised on the notion that an adversary perceives a strategic opportunity to act and chooses to not act because the benefits do not outweigh the costs. Preclusion, alternatively, reduces the number of potential strategic opportunities.
8. Martin C. Libicki, "Norms and Normalization," *Cyber Defense Review* (Spring 2020): 41–52, https://cyberdefensereview.army.mil/Portals/6/CDR%20V5N1%20-%2004_Libicki_WEB.pdf.

Chapter 7

1. Our goal here is not to produce a comprehensive history, but to provide an examination of the theory to policy linkage during this period under review to assess the alignment and misalignment between the expectations of cyber persistence theory and the policies of the United

States. In this chapter, we are adopting the frame of cyber persistence from the perspective of academic researchers and policy analysts. We acknowledge that all three authors played central roles in the policy application of the principles of cyber persistence. We attempt here not to tell that story, but rather remain focused on the theory-policy gap that was initially bridged and what it tells us about the cyber strategic environment and its logic. For a summary of the first ten years by USCYBERCOM's Command Historian, see Michael Warner, "US Cyber Command's First Decade," Hoover Institution/Aegis Series Paper No. 2008, November 2020, https://www.hoover.org/sites/default/files/research/docs/warner_webready.pdf.

2. The White House, *International Strategy for Cyberspace: Prosperity, Security, and Openness in a Networked World*, May 2011, https://obamawhitehouse.archives.gov/sites/default/files/rss_viewer/international_strategy_for_cyberspace.pdf; US Department of Defense, *Department of Defense Strategy for Operating in Cyberspace*, July 2011, https://csrc.nist.gov/CSRC/media/Projects/ISPAB/documents/DOD-Strategy-for-Operating-in-Cyberspace.pdf.
3. See The White House, "White House Report to Congress on Cyber Deterrence Policy," December 29, 2015, https://insidecybersecurity.com/sites/insidecybersecurity.com/files/documents/dec2015/cs2015_0133.pdf. And US Department of Defense, *The DoD Cyber Strategy*, April 2015, https://archive.defense.gov/home/features/2015/0415_cyberstrategy/final_2015_dod_cyber_strategy_for_web.pdf.
4. The White House, *National Cyber Strategy of the United States of America*, September 2018, https://www.whitehouse.gov/wp-content/uploads/2018/09/National-Cyber-Strategy.pdf. As of this writing, the Biden-Harris administration had not yet produced a national cyber strategy.
5. US Department of Defense, *Department of Defense Cyber Strategy: Summary (2018)*, https://media.defense.gov/2018/Sep/18/2002041658/-1/-1/1/CYBER_STRATEGY_SUMMARY_FINAL.PDF.
6. *Achieve and Maintain Cyberspace Superiority: Command Vision for U.S. Cyber Command, April 2018*, https://www.cybercom.mil/Portals/56/Documents/USCYBERCOM%20Vision%20April%202018.pdf. More recently, General Paul Nakasone, commander of USCYBERCOM, described persistent engagement as CYBERCOM's "doctrine." See Paul M. Nakasone and Michael Sulmeyer, "How to Compete in Cyberspace: Cyber Command's New Approach," *Foreign Affairs*, August 25, 2020, https://www.foreignaffairs.com/articles/united-states/2020-08-25/cybersecurity.
7. This is not to suggest that the "defend forward/persistent engagement" strategy is the ideal type of cyber persistence theory's prescriptions; rather, it is better understood as an archetype for States characterized by similar institutional constraints, perceptions of international constraints, and operational capacity.
8. Lynn, "Defending a New Domain: The Pentagon's Cyberstrategy."
9. For a brief history of the formation of USCYBERCOM as a sub-unified command, see Michael Warner, "US Cyber Command's Road to Full Operational Capability," in *Stand Up and Fight! The Creation of US Security Organizations, 1942–2005*, ed. Ty Seidule and Jacqueline E. Whitt (Carlisle, PA: US Army War College, 2015), https://press.armywarcollege.edu/monographs/11/.
10. The White House, *International Strategy for Cyberspace*.
11. *Fact Sheet on Presidential Policy Directive 20*, https://epic.org/privacy/cybersecurity/Pres-Policy-Dir-20-FactSheet.pdf.
12. Hearing of the House (Select) Intelligence Committee Subject, "Cybersecurity Threats: The Way Forward," November 20, 2014, https://www.nsa.gov/news-features/speeches-testimonies/Article/1620360/hearing-of-the-house-select-intelligence-committee-subject-cybersecurity-threat/. Admiral Rogers also noted, "I would argue clearly that approach is not achieving the results that we want."
13. Active cyber defense is described in the DoD strategy as "DoD's synchronized, real-time capability to discover, detect, analyze, and mitigate threats and vulnerabilities."
14. US Department of Defense, *Department of Defense Strategy for Operating in Cyberspace*, July 2011.
15. Leon E. Panetta, "Defending the Nation from Cyber Attack," Speech to Business Executives for National Security, October 11, 2012, https://www.bens.org/document.doc?id=188.

16. Garrett M. Graff, "The Man Who Speaks Softly—and Commands a Big Cyber Army," *Wired*, October 13, 2020, https://www.wired.com/story/general-paul-nakasone-cyber-command-nsa/.
17. The closest approximation may be the 2011 International Strategy for Cyberspace, an initiative spearheaded by Christopher Painter, who was named the DOS cyber coordinator in 2011 and subsequently established the Office of the Coordinator for Cyber Issues. See Emily O. Goldman, "From Reaction to Action: Adopting a Competitive Posture in Cyber Diplomacy," *Texas National Security Review, Special Issue: Cyber Competition* (Fall 2020), https://tnsr.org/2020/09/from-reaction-to-action-adopting-a-competitive-posture-in-cyber-diplomacy/#_ftn22.
18. US Department of Justice, "U.S. Charges Five Chinese Military Hackers for Cyber Espionage Against U.S. Corporations and a Labor Organization for Commercial Advantage: First Time Criminal Charges Are Filed Against Known State Actors for Hacking," May 19, 2014, https://www.justice.gov/opa/pr/us-charges-five-chinese-military-hackers-cyber-espionage-against-us-corporations-and-labor.
19. The White House, *International Strategy for Cyberspace*, 3.
20. See Michael Warner, *The Rise and Fall of Intelligence: An International Security History* (Washington, DC: Georgetown University Press, 2014); and Michael Warner, "Invisible Battlegrounds: On Force and Revolutions, Military and Otherwise," in *The Palgrave Handbook of Security, Risk and Intelligence*, ed. Robert Dover, Huw Dylan, and Michael S. Goodman (London: Palgrave Macmillan, 2017), 247–264.
21. The White House, Presidential Policy Directive 41—United States Cyber Incident Coordination, July 26, 2016, https://obamawhitehouse.archives.gov/the-press-office/2016/07/26/presidential-policy-directive-united-states-cyber-incident. A *significant cyber incident* is defined as a cyber incident (or group of related cyber incidents) that would likely result in demonstrable harm to the national security interests, foreign relations, or economy of the United States or to the public confidence, civil liberties, or public health and safety of the American people.
22. Statement of Admiral Michael S. Rogers, Commander United States Cyber Command, Before the Senate Armed Services Committee, April 5, 2016, https://www.armed-services.senate.gov/imo/media/doc/Rogers_04-05-16.pdf.
23. For example, in October 2015, Secretary of Defense Ashton Carter recommended to the US president using cyber operations against the Islamic State. See United States Senate, Hearing to Receive Testimony on Counter-ISIL (Islamic State of Iraq and the Levant) Operations and Middle East Strategy, April 28, 2016, https://www.armed-services.senate.gov/imo/media/doc/16-51_04-28-16.pdf.
24. US Department of Defense, *The DoD Cyber Strategy*, April 2015.
25. Ibid. (emphasis added)
26. The White House, "White House Report to Congress on Cyber Deterrence Policy."
27. Ibid.
28. Egan, "Remarks on International Law and Stability in Cyberspace."
29. For a compilation on unsealed indictments from 2014 through 2018, see Garrett Hinck and Tim Maurer, "Persistent Enforcement: Criminal Charges as a Response to Nation-State Malicious Cyber Activity," *Journal of National Security Law and Policy* 10, no. 3 (2020), https://jnslp.com/2020/01/23/persistent-enforcement-criminal-charges-as-a-response-to-nation-state-malicious-cyber-activity/.
30. Schwartz, "Dridex Malware Campaign Disrupted."
31. See Executive Order 13694, "Blocking the Property of Certain Persons Engaging in Significant Malicious Cyber-Enabled Activities" (Washington, DC: Federal Register, 2015), https://www.treasury.gov/resourcecenter/sanctions/Programs/Documents/cyber_eo.pdf; and Executive Order 13757, "Taking Additional Steps to Address the National Emergency with Respect to Significant Malicious Cyber-Enabled Activities" (Washington, DC: Federal Register, December 2016), https://www.treasury.gov/resourcecenter/sanctions/Programs/Documents/cyber2_eo.pdf.
32. For a critique of the effectiveness of the indictment approach in stemming China's cyber-enabled theft of US intellectual property, see Jack Goldsmith and Robert D. Williams, "The

Failure of the United States' Chinese-Hacking Indictment Strategy," *Lawfare*, December 28, 2018, https://www.lawfareblog.com/failure-united-states-chinese-hacking-indictment-strategy. Goldsmith and Williams argued that "on the basis of the public record in light of its publicly stated aims, the indictment strategy appears to be a magnificent failure."

33. Andrew Blake, "John McCain Says White House's Cyber Deterrence Policy Comes Up Short," *Washington Times*, January 15, 2016, https://www.washingtontimes.com/news/2016/jan/15/john-mccain-says-white-houses-cyber-deterrence-pol/.
34. Joseph Marks, "McCain Leaves a Rich Cyber Legacy," *Nextgov*, August 27, 2018, https://www.nextgov.com/cybersecurity/2018/08/mccain-leaves-rich-cyber-legacy/150847/.
35. Morgan Chalfant, "McCain Hits Trump over Lack of Cyber Policy," *The Hill*, August 23, 2017, https://thehill.com/policy/cybersecurity/347660-mccain-hits-trump-over-lack-of-cyber-policy.
36. Senator McCain more rightly should have been asking for a cyber security strategy vice a cyber deterrence strategy. That he did not shows that he, and many policymakers, assumed deterrence and security were synonymous. That may have been the case in the nuclear and conventional strategic environments, but, as we argue in this book, it is not the case in the cyberspace strategic environment.
37. The White House, *National Cyber Strategy of the United States of America*, September 2018.
38. Ibid.
39. The White House, *National Security Strategy of the United States of America*, November 2017. Notably, the Strategy also argues that "the United States often views the world in binary terms, with states being either 'at peace' or 'at war,' when it is actually an arena of continuous competition" (28). This binary worldview harkens back to a point we made in Chapter 4 that coercive environments are characterized by definable states of peace, militarized crisis, and war and that periods of non-aggression are typically the norm and are punctuated with aggression on an episodic basis—we declare war and live in peace.
40. The White House, *National Cyber Strategy of the United States of America*, September 2018.
41. US Department of Defense, *Department of Defense Cyber Strategy: Summary (2018)*, 4.
42. US Cyber Command, *Achieve and Maintain Cyberspace Superiority*.
43. See the discussion of Operation Glowing Symphony in Chapter 4.
44. "An Interview with Paul M. Nakasone," *Joint Force Quarterly* 92 (1st Quarter 2019), https://ndupress.ndu.edu/JFQ/Joint-Force-Quarterly-92.aspx.
45. Nakasone would later explicitly connect persistent engagement with the DoD strategy. See Statement of General Paul M. Nakasone, Commander, United States Cyber Command, Before the House Committee on Armed Services, Subcommittee on Intelligence, Emergent Threats, and Capabilities, March 3, 2019, https://armedservices.house.gov/_cache/files/e/d/ed0549b9-c479-4ae0-943d-66cf8fd933c1/AEDF855100875FF9DBB6F5E7472F6E36.nakasone-cybercom-hasc-posture-statement-final-3-13-19.pdf; and Paul M. Nakasone and Michael Sulmeyer, "How to Compete in Cyberspace: Cyber Command's New Approach," *Foreign Affairs*, August 25, 2020, https://www.foreignaffairs.com/articles/united-states/2020-08-25/cybersecurity.
46. Paul M. Nakasone, "A Cyber Force for Persistent Operations," *Joint Force Quarterly* 92 (1st Quarter 2019), https://ndupress.ndu.edu/JFQ/Joint-Force-Quarterly-92.aspx; and US Senate Committee on Armed Services, "Hearing to Review Testimony on United States Special Operations Command and United States Cyber Command in Review of the Defense Authorization Request for Fiscal Year 2020 and the Future Years Defense Program," February 14, 2019, https://www.armed-services.senate.gov/imo/media/doc/19-13_02-14-19.pdf. Additionally, Kenneth Rapuano, assistant secretary of defense for homeland defense and global security and principal cyber advisor to the secretary of defense, noted that "[i]n order to be successful, we must be in malicious actors' networks and systems." Statement of Mr. Kenneth Rapuano, Assistant Secretary of Defense for Homeland Defense and Global Security and Principal Cyber Advisor, Testimony before the House Armed Services Committee Subcommittee on Intelligence and Emerging Threats and Capabilities, March 4, 2020, https://www.congress.gov/116/meeting/house/110592/witnesses/HHRG-116-AS26-Wstate-RapuanoK-20200304.pdf.
47. Nakasone, "A Cyber Force for Persistent Operations."

48. James N. Miller and Robert J. Butler, *National Cyber Defense Center: A Key Next Step toward a Whole-of-Nation Approach to Cyber Security* (Laurel, MD: Johns Hopkins Applied Physics Laboratory, 2021), https://www.jhuapl.edu/Content/documents/NationalCyberDefenseCenter.pdf.
49. Ben Watson, "Is the U.S. Ready to Escalate in Cyberspace?," *Defense One*, November 21, 2018, https://www.defenseone.com/feature/is-the-us-ready-to-escalate-in-cyberspace.
50. Eric Geller and Christian Vasquez, "Beware Cyber Command's Twitter Account," *Politico*, November 9, 2018, https://www.politico.com/newsletters/morning-cybersecurity/2018/11/09/beware-cyber-commands-twitter-account-407006.
51. National Security Advisor John Bolton on Cyber Strategy, Off-camera, on-record press conference, September 20, 2020, https://www.c-span.org/video/?451807-1/national-security-adviser-bolton-briefs-cyber-strategy-audio-only.
52. Robert Chesney, "CYBERCOM's Out-of-Network Operations." The relevant section of the National Defense Authorization Act is available at https://casetext.com/statute/united-states-code/title-10-armed-forces/subtitle-a-general-military-law/part-i-organization-and-general-military-powers/chapter-19-cyber-matters/section-394-authorities-concerning-military-cyber-operations.
53. https://casetext.com/statute/united-states-code/title-10-armed-forces/subtitle-a-general-military-law/part-i-organization-and-general-military-powers/chapter-19-cyber-matters/section-394-authorities-concerning-military-cyber-operations (emphasis added).
54. US Department of Defense, *Department of Defense Cyber Strategy: Summary* (2018).
55. Ibid.
56. For a review of all recent changes in US domestic law supporting the shift in US cyber strategy, see Robert Chesney, "The Domestic Legal Framework for U.S. Military Cyber Operations," *Lawfare*, August 5, 2020, https://www.lawfareblog.com/domestic-legal-framework-us-military-cyber-operations.
57. Ney, "DoD General Counsel Remarks at U.S. Cyber Command Legal Conference." As we noted in Chapter 5, published opinions of government legal advisors may shed light on a State's legal position, though not if the State declined to follow the advice.
58. Lauren C. Williams, "NSA Chief Explains New Cyber Directorate," October 10, 2019, https://fcw.com/articles/2019/10/10/nsa-nagasone-cyber-group-williams.aspx.
59. The same can be said for US allies and partners. For example, in 2020, the European Union announced its first ever sanctions regime against cyber operators affiliated with Russia to "prevent, discourage, deter and respond to such malicious behaviour in cyberspace." The sanctions include a travel ban and asset freeze to natural persons and an asset freeze to entities or bodies. See Council of the European Union Press Release, "Declaration by the High Representative Josep Borrell on Behalf of the EU: European Union Response to Promote International Security and Stability in Cyberspace," July 30, 2020, https://www.consilium.europa.eu/en/press/press-releases/2020/07/30/declaration-by-the-high-representative-josep-borrell-on-behalf-of-the-eu-european-union-response-to-promote-international-security-and-stability-in-cyberspace/.
60. The White House, *National Cyber Strategy of the United States of America*, September 2018. The origin of the CDI is the Presidential Executive Order on Strengthening the Cybersecurity of Federal Networks and Critical Infrastructure, May 11, 2017, https://www.whitehouse.gov/presidential-actions/presidential-executive-order-strengthening-cybersecurity-federal-networks-critical-infrastructure/.
61. The White House, *National Cyber Strategy of the United States of America*, September 2018.
62. Christopher A. Ford, "International Security for Cyberspace: New Models for Reducing Risk," *Arms Control and International Security Papers* 1, no. 20 (2020), https://www.state.gov/wp-content/uploads/2020/10/T-paper-series-Cybersecurity-Format-508.pdf.
63. Ibid. Potential consequences from the actions of "interagency colleagues" presumably speak to DOJ and USDT capabilities. These are addressed in upcoming paragraphs.
64. The White House, *National Cyber Strategy of the United States of America*, September 2018.
65. Ford, "International Security for Cyberspace."
66. US Department of Justice, "North Korean Regime-Backed Programmer Charged with Conspiracy to Conduct Multiple Cyber Attacks and Intrusions," September 6, 2018, https://

www.justice.gov/opa/pr/north-korean-regime-backed-programmer-charged-conspiracy-conduct-multiple-cyber-attacks-and.

67. *United States of America v. Viktor Borisovich Netyksho, Boris Alekseyevich Antonov, Dmitriy Sergeyevich Badin, Ivan Sergeyevich Yermakov, Aleksey Viktorovich Lukasheb, Sergey Aleksandrovich Morgachev, Nikolay Yuryvich Kozachek, Pavel Vyacheslavovich Yershov, Artem Andreyevich Malyshev, Aleksandr Vladimirovich Osadchuk, Aleksey Aleksandrovich Potemkin, and Anatoliy Sergeyovich Kovalev*, Filed July 13, 2018, https://www.justice.gov/file/1080281.
68. Patrick Howell O'Neill, "DOJ Drops Massive Report on Its Efforts to Protect U.S. from Cyberattacks," *CyberScoop*, July 19, 2018, https://www.cyberscoop.com/department-of-justice-internal-cyber-digital-task-force-report/.
69. US Department of Justice, *Report of the Attorney General's Cyber Digital Task Force*, July 2, 2018, https://www.justice.gov/ag/page/file/1076696/download.
70. *U.S. House Resolution 5576—Cyber Deterrence and Response Act of 2018 (115th Congress 2017–2018)*, https://www.congress.gov/bill/115th-congress/house-bill/5576/text.
71. "U.S. Treasury Sanctions North Korean Hacker, Company for Cyber Attacks," *Reuters*, September 6, 2018, https://www.reuters.com/article/us-cyber-northkorea-sanctions/u-s-treasury-sanctions-north-korean-hacker-company-for-cyber-attacks-idUSKCN1LM2I7.
72. Kuhn, *The Structure of Scientific Revolutions*, 4th ed.
73. Michael P. Fischerkeller, Richard J. Harknett, and Jelena Vićić, "The Limits of Deterrence and the Need for Persistence," in *The Cyber Deterrence Problem*, ed. Aaron Brantly (London: Rowman & Littlefield Ltd., 2020), 21–38.
74. See Senator Nelson: "Are there examples of a U.S. response to a cyber attack that deterred the adversary from future attacks, a new 'mutual assured destruction' deterrence policy?" Senate Armed Services Committee Hearing on Encryption & Cyber Matters, September 13, 2016, https://www.armed-services.senate.gov/imo/media/doc/16-68_09-13-16.pdf. See also Senator McCain: "I remain concerned that the administration's cyber policy as a whole remains detached from reality. . . . In December, the administration provided its response, nearly a year and a half late to this committee's requirement for a cyber deterrence policy. The response reflected a troubling lack of seriousness and focus, as it simply reiterated many of the same pronouncements from years past that failed to provide any deterrent value or decrease the vulnerability of our nation in cyberspace." Senate Armed Service Committee Hearing on U.S. Cyber Command, April 5, 2016, https://www.armed-services.senate.gov/imo/media/doc/16-35_4-05-16.pdf.
75. Warner, "A Brief History of Cyber Conflict."
76. The White House, *National Security Decision Directive Number 145*.
77. Arquilla and Rondfeldt, "Cyberwar Is Coming!"
78. US Department of Defense, *Joint Publication 3-13, Information Operations*.
79. Lynn, "Defending a New Domain: The Pentagon's Cyberstrategy."
80. Seminal works include Schelling, *Strategy of Conflict* and *Arms and Influence*; and Alexander L. George, David K. Hall, and William R. Simons, *Limits of Coercive Diplomacy* (Boston: Little, Brown & Company, 1971).
81. Harknett and Smeets, "Cyber Campaigns and Strategic Outcomes."
82. Brendan Rittenhouse Green, *The Revolution That Failed: Nuclear Competition, Arms Control, and the Cold War* (Cambridge: Cambridge University Press, 2020).
83. A cyber operations tracker created by the Council on Foreign Relations (CFR) Digital and Cyberspace Policy program identifies 426 publicly known State-sponsored incidents that have occurred since 2005. None of the events rises to the level of an armed-attack equivalent or a "significant cyber incident." https://www.cfr.org/cyber-operations/#CyberOperations.
84. The most recent example is the Cyberspace Solarium Commission, https://www.solarium.gov/.
85. Kuhn, *The Structure of Scientific Revolutions*.
86. Ibid., 78.
87. See David Dollar and Peter Petri, "Why It's Time to End the Tit-for-Tat Tariffs in the U.S.-China Trade War," *Brookings*, October 5, 2018, https://www.brookings.edu/blog/order-from-chaos/2018/10/05/why-its-time-to-end-the-tit-for-tat-tariffs-in-the-u-s-china-trade-war/; "Next China: Tit for Tat," *BloombergNews*, July 16, 2020, https://www.bloomb

erg.com/news/newsletters/2020-07-17/next-china-tit-for-tat; Lucas Tcheyan and Sam Bresnick, "Reciprocity Is a Tool, Not a Strategy, against China," *Foreign Policy*, August 20, 2020, https://foreignpolicy.com/2020/08/20/china-reciprocity-trump-tit-for-tat-strat egy-trade-war/; Rajaram Panda, "Sino-US Ties: From Ping Pong Diplomacy to Tit-for-Tat Diplomacy," *Modern Diplomacy*, July 7, 2020, https://moderndiplomacy.eu/2020/07/27/sino-us-ties-from-ping-pong-diplomacy-to-tit-for-tat-diplomacy/; and Chun Han Wong, "China Restricts U.S. Diplomats' Movements in Latest Tit-for-Tat," *Wall Street Journal*, September 11, 2020, https://www.wsj.com/amp/articles/china-restricts-u-s-diplomats-movements-in-latest-tit-for-tat-11599834078.

88. Office of the United States Trade Representative, "USTR Issues Tariffs on Chinese Products in Response to Unfair Trade Practices," June 15, 2018, https://ustr.gov/about-us/policy-offi ces/press-office/press-releases/2018/june/ustr-issues-tariffs-chinese-products.
89. See Colum Lynch, "U.S. State Department Appoints Envoy to Counter Chinese Influence at the U.N.," *Foreign Policy*, January 22, 2020, https://foreignpolicy.com/2020/01/22/us-state-department-appoints-envoy-counter-chinese-influence-un-trump/; and Nick Cumming-Bruce, "U.S.-Backed Candidate for Global Tech Post Beats China's Nominee," *New York Times*, March 4, 2020, https://www.nytimes.com/2020/03/04/business/economy/un-world-intel lectual-property-organization.html.
90. Billy Mitchell, "State Department Seeks 'Clean Path' 5G Networks," *FedScoop*, June 9, 2020, https://www.fedscoop.com/state-department-seeks-clean-path-5g-networks/.
91. See US Department of State, "The Clean Network," https://www.state.gov/the-clean-netw ork/#:~:text=The%20Clean%20Network%20program%20is,as%20the%20Chinese%20Co mmunist%20Party; and Roslyn Layton, "State Department's 5G Clean Network Club Gains Members Quickly," *Forbes*, September 4, 2020, https://www.forbes.com/sites/roslynlayton/2020/09/04/state-departments-5g-clean-network-club-gains-members-quickly/?sh=49119 e7b7536.
92. Shannon Vavra, "Trump Administration Expands Economic Restrictions on Huawei," *CyberScoop*, August 17, 2020, https://www.cyberscoop.com/huawei-entity-list-commerce-trump-expanded-38-affiliates/.
93. US Department of Commerce, "Commerce Department Prohibits WeChat and TikTok Transactions to Protect the National Security of the United States," September 18, 2020, https://www.commerce.gov/news/press-releases/2020/09/commerce-department-prohibits-wechat-and-tiktok-transactions-protect. This concern was informed by China's cyber campaigns focusing on harvesting PII from the US Office of Personnel Management, Anthem, Marriott Corporation, Equifax Corporation, and others.
94. Florian J. Egloff, "Public Attribution of Cyber Intrusions," *Journal of Cybersecurity* 6 (2020), https://doi.org/10.1093/cybsec/tyaa012.
95. "Full Transcript: President Obama's Final End-of-Year Press Conference," *Politico*, December 16, 2016, https://www.politico.com/story/2016/12/obama-press-conference-transcript-232763.
96. Hinck and Maurer, "Persistent Enforcement."
97. See, respectively, Jack Goldsmith, "The Puzzle of the GRU Indictment," *Lawfare*, October 21, 2020, https://www.lawfareblog.com/puzzle-gru-indictment; and Peter Machtiger, "The Latest GRU Indictment: A Failed Exercise in Deterrence," *JustSecurity*, October 29, 2020, https://www.justsecurity.org/73071/the-latest-gru-indictment-a-failed-exercise-in-deterre nce/. Additionally, two former NSA analysts, Blake Darché and Mark Kuhr, agree that Russian indictments will have little to no impact on the pace and volume of Russian cyberattacks on the United States. See Chris Bing, "Former NSA Hackers: Yahoo Indictments Won't Slow Down Russian Cyberattacks," *CyberScoop*, March 17, 2017, https://perma.cc/M7G7-ANKP.
98. Elias Groll, "The U.S. Hoped Indicting 5 Chinese Hackers Would Deter Beijing's Cyberwarriors. It Hasn't Worked," *Foreign Policy*, September 2, 2015, https://foreignpolicy.com/2015/09/02/the-u-s-hoped-indicting-5-chinese-hackers-would-deter-beijings-cyberw arriors-it-hasnt-worked/. US Department of State, *Bureau for International Narcotics and Law Enforcement Affairs International Narcotics Control Strategy Report, Volume I*, March 2020, https://www.state.gov/wp-content/uploads/2020/06/Tab-1-INCSR-Vol.-I-Final-for-Print ing-1-29-20-508-4.pdf.

99. An implicit admission of this is commentary by DOJ officials after an October 2020 indictment of five GRU cyber operators. The officials claimed that although the indictment was not a specific warning to Moscow to avoid interfering in this year's election, it serves as a "general" warning that such activities are not deniable. See Ellen Nakashima and Devlin Barrett, "U.S. Charges Russian Intelligence Officers in Several High-Profile Cyberattacks," *Washington Post*, October 19, 2020, https://www.washingtonpost.com/national-security/russia-cyberattacks-election-interference/2020/10/19/51a84208-1208-11eb-bc10-40b25382f1be_story.html.
100. *United States of America v. Li Xiaoyou (a/k/a "Oro0lxy") and Dong Jiazhi*, filed July 7, 2020, https://www.justice.gov/opa/press-release/file/1295981/download.
101. Stefan Soesanto, "Europe's Incertitude in Cyberspace," *Lawfare*, August 3, 2020, https://www.lawfareblog.com/europes-incertitude-cyberspace.
102. Ibid.
103. Jamie Collier, "Europe's New Sanction Regime Suggests a Growing Cyber Diplomacy Presence," *FireEye*, August 06, 2020, https://www.fireeye.com/blog/executive-perspective/2020/08/europe-new-sanction-regime-suggest-a-growing-cyber-diplomacy-presence.html. For a review of the weakness of such sanctions in achieving foreign policy objectives, including coercive objectives, see: Gary Hufbauer, Jeffrey Schott, Kimberly Elliott, and Barbara Oegg, *Economic Sanctions Reconsidered*, 3rd ed. (Washington, DC: Peterson Institute for International Economics, 2007); Clifton Morgan, Navin Bapat, and Valentina Krustev, "The Threat and Imposition of Economic Sanctions, 1971–2000," *Conflict Management and Peace Science* 28, no. 1 (2008): 92–110, https://doi.org/10.1177%2F0738894208097668; and Thomas Bierstecker, Zuzana Hudokova, and Marcos Tourinho, "The Effectiveness of UN Targeted Sanctions: Findings from the Targeted Sanctions Consortium," https://www.academia.edu/8406764/The_Effectiveness_of_UN_Targeted_Sanctions_Findings_from_the_Targeted_Sanctions_Consortium_TSC_.
104. US Department of the Treasury, "Treasury Sanctions North Korean State-Sponsored Malicious Cyber Groups," September 13, 2019, https://home.treasury.gov/news/press-releases/sm774.
105. Interestingly, USDT assesses the *impact* but not the *effectiveness* of its sanctions. The two are distinguished as follows: impact is assessed by analyzing the potential or observed effect of the sanction on the target, whereas effectiveness is assessed based on the extent to which a sanctions program is achieving the overall broader policy goals of the sanctions program, such as ultimately altering or deterring specific behaviors of the target. See United States Government Accountability Office, "Economic Sanctions: Agencies Assess Impacts on Targets, and Studies Suggest Several Factors Contribute to Sanctions' Effectiveness," October 2019, https://www.gao.gov/assets/710/701891.pdf.
106. Additionally, because the sanctioned actor continues to operate in and through cyberspace, the deterrer continues to accrue losses. See Michael P. Fischerkeller and Richard J. Harknett, "Initiative Persistence as the Central Approach for US Cyber Strategy," *Kybernao* 1, no. 1 (August 2021), https://www.artsci.uc.edu/content/dam/refresh/artsandsciences-62/departments/political-science/ccsp/pdf_downloadableflyers/Kybernao_PaperSeries_Issue1_Final.pdf.
107. Hinck and Maurer, for example, consider several potential contributions that unsealed indictments may make to a national cyber security strategy and to cyber stability. Hinck and Maurer, "Persistent Enforcement."
108. Quoted in Aaron Blake, "Trump's Policy of Friendly Deterrence Toward Russia Suffers Another Setback," *Washington Post*, December 16, 2020, https://www.washingtonpost.com/politics/2020/12/16/trumps-policy-friendly-deterrence-toward-russia-suffers-another-setback/. In spite of this acknowledgment, O'Brien concludes with the comment that "But, we're looking at every potential deterrence [*sic*] we can on those countries, as well as others." This is an exemplar of Kuhn's notion of paradigmatic crisis (i.e., in the face of overwhelming evidence adherents to a paradigm nonetheless anchor their thinking to the failing paradigm).
109. The White House, *Interim National Security Strategic Guidance*, March 2021, https://www.whitehouse.gov/wp-content/uploads/2021/03/NSC-1v2.pdf.

110. Secretary of Defense Remarks for the US INDOPACOM Change of Command, April 30, 2021, https://www.defense.gov/Newsroom/Speeches/Speech/Article/2592093/secretary-of-defense-remarks-for-the-us-indopacom-change-of-command/.
111. Eric Geller, "White House Announces Ransomware Task Force—and Hacking Back Is One Option," *Politico*, July 14, 2021. https://www.politico.com/news/2021/07/14/white-house-ransomware-task-force-499723.
112. Many in favor of an explicit bargaining approach claim support for a deterrence strategy as their rationale. Explicit bargaining signals and creates red lines, they say, which are foundational elements of any strategy of deterrence. Although that logic is sound in the nuclear and conventional contexts, we argue that the broad application of deterrence theory to the cyber strategic environment is not sound. Additionally, evoking deterrence implies episodic behavior in cyberspace whereas, we argue, States are primarily pursuing long-term cyber campaigns in pursuit of strategic ends. As we discussed in Chapter 5, treating adversary State behavior as episodic instead of as ongoing campaigns precludes the option of using countermeasures as an internationally lawful remedy to such campaigns.
113. See David E. Sanger and Julian E. Barnes, "U.S. Tried a More Aggressive Cyberstrategy, and the Feared Attacks Never Came," *New York Times*, November 9, 2020, https://www.nytimes.com/2020/11/09/us/politics/cyberattacks-2020-election.html; and Ellen Nakashima, "U.S. Undertook Cyber Operation against Iran as Part of Effort to Secure the 2020 Election," *Washington Post*, November 3, 2020, https://www.washingtonpost.com/national-security/cybercom-targets-iran-election-interference/2020/11/03/aa0c9790-1e11-11eb-ba21-f2f001f0554b_story.html.
114. B. H. Liddell Hart, *Thoughts on War* (London: Farber, 1944).
115. It is important to re-emphasize that one of the challenges of accepting the cyber persistence paradigm is that the deterrence paradigm remains central to the nuclear strategic environment and conventional strategic environment. Thus, the need to simultaneously apply the right logic to the right environment is complicated when introducing a new logic alongside a logic that remains salient to another strategic environment.
116. Finnemore and Hollis argued, "A more nuanced understanding of how accusations [public attributions and unsealed indictments] work could help states better construct them for desired ends, including improved enforcement of international law *and constituting the contents of the law itself*. To date, cyber accusations have emphasized the former, with little to no attention to the latter possibility. However, careful crafting of accusations [public attributions or unsealed indictments], with attention to their structure and function, could enhance their effectiveness for either end" (emphasis added). Finnemore and Hollis, "Beyond Naming and Shaming."
117. "Georgia Hit by Massive Cyber-attack," *BBC News*, October 28, 2019, https://www.bbc.com/news/technology-50207192.
118. See, respectively, "Attribution of Malicious Cyber Activity in Georgia by Russian Military Intelligence," February 21, 2020, https://www.foreignminister.gov.au/minister/marise-payne/media-release/attribution-malicious-cyber-activity-georgia-russian-military-intelligence; "Canada Condemns Russia's Malicious Cyber-activity Targeting Georgia," February 20, 2020, https://www.canada.ca/en/global-affairs/news/2020/02/canada-condemns-russias-malicious-cyber-activity-targeting-georgia.html; and "Latvia Condemns Cyber-attack against Georgia," February 21, 2020, https://www.mfa.gov.lv/en/news/latest-news/65504-latvia-condemns-cyber-attack-against-georgia.
119. Przemysław Roguski, "Russian Cyber Attacks against Georgia, Public Attributions and Sovereignty in Cyberspace," *JustSecurity*, March 6, 2020, https://www.justsecurity.org/69019/russian-cyber-attacks-against-georgia-public-attributions-and-sovereignty-in-cyberspace/.
120. The conduct of any person or entity empowered by the law of the State to exercise elements of governmental authority may be considered evidence of State practice provided the person or entity is acting in that capacity in the particular instance of conduct. See *United Nations Report of the International Law Commission, Sixty-eighth session (2 May–10 June and 4 July–12 August 2016), A/71/10.*

121. Michael R. Pompeo, "The United States Condemns Russian Cyber Attack against the Country of Georgia," February 20, 2020, https://www.state.gov/the-united-states-condemns-russian-cyber-attack-against-the-country-of-georgia/.
122. Dominic Raab, "UK Condemns Russia's GRU over Georgia Cyber-attacks," February 20, 2020, https://www.gov.uk/government/news/uk-condemns-russias-gru-over-georgia-cyber-attacks.
123. "The Netherlands Considers Russia's GRU Responsible for Cyber Attacks against Georgia," February 20, 2020, https://www.government.nl/ministries/ministry-of-foreign-affairs/documents/diplomatic-statements/2020/02/20/the-netherlands-considers-russia%E2%80%99s-gru-responsible-for-cyber-attacks-against-georgia.
124. "New Zealand Condemns Malicious Cyber Activity against Georgia," February 21, 2020, https://www.gcsb.govt.nz/news/gcsb-media-statement/.
125. Aaron Mehta, "Esper Proclaims Election Security an 'Enduring Mission' for Pentagon," *C4ISRNET*, September 19, 2019, https://www.c4isrnet.com/cyber/2019/09/19/esper-proclaims-election-security-an-enduring-mission-for-pentagon/.
126. Julian E. Barnes, "U.S. Cyber Command Expands Operations to Hunt Hackers from Russia, Iran and China," *New York Times*, November 2, 2020, https://www.nytimes.com/2020/11/02/us/politics/cyber-command-hackers-russia.html.
127. DoD's general counsel stated that "'a cyber operation by a State that interferes with another country's ability to hold an election' or that tampers with 'another country's election results would be a clear violation of the rule of non-intervention.'" Ney, "DoD General Counsel Remarks at U.S. Cyber Command Legal Conference."
128. Since 2015, for example, the DOJ has filed numerous unsealed indictments against Chinese cyber operators charged with crimes arguably representing violations of a prohibition on non-intervention in the "economic system" of the United States (by reducing US competitive economic advantage through cyber-enabled intellectual property theft). Additionally, in 2020, the DOJ publicly announced a complaint to seize virtual currency illicitly acquired by DPRK-sponsored cyber operators. See, respectively, US Department of Justice, "China Initiative," September 1, 2020, https://www.justice.gov/usao-edtx/china-initiative; and US Department of Justice, "United States Files Complaint to Forfeit 280 Cryptocurrency Accounts Tied to Hacks of Two Exchanges by North Korean Actors," August 27, 2020, https://www.justice.gov/opa/pr/united-states-files-complaint-forfeit-280-cryptocurrency-accounts-tied-hacks-two-exchanges. Additionally, in 2020, Germany's Federal Prosecutor issued an arrest warrant against Russian citizen Dmitry Badin (an assumed member of Russia's GRU elite cyber unit 26165 [APT 28]), the main suspect in the 2015 hacking of the German Bundestag, which resulted in disruption of the functioning of a legislative body (a violation of the non-intervention principle). See "Germany Issues Arrest Warrant for Russian Suspect in Parliament Hack: Newspaper," *Reuters*, May 5, 2020, https://www.reuters.com/article/us-russia-germany-warrant/germany-issues-arrest-warrant-for-russian-suspect-in-parliament-hack-newspaper-idUSKBN22H0TB.
129. *United States of America v. Internet Research Agency*, February 16, 2018, https://www.justice.gov/file/1035477/download.
130. See Steven J. Barela, "Cross-Border Cyber Ops to Erode Legitimacy: An Act of Coercion," *JustSecurity*, January 12, 2017, https://www.justsecurity.org/36212/cross-border-cyber-ops-erode-legitimacy-act-coercion/; and David Ohlin, "Did Russian Cyber Interference in the 2016 Election Violate International Law?," respectively.
131. Eric Talbot Jensen and Sean Watts, "A Cyber Duty of Due Diligence: Gentle Civilizer or Crude Destabilizer," *Texas Law Review* 95 (2017): 1555–1577, https://texaslawreview.org/wp-content/uploads/2017/11/Jensen.Watts_.pdf.
132. James Crawford, *The International Law Commission's Articles on State Responsibility: Introduction, Text and Commentaries* (London: Cambridge University Press, 2002), 285. "A State that is unable to establish attribution to a reliably certain level thus accepts the risk that its countermeasures will themselves amount to an internationally wrongful act." *International Law Commission Draft Articles on Responsibility of States for Internationally Wrongful Acts, U.N. Doc. A/56/10,* 2001.

133. Decisions of national courts at all levels may count as evidence of State practice (both of the defending and the indicted State) and may also be considered evidence of expressions of *opinio juris*. Public indictments could also serve to "establish a solid narrative for . . . what is appropriate or completely out of line" in support of a tacit bargaining strategy. See Arielle Waldman, "Nation-state Hacker Indictments: Do They Help or Hinder?," *SearchSecurity*, April 2021, https://searchsecurity.techtarget.com/feature/Nation-state-hacker-indictments-Do-they-help-or-hinder.
134. Even private company accusations may have constitutive value if done at the behest of a State or when other States accept and adopt the behavioral lines drawn by the accusation as an existing or developing customary legal norm. Finnemore and Hollis, "Beyond Naming and Shaming."
135. Ibid. Finnemore and Hollis argue that "Today's accusations may serve as early evidence of a 'usage'—that is, a habitual practice followed without any sense of legal obligation. If such accusations persist and spread over time, states may come to assume that these accusations are evidence of *opinio juris*, delineating which acts are either appropriate or wrongful as a matter of international law. In other words, accusations can directly contribute to the formation of customary international law."
136. Finnemore and Hollis, "Constructing Norms for Global Cybersecurity."
137. A useful volume for US cyber policymakers seeking to better understand the US DoD strategic approach of defend forward/persistent engagement in light of international law is Goldsmith, ed., *The United States' Defend Forward Cyber Strategy*.
138. Fischerkeller and Harknett, "Initiative Persistence as the Central Approach for US Cyber Strategy."
139. See Michael P. Fischerkeller and Richard J. Harknett, "A Response on Persistent Engagement and Agreed Competition," *Lawfare*, June 27, 2019, https://www.lawfareblog.com/response-persistent-engagement-and-agreed-competition; Michael P. Fischerkeller, "The Cyberspace Solarium Commission Report and Persistent Engagement," *Lawfare*, March 23, 2020, https://www.lawfareblog.com/cyberspace-solarium-commission-report-and-persistent-engagement; and Richard J. Harknett, "SolarWinds: The Need for Persistent Engagement," *Lawfare*, December 23, 2020, https://www.lawfareblog.com/solarwinds-need-persistent-engagement.
140. The National Cyber Security Centre, https://www.ncsc.gov.uk/. Israel National Cyber Directorate, https://www.gov.il/en/departments/israel_national_cyber_directorate.
141. United States Department of Justice, "Information about the Department of Justice's China Initiative and a Compilation of China-related Prosecutions since 2018," November 12, 2020, https://www.justice.gov/opa/information-about-department-justice-s-china-initiative-and-compilation-china-related.
142. US Cyber Command, "New CNMF Initiative Shares Malware Samples with Cybersecurity Industry," November 5, 2018, https://www.cybercom.mil/Media/News/News-Display/Article/1681533/new-cnmf-initiative-shares-malware-samples-with-cybersecurity-industry/. USCYBERCOM's press release notes that "Recognizing the value of collaboration with the public sector, the CNMF has initiated an effort to share unclassified malware samples it has discovered that it believes will have the greatest impact on improving global cybersecurity. For members of the security community, CNMF-discovered malware samples will be logged at this website."
143. Nakasone, "A Cyber Force for Persistent Operations."
144. The existence of the group was first reported by the *Washington Post* and later confirmed by General Nakasone. See Ellen Nakashima, "NSA and Cyber Command to Coordinate Actions to Counter Russian Election Interference in 2018 amid Absence of White House Guidance," *Washington Post*, July 17, 2018, https://www.washingtonpost.com/world/national-security/nsa-and-cyber-command-to-coordinate-actions-to-counter-russian-election-interference-in-2018-amid-absence-of-white-house-guidance/2018/07/17/baac95b2-8900-11e8-85ae-511bc1146b0b_story.html; and Sean Lyngaas, "NSA Chief Confirms He Set Up Task Force to Counter Russian Hackers," July 23, 2018, https://www.cyberscoop.com/russia-small-group-paul-nakasone-nsa-aspen/.

145. Shannon Vavra, "Cyber Command Has Redeployed Overseas in Effort to Protect 2020 Elections," *CyberScoop*, May 7, 2019, https://www.cyberscoop.com/cyber-command-redeployed-overseas-effort-protect-2020-elections/. In the fall of 2017, the Federal Bureau of Investigation's director Christopher Wray established the Foreign Influence Task Force to identify and counteract malign foreign influence operations targeting the United States. In May 2018, the DHS (through its Cybersecurity and Infrastructure Security Agency) established the Countering Foreign Influence Task Force (CFITF), which was "charged with building national resilience to foreign influence activities" by "helping the American people and the Department of Homeland Security (DHS) stakeholders understand the scope and scale of influence activities targeting elections and critical infrastructure, and by enabling them to take actions to mitigate risks associated with foreign influence operations." See, respectively, FBI National Press Office, "The FBI Launches a Combating Foreign Influence Webpage," August 30, 2018, https://www.fbi.gov/news/pressrel/press-releases/the-fbi-launches-a-combating-foreign-influence-webpage; and Cybersecurity and Infrastructure Security Agency, "Countering Foreign Influence Task Force," https://www.cisa.gov/cfi-task-force.
146. US Department of the Treasury, "Treasury Sanctions North Korean State-Sponsored Malicious Cyber Groups," September 13, 2019, https://home.treasury.gov/news/press-releases/sm774.
147. Ibid.
148. Spencer S. Hsu, "U.S. Cyber Command Helps Prosecutors Seize Stolen Cryptocurrency Traced to Illicit N. Korea Nuclear Weapons Program," *Washington Post*, August 28, 2020, https://www.washingtonpost.com/local/legal-issues/us-cyber-command-helps-prosecutors-seize-stolen-cryptocurrency-traced-to-illicit-n-korea-nuclear-weapons-program/2020/08/28/12e7959c-e886-11ea-970a-64c73a1c2392_story.html.
149. CISA uses the Traffic Light Protocol (TLP) to facilitate greater information sharing. "Unlimited distribution," referred to as "TLP: White," by definition, has no associated constraints. TLP: Green, TLP: Amber, and TLP: Red are designations representing increasing limits on distribution. "Cybersecurity and Infrastructure Security Agency, Traffic Light Protocol Definitions and Usage," https://www.cisa.gov/tlp.
150. Shannon Vavra, "Why Cyber Command's Latest Warning Is a Win for the Government's Information Sharing Efforts," *CyberScoop*, July 10, 2019, https://www.cyberscoop.com/cyber-command-information-sharing-virustotal-iran-russia/.
151. Cybersecurity and Infrastructure Security Agency, "Malware Analysis Report (AR20-045A): MAR-10265965-1.v1—North Korean Trojan: BISTROMATH," February 14, 2020, https://us-cert.cisa.gov/ncas/analysis-reports/ar20-045a.
152. Cybersecurity and Infrastructure Security Agency, "Alert (AA20-205A) NSA and CISA Recommend Immediate Actions to Reduce Exposure across Operational Technologies and Control Systems," July 23, 2020, https://us-cert.cisa.gov/ncas/alerts/aa20-205a.
153. Cybersecurity and Infrastructure Security Agency, "Alert (AA20-301A) North Korean Advanced Persistent Threat Focus: Kimsuky," October 27, 2020, https://us-cert.cisa.gov/ncas/alerts/aa20-301a.
154. Cybersecurity and Infrastructure Security Agency, "Malware Analysis Report (AR20-303B): MAR-10310246-1.v1—ZEBROCY Backdoor," October 29, 2020, https://us-cert.cisa.gov/ncas/analysis-reports/ar20-303b.
155. Kevin Randolph, "U.S. Energy, Homeland Security, Defense Departments to Partner on Initiative to Protect Energy Infrastructure from Cyberthreats," *Daily Energy Insider*, February 5, 2020, https://dailyenergyinsider.com/news/24113-u-s-energy-homeland-security-defense-departments-to-partner-on-initiative-to-protect-energy-infrastructure-from-cyberthreats/.
156. David McLaughlin, Saleha Mohsin, and Jacob Rund, "All about Cfius, Trump's Watchdog on China Dealmaking," *Washington Post*, September 15, 2020, https://www.washingtonpost.com/business/energy/all-about-cfius-trumps-watchdog-on-china-dealmaking/2020/09/15/1fdb46fa-f762-11ea-85f7-5941188a98cd_story.html.

157. US Department of the Treasury, "The Committee on Foreign Investment in the United States (CFIUS)," https://home.treasury.gov/policy-issues/international/the-committee-on-foreign-investment-in-the-united-states-cfius.
158. Ibid.
159. See Cybersecurity and Infrastructure Security Agency, https://www.cisa.gov/news/2021/08/05/cisa-launches-new-joint-cyber-defense-collaborative; and Cybersecurity and Infrastructure Security Agency, https://www.cisa.gov/publication/jcdc-fact-sheet.
160. We choose the term "alignment" rather than "partnership," which has been the concept for twenty-five years, to emphasize the fact that the primary private sector profit-making interest and the primary security-seeking interest of the State are different starting points. Partnerships are based on a shared primary interest; alignments are not.
161. For an overview, see Megan Brown, "Cyber Imperative: Preserve and Strengthen Public-Private Partnerships" (The National Security Institute at George Mason University's Antonin Scalia Law School, 2018), https://nationalsecurity.gmu.edu/2018/10/nsi-policy-paper-cyber-imperative-preserve-and-strengthen-public-private-partnerships/.
162. "The Cybersecurity 202: Top Cybersecurity Companies Are Pooling Their Intel to Stop Cyberattacks," *Washington Post*, May 23, 2019, https://www.washingtonpost.com/news/powerpost/paloma/the-cybersecurity-202/2019/05/23/the-cybersecurity-202-top-cybersecurity-companies-are-pooling-their-intel-to-stop-cyberattacks/5ce5ef73a7a0a46b92a3fd95/.
163. Salvador Rodriguez, "The FBI Visits Facebook to Talk about 2020 Election Security, with Google, Microsoft and Twitter Joining," *CNBC*, September 24, 2019, https://www.cnbc.com/amp/2019/09/04/facebook-twitter-google-are-meeting-with-us-officials-to-discuss-2020-election-security.html.
164. Ibid.
165. US Department of Justice, "United States Seizes Domain Names Used by Iran's Islamic Revolutionary Guard Corps," October 7, 2020, https://www.justice.gov/usao-ndca/pr/united-states-seizes-domain-names-used-iran-s-islamic-revolutionary-guard-corps.
166. National Security Agency | Central Security Service. "The NSA's Cybersecurity Collaboration Center." https://www.nsa.gov/what-we-do/cybersecurity/cybersecurity-collaboration-center/.
167. Andy Greenberg, "Chinese Hacking Spree Hit an 'Astronomical' Number of Victims," *Wired*, March 5, 2021, https://www.wired.com/story/china-microsoft-exchange-server-hack-victims/.
168. Microsoft Security, "HAFNIUM Targeting Exchange Servers with 0-day Exploits," March 2, 2021, https://www.microsoft.com/security/blog/2021/03/02/hafnium-targeting-exchange-servers/.
169. United States Department of Justice, "Justice Department Announces Court-authorized Effort to Disrupt Exploitation of Microsoft Exchange Server Vulnerabilities," April 13, 2021, https://www.justice.gov/usao-sdtx/pr/justice-department-announces-court-authorized-effort-disrupt-exploitation-microsoft.
170. National Cybersecurity Alliance, https://staysafeonline.org/cybersecurity-awareness-month/.
171. By 2021, the notion of shared responsibility was the context (and policy) of the private sector, which adopted the terminology of "shared responsibility" for cloud computing security. See Chris Tozzi, "Avoiding the Pitfalls of the Shared Responsibility Model for Cloud Security," *PaloAlto Networks*, September 24, 2020, https://blog.paloaltonetworks.com/prisma-cloud/pitfalls-shared-responsibility-cloud-security/.
172. The 2009 White House *Cybersecurity Policy Review* devoted a few lines that call for an education campaign that "should focus on public messages to promote responsible use of the Internet and awareness of fraud, identity theft, cyber predators, and cyber ethics." The White House, *Cybersecurity Policy Review*, 14, https://obamawhitehouse.archives.gov/cyberreview/documents/.
173. https://staysafeonline.org/cybersecurity-awareness-month/theme/.
174. Gregory J. Rattray and Jason Healey, "Non-State Actors in Cyber Conflict," in *America's Cyber Future: Security and Prosperity in the Information Age, Volume II*, ed. Kristen M. Lord

and Travis Sharp (Washington, DC: Center for a New American Security, June 2011), https://www.files.ethz.ch/isn/129907/CNAS_Cyber_Volume%20II_2.pdf.

175. Tim Maurer and Arthur Nelson, *International Strategy to Better Protect the Financial System against Cyber Threats* (Washington, DC: Carnegie Endowment for International Peace, 2020), https://carnegieendowment.org/2020/11/18/international-strategy-to-better-protect-financial-system-against-cyber-threats-pub-83105.
176. See https://www.honeynet.org/about/.
177. See https://www.stopbadware.org/. StopBadware started as a project of the Berkman Center for Internet & Society at Harvard University, spun off as an independent nonprofit organization in 2010, and then turned back into a university research project in 2015.
178. See https://www.cyberthreatalliance.org/.
179. "Financial Systemic Analysis & Resilience Center Appoints Scott DePasquale as President," *Cision PRNewswire*, March 15, 2017, https://www.prnewswire.com/news-releases/financial-systemic-analysis--resilience-center-appoints-scott-depasquale-as-president-300423909.html. The frameworks and models developed by FSARC are now embodied in the Analysis and Resilience Center for Systemic Risk. See https://systemicrisk.org/.
180. See, Maggie Miller, "New Coalition Aims to Combat Growing Wave of Ransomware Attacks," *The Hill*, January 17, 2021, https://thehill.com/policy/cybersecurity/534544-new-coalition-aims-to-combat-growing-wave-of-ransomware-attacks; and https://securityandtechnology.org/ransomwaretaskforce/. Formal task force members include, but are not limited to, Andreessen Horowitz, Aspen Digital, BlackBaud, BlueVoyant, The Center for Internet Security, CFC Underwriting, Chainalysis, Citrix, FireEye, Microsoft, Third Way, and Ernst & Young.
181. See Cybersecurity and Infrastructure Security Agency, *CISA Global*, February 2021, https://www.cisa.gov/sites/default/files/publications/CISA%20Global_2.1.21_508.pdf; and Mariam Baksh, "International Cybersecurity Work of Both the State and Homeland Security Departments Will Rely on Support from Congress," *Nextgov*, February 19, 2021, https://www.nextgov.com/cybersecurity/2021/02/cisa-chief-says-agencys-global-initiative-support-state-department/172184/.
182. See Shannon Vavra, "Cyber Command Deploys Abroad to Fend Off Foreign Hacking Ahead of the 2020 Election," *CyberScoop*, August 25, 2020, https://www.cyberscoop.com/2020-presidential-election-cyber-command-nakasone-deployed-protect-interference-hacking/; and US Cyber Command, "Hunt Forward Estonia: Estonia, US Strengthen Partnership in Cyber Domain with Joint Operation," December 3, 2020, https://www.cybercom.mil/Media/News/Article/2433245/hunt-forward-estonia-estonia-us-strengthen-partnership-in-cyber-domain-with-joi/.
183. Emily O. Goldman, "The Cyber Paradigm Shift," in *Ten Years In: Implementing Strategic Approaches to Cyberspace*, ed. Jacquelyn G. Schneider, Emily O. Goldman, and Michael Warner (Newport, RI: Naval War College Pres 2020), 31–46, https://digital-commons.usnwc.edu/usnwc-newport-papers/45/.
184. Kuhn, *The Structure of Scientific Revolutions*, 2nd ed., 150–151.
185. Ibid., 157.
186. Goldman has offered a blueprint for a DOS cyber strategy. See Goldman, "From Reaction to Action: Adopting a Competitive Posture in Cyber Diplomacy."

BIBLIOGRAPHY

Achen, Christopher H. *The Statistical Analysis of Quasi-Experiments.* Berkeley: University of California Press, 1986.

Achen, Christopher H., and Duncan Snidal. "Rational Deterrence Theory and Comparative Case Studies." *World Politics: A Quarterly Journal of International Relations* 41, no. 2 (January 1989): 143–169. https://doi.org/10.2307/2010405.

Achieve and Maintain Cyberspace Superiority: Command Vision for U.S. Cyber Command. April 2018. https://www.cybercom.mil/Portals/56/Documents/USCYBERCOM%20Vision%20April%202018.pdf.

Adamson, Liisi. "International Law and Cyber Norms: A Continuum?" In *Governing Cyberspace: Behavior, Power and Diplomacy,* edited by Dennis Broeders and Bibi van den Berg, 19–44. Rowman & Littlefield, 2020. https://rowman.com/WebDocs/Open_Access_Governing_Cyberspace_Broeders_and_van_den_Berg.pdf.

Adelman, Kenneth L. "Arms Control: Arms Control with and without Agreements." *Foreign Affairs* 63, no. 2 (Winter 1984/85): 240–263. https://www.foreignaffairs.com/articles/1984-12-01/arms-control-arms-control-and-without-agreements.

Agreement Between the Government of The United States of America and the Government of The Union of Soviet Socialist Republics on the Prevention of Incidents On and Over the High Seas. May 25, 1972. https://2009-2017.state.gov/t/isn/4791.htm.

Alperavitch, Dmitri. "Revealed: Operation Shady RAT." 2011. http://www.a51.nl/sites/default/files/pdf/wp_operation_shady_rat.pdf.

Altman, Dan. "Advancing without Attacking: The Strategic Game around the Use of Force." *Security Studies* 27, no. 1 (2018): 58–88. https://doi.org/10.1080/09636412.2017.1360074.

Altman, Dan. "By *Fait Accompli*, Not Coercion: How States Wrest Territory from Their Adversaries." *International Studies Quarterly* 61, no. 4 (2017): 881–891. https://doi.org/10.1093/isq/sqx049.

Altman, Dan. "Red Lines and *Faits Accomplis* in Interstate Coercion and Crisis." PhD diss., Massachusetts Institute of Technology, June 2015. https://dspace.mit.edu/bitstream/handle/1721.1/99775/927329080-MIT.pdf?sequence=1&isAllowed=y.

Analysis and Resilience Center for Systemic Risk. https://systemicrisk.org/.

Application Security Services. "Murder of Cybersecurity by Legacy Applications." *ImmuniWeb,* September 21, 2018. https://www.immuniweb.com/blog/murder-of-cybersecurity-by-legacy-applications.html.

Arquilla, John. "Cyberwar Is Already Upon Us." *Foreign Policy,* February 27, 2012. https://foreignpolicy.com/2012/02/27/cyberwar-is-already-upon-us/.

Arquilla, John. "Rebuttal Cyberwar Is Already Upon Us." *Foreign Policy,* March/April 2012.

Arquilla, John, and David Ronfeldt. "Cyberwar Is Coming." *Comparative Strategy* 12 (April–June 1993): 141–165. https://doi.org/10.1080/01495939308402915.

Arquilla, John, and David Ronfeldt. *In Athena's Camp: Preparing for Conflict in the Information Age*. Santa Monica, CA: Rand Corporation, 1997. https://www.rand.org/pubs/monograph_reports/MR880.html.

Arquilla, John, and David Ronfeldt. *Networks and Netwars: The Future of Terror, Crime, and Militancy*. Santa Monica, CA: Rand Corporation, 2001. https://www.rand.org/pubs/monograph_reports/MR1382.html.

Arsene, Liviu, and Radu Tudorica. "Trickbot Is Dead: Long Live Trickbot!" *BitDefender*, November 23, 2020. https://labs.bitdefender.com/2020/11/trickbot-is-dead-long-live-trickbot/.

Art, Robert, and Kelly M. Greenhill. "Coercion: An Analytical Overview." In *Coercion: The Power to Hurt in International Politics*, edited by Kelly M. Greenhill and Peter Krause, 3–32. New York: Oxford University Press, 2018.

"Attribution of Malicious Cyber Activity in Georgia by Russian Military Intelligence." February 21, 2020. https://www.foreignminister.gov.au/minister/marise-payne/media-release/attribution-malicious-cyber-activity-georgia-russian-military-intelligence.

Australia's International Cyber Engagement Strategy: 2019 International Law Supplement. 2019. https://www.dfat.gov.au/publications/international-relations/international-cyber-engagement-strategy/aices/chapters/2019_international_law_supplement.html.

Austria. "Pre-Draft Report of the OEWG–ICT: Comments by Austria." March 13, 2020. https://front.un-arm.org/wp-content/uploads/2020/04/comments-by-austria.pdf.

Avgerinos, Thanassis, Alexandre Rebert, Sang Kil Cha, and David Brumley. "Enhancing Symbolic Execution with Veritesting." 2016. https://users.ece.cmu.edu/~aavgerin/papers/veritesting-icse-2014.pdf.

Ayoub, Kareem, and Kenneth Payne. "Strategy in the Age of Artificial Intelligence." *Journal of Strategic Studies* 39 (2015): 5–6. https://doi.org/10.1080/01402390.2015.1088838.

Baksh, Mariam. "International Cybersecurity Work of Both the State and Homeland Security Departments Will Rely on Support from Congress." *Nextgov*, February 19, 2021. https://www.nextgov.com/cybersecurity/2021/02/cisa-chief-says-agencys-global-initiative-support-state-department/172184/.

Barela, Steven J. "Cross-Border Cyber Ops to Erode Legitimacy: An Act of Coercion." *JustSecurity*, January 12, 2017. https://www.justsecurity.org/36212/cross-border-cyber-ops-erode-legitimacy-act-coercion/.

Barnes, Julian E. "U.S. Cyber Command Expands Operations to Hunt Hackers from Russia, Iran and China." *New York Times*, November 2, 2020. https://www.nytimes.com/2020/11/02/us/politics/cyber-command-hackers-russia.html.

Barnsby, Robert E., and Shane R. Reeves. "Give Them an Inch, They'll Take a Terabyte: How States May Interpret Tallinn Manual 2.0's International Human Rights Law Chapter." *Texas Law Review* 95, no. 1515 (2017). https://texaslawreview.org/wp-content/uploads/2017/11/Barnsby.Reeves.pdf.

Barry, Rob, and Dustin Volz. "Ghosts in the Clouds: Inside China's Major Corporate Hack." *Wall Street Journal*, December 30, 2019. https://www.wsj.com/articles/ghosts-in-the-clouds-inside-chinas-major-corporate-hack-11577729061.

Basu, Arindrajit, Irene Poetranto, and Justin Lau. "The UN Struggles to Make Progress on Securing Cyberspace." *Carnegie Endowment for International Peace*. May 19, 2021. https://carnegieendowment.org/2021/05/19/un-struggles-to-make-progress-on-securing-cyberspace-pub-84491.

Bejtlich, Richard. "What Is APT and What Does It Want?" *TaoSecurity*, January 16, 2010. https://taosecurity.blogspot.com/2010/01/what-is-apt-and-what-does-it-want.html.

Bekerman, Dina, and Sharit Yarushalmi. "The State of Vulnerabilities in 2019." January 23, 2020. https://www.imperva.com/blog/the-state-of-vulnerabilities-in-2019/.

Bendrath, Ralf. "The American Cyber-Angst and the Real World." In *Bombs and Bandwidth: The Emerging Relationship between Information Technology and Security*, edited by Robert Latham, 49–73. New York: The New Press, 2003.

Bendrath, Ralf. "The Cyberwar Debate: Perception and Politics in US Critical Infrastructure Protection." *Information & Security: An International Journal* 47 (2001): 80–103. https://infosec-journal.com/article/cyberwar-debate-perception-and-politics-us-critical-infrastructure-protection.

Betz, David J., and Tim Stevens. *Cyberspace and the State: Towards a Strategy for Cyber-Power*. London: Routledge, 2011.

Biersteker, Thomas, Zuzana Hudokova, and Marcos Tourinho. "The Effectiveness of UN Targeted Sanctions: Findings from the Targeted Sanctions Consortium." 2013. https://www.academia.edu/8406764/The_Effectiveness_of_UN_Targeted_Sanctions_Findings_from_the_Targeted_Sanctions_Consortium_TSC_.

Bing, Chris. "Former NSA Hackers: Yahoo Indictments Won't Slow Down Russian Cyberattacks." *CyberScoop*, March 17, 2017. https://perma.cc/M7G7-ANKP.

Bing, Chris. "U.S. Cyberwarriors Are Getting Better at Fighting ISIS Online, Says Top General." *CyberScoop*, May 23, 2017. https://www.cyberscoop.com/paul-nakasone-isis-cyber-attacks-army-cyber-command/.

Blake, Aaron. "Trump's Policy of Friendly Deterrence toward Russia Suffers Another Setback." *Washington Post*, December 16, 2020. https://www.washingtonpost.com/politics/2020/12/16/trumps-policy-friendly-deterrence-toward-russia-suffers-another-setback/.

Blake, Andrew. "John McCain Says White House's Cyber Deterrence Policy Comes Up Short." *Washington Times*, January 15, 2016. https://www.washingtontimes.com/news/2016/jan/15/john-mccain-says-white-houses-cyber-deterrence-pol/.

Borghard, Erica D., and Shawn W. Lonergan. "Cyber Operations as Imperfect Tools of Escalation." *Strategic Studies Quarterly* 13, no. 3 (Fall 2019): 122–145. https://www.jstor.org/stable/26760131.

Borghard, Erica D., and Shawn W. Lonergan, "The Logic of Coercion in Cyberspace." *Security Studies* 26, no. 3 (2017): 452–481. https://www.tandfonline.com/doi/full/10.1080/09636412.2017.1306396.

Borys, Stephanie. "Licence to Hack: Using a Keyboard to Fight Islamic State." *ABC News Australia*, December 17, 2019. https://www.abc.net.au/news/2019-12-18/inside-the-islamic-state-hack-that-crippled-the-terror-group/11792958?nw=0.

Bracken, Paul. "National Security Organization and Conflict Termination." Presented at Haverford College, April 27, 1990.

Brodie, Bernard. *The Absolute Weapon: Atomic Power and World Order*. New York: Harcourt, Brace, 1946.

Brodie, Bernard. "The Continuing Relevance of *On War*." In Carl von Clausewitz, *On War*, edited and translated by Michael Howard and Peter Paret, 45–58. Princeton, NJ: Princeton University Press, 1976.

Broeders, Dennis. "Aligning the International Protection of 'The Public Core of the Internet' with State Sovereignty and National Security." *Journal of Cyber Policy* 2, no. 3 (November 2017): 366–376. https://doi.org/10.1080/23738871.2017.1403640.

Brown, Megan. "Cyber Imperative: Preserve and Strengthen Public-Private Partnerships." The National Security Institute at George Mason University's Antonin Scalia Law School, 2018. https://nationalsecurity.gmu.edu/2018/10/nsi-policy-paper-cyber-imperative-preserve-and-strengthen-public-private-partnerships/.

Brown, Michael, and Pavneet Singh. "China's Technology Transfer Strategy: How Chinese Investments in Emerging Technology Enable a Strategic Competitor to Access the Crown Jewels of U.S. Innovation." *Defense Innovation Unit Experimental (DIUx)*. January 2018. https://admin.govexec.com/media/diux_chinatechnologytransferstudy_jan_2018_(1).pdf.

Brundage, Miles, Shahar Avin, Jack Clark, Helen Toner, Peter Eckersley, Ben Garfinkel, Allan Dafoe, Paul Scharre, Thomas Zeitzoff, Bobby Filar, Hyrum Anderson, Heather Roff, Gregory C. Allen, Jacob Steinhardt, Carrick Flynn, Seán Ó hÉigeartaigh, Simon Beard, Haydn Belfield, Sebastian Farquhar, Clare Lyle, Rebecca Crootof, Owain Evans, Michael Page, Joanna Bryson, Roman Yampolskiy, and Dario Amodei. "The Malicious Use of Artificial Intelligence: Forecasting, Prevention, and Mitigation." February 2018. https://img1.wsimg.com/blobby/go/3d82daa4-97fe-4096-9c6b-376b92c619de/downloads/MaliciousUseofAI.pdf?ver=1553030594217.

Brussels Summit Communiqué. June 14, 2021. https://www.nato.int/cps/en/natohq/news_185000.htm.

Buchanan, Ben. *The Cyber Security Dilemma: Hacking, Trust, and Fear between Nations*. London: Oxford University Press, 2016.

Buchanan, Ben. *The Hacker and the State: Cyber Attacks and the New Normal of Geopolitics*. Cambridge, MA: Harvard University Press, 2020.

Buckley, Chris, and Keith Bradsher. "Xi Jinping's Marathon Speech: Five Takeaways." *New York Times*, October 18, 2017. https://www.nytimes.com/2017/10/18/world/asia/china-xi-jinping-party-congress.html.

Bulazel, Alexei, Sophia d'Antoine, Perri Adams, and Dave Aitel. "The Risks of Huawei Risk Mitigation." *Lawfare*, April 24, 2019. https://www.lawfareblog.com/risks-huawei-risk-mitigation.

Burt, Tom. "Microsoft Report Shows Increasing Sophistication of Cyber Threats." *Microsoft*. September 29, 2020. https://blogs.microsoft.com/on-the-issues/2020/09/29/microsoft-digital-defense-report-cyber-threats/.

Burt, Tom. "An Update on Disruption of TrickBot." *Microsoft on the Issues*. October 20, 2020. https://blogs.microsoft.com/on-the-issues/2020/10/20/trickbot-ransomware-disruption-update/.

Burton, Joe, and Simona R. Soare. "Understanding the Strategic Implications of the Weaponization of Artificial Intelligence." *IEEE Xplore*, July 11, 2019. https://ieeexplore.ieee.org/document/8756866.

Busselen, Michael. "CrowdStrike CTO Explains 'Breakout Time'—A Critical Metric in Stopping Breaches." *CrowdStrike Blog*, June 6, 2018. https://www.crowdstrike.com/blog/crowdstrike-cto-explains-breakout-time-a-critical-metric-in-stopping-breaches/.

Caesar, Ed. "The Incredible Rise of North Korea's Hacking Army." *The New Yorker*, April 19, 2021. https://www.newyorker.com/magazine/2021/04/26/the-incredible-rise-of-north-koreas-hacking-army.

"Canada Condemns Russia's Malicious Cyber-activity Targeting Georgia." February 20, 2020. https://www.canada.ca/en/global-affairs/news/2020/02/canada-condemns-russias-malicious-cyber-activity-targeting-georgia.html.

Carson, Austin. *Secret Wars: Covert Conflict in International Politics*. Princeton, NJ: Princeton University Press, 2018.

Cartwright, Mark. "The Art of War." 2013. https://www.ancient.eu/The_Art_of_War/.

Cavaiola, Lawrence J., David C. Gompert, and Martin Libicki. "Cyber House Rules: On War, Retaliation and Escalation." *Survival* 57, no. 1 (2015): 81–104. https://doi.org/10.1080/00396338.2015.1008300.

Cavelty, Myriam Dunn. *Cyber-Security and Threat Politics: US Efforts to Secure in the Information Age*. Abingdon: Routledge, 2007. https://www.researchgate.net/publication/277714726_Cyber-Security_and_Threat_Politics_US_Efforts_to_Secure_the_Information_Age.

Chalfant, Morgan. "McCain Hits Trump over Lack of Cyber Policy." *The Hill*, August 23, 2017. https://thehill.com/policy/cybersecurity/347660-mccain-hits-trump-over-lack-of-cyber-policy.

Chen, Wi, Xilu Chen, Chang-Tai Hsieh, and Zheng (Michael) Song. "A Forensic Examination of China's National Accounts." *Brookings Papers on Economic Activity*. 2019. https://www.brookings.edu/wp-content/uploads/2019/03/bpea-2019-forensic-analysis-china.pdf.

Chesney, Robert. "Crossing a Cyber Rubicon? Overreactions to the IDF's Strike on the Hamas Cyber Facility." *Lawfare,* May 6, 2019. https://www.lawfareblog.com/crossing-cyber-rubicon-overreactions-idfs-strike-hamas-cyber-facility.

Chesney, Robert. "CYBERCOM's Out-of-Network Operations: What Has and Has Not Changed over the Past Year?" *Lawfare,* May 9, 2019. https://www.lawfareblog.com/cybercoms-out-network-operations-what-has-and-has-not-changed-over-past-year.

Chesney, Robert. "The Domestic Legal Framework for U.S. Military Cyber Operations." *Lawfare,* August 5, 2020. https://www.lawfareblog.com/domestic-legal-framework-us-military-cyber-operations.

China General Nuclear Power Group. "China General Nuclear Power Group's Fangchenggang-3 Begins Construction with First Concrete Pour." January 8, 2016. https://electricenergyonline.com/article/organization/29681/559216/China-General-Nuclear-Power-Group-s-Fangchenggang-3-begins-construction-with-first-concrete-pour.htm.

"China May Be Running out of Time to Escape the Middle-Income Trap." *Asia Society.* October 2017. https://asiasociety.org/new-york/china-may-be-running-out-time-escape-middle-income-trap.

"China's Top Political Stresses Indigenous Innovation." *Xinhua English,* April 19, 2011. http://english.sina.com/china/p/2011/0419/369456.html.

Cimpanu, Catalin. "Backdoor Accounts Discovered in 29 FTTH Devices from Chinese Vendor C-Data." *ZDNet,* July 10, 2020. https://www.zdnet.com/google-amp/article/backdoor-accounts-discovered-in-29-ftth-devices-from-chinese-vendor-c-data/.

Cimpanu, Catalin. "Centreon Says Only 15 Entities Were Targeted in Recent Russian Hacking Spree." *ZDNet,* February 16, 2021. https://www.zdnet.com/google-amp/article/centreon-says-only-15-entitites-were-targeted-in-recent-russian-hacking-spree/.

Cimpanu, Catalin. "FBI Says an Iranian Hacking Group Is Attacking F5 Networking Devices." *ZDNet,* August 10, 2020. https://www.zdnet.com/article/fbi-says-an-iranian-hacking-group-is-attacking-f5-networking-devices/.

Cimpanu, Catalin. "Hackers Are Trying to Steal Admin Passwords from F5 BIG-IP Devices." *ZDNet,* July 4, 2020. https://www.zdnet.com/article/hackers-are-trying-to-steal-admin-passwords-from-f5-big-ip-devices/#ftag=CAD-00-10aag7e.

Clarke, Richard A., and Robert K. Knake. *Cyber War.* New York: Ecco, 2010.

Clover, Charles. "Xi Jinping Signals Departure from Low-Profile Policy." *Financial Times,* October 20, 2017. https://www.ft.com/content/05cd86a6-b552-11e7-a398-73d59db9e399.

Collier, Jamie. "Europe's New Sanction Regime Suggests a Growing Cyber Diplomacy Presence." *FireEye,* August 06, 2020. https://www.fireeye.com/blog/executive-perspective/2020/08/europe-new-sanction-regime-suggest-a-growing-cyber-diplomacy-presence.html.

Committee on Armed Services, US House of Representatives. "Cyber Warfare in the 21st Century: Threats, Challenges, and Opportunities." March 1, 2017. https://www.gpo.gov/fdsys/pkg/CHRG-115hhrg24680/pdf/CHRG-115hhrg24680.pdf.

Corn, Colonel (Retired) Gary. "Punching on the Edges of the Grey Zone: Iranian Cyber Threats and State Cyber Responses Chatham House Report on Cyber Sovereignty and Non-Intervention—Helpful or Off Target?" *JustSecurity,* February 11, 2020. https://www.justsecurity.org/68622/punching-on-the-edges-of-the-grey-zone-iranian-cyber-threats-and-state-cyber-responses/.

Corn, Colonel (Retired) Gary. "Tallinn Manual 2.0—Advancing the Conversation." *JustSecurity,* February 15, 2017. https://www.justsecurity.org/37812/tallinn-manual-2-0-advancing-conversation/#more37812.

Corn, Colonel (Retired) Gary, and Eric Jensen. "The Technicolor Zone of Cyberspace—Part I: Analyzing the Major U.K. Speech on International Law of Cyber." *JustSecurity,* May 30, 2018. https://www.justsecurity.org/57217/technicolor-zone-cyberspace-part/.

Corn, Colonel (Retired) Gary, and Eric Jensen. "The Technicolor Zone of Cyberspace—Part 2." *JustSecurity,* June 8, 2018. https://www.justsecurity.org/57545/technicolor-zone-cyberspace-part-2/.

Corn, Gary, and Eric Talbot Jensen. "The Use of Force and Cyber Countermeasures." *Temple International & Comparative Law Journal* 127 (2018). https://papers.ssrn.com/sol3/papers.cfm?abstract_id=3190253&download=yes.

Cornell University, INSEAD, and WIPO. "Global Innovation Index 2019." Ithaca, Fontainebleau, and Geneva, 2019. https://www.globalinnovationindex.org/userfiles/file/reportpdf/GII2019-keyfinding-E-Web3.pdf.

Council of Europe. "Convention on Cybercrime." *European Treaty Series No. 185*. November 23, 2001. https://www.europarl.europa.eu/meetdocs/2014_2019/documents/libe/dv/7_conv_budapest_/7_conv_budapest_en.pdf.

Council of the European Union Press Release. "Declaration by the High Representative Josep Borrell on Behalf of the EU: European Union Response to Promote International Security and Stability in Cyberspace." July 30, 2020. https://www.consilium.europa.eu/en/press/press-releases/2020/07/30/declaration-by-the-high-representative-josep-borrell-on-behalf-of-the-eu-european-union-response-to-promote-international-security-and-stability-in-cyberspace/.

Council on Foreign Relations Digital and Cyberspace Policy Program. Accessed on February 1, 2021. https://www.cfr.org/cyber-operations/#CyberOperations.

Cox, Joseph. "How U.S. Military Hackers Prepared to Hack the Islamic State." *Vice*, August 1, 2018. https://www.vice.com/en_us/article/ne5d5g/how-us-military-cybercom-hackers-hacked-islamic-state-documents.

Crane, Keith, Jill E. Luoto, Scott Warren Harold, David Yang, Samuel K. Berkowitz, and Xiao Wang. *The Effectiveness of China's Industrial Policies in Commercial Aviation Manufacturing*. Santa Monica, CA: Rand Corporation, 2014. https://www.rand.org/content/dam/rand/pubs/research_reports/RR200/RR245/RAND_RR245.pdf.

Crawford, James. *The International Law Commission's Articles on State Responsibility: Introduction, Text and Commentaries*. London: Cambridge University Press, 2002.

CrowdStrike. "What Is an Advanced Persistent Threat (APT)?" November 18, 2019. https://www.crowdstrike.com/epp-101/advanced-persistent-threat-apt/.

CrowdStrike. "Which Adversary Had the Fastest Breakout Time in 2018?" June 6, 2018. https://www.crowdstrike.com/blog/crowdstrike-cto-explains-breakout-time-a-critical-metric-in-stopping-breaches/.

Cumming-Bruce, Nick. "U.S.-backed Candidate for Global Tech Post Beats China's Nominee." *New York Times*, March 4, 2020. https://www.nytimes.com/2020/03/04/business/economy/un-world-intellectual-property-organization.html.

Cyber Threat Alliance. Accessed on January 29, 2022. https://www.cyberthreatalliance.org/.

"The Cybersecurity 202: Top Cybersecurity Companies Are Pooling Their Intel to Stop Cyberattacks." *Washington Post*, May 23, 2019. https://www.washingtonpost.com/news/powerpost/paloma/the-cybersecurity-202/2019/05/23/the-cybersecurity-202-top-cybersecurity-companies-are-pooling-their-intel-to-stop-cyberattacks/5ce5ef73a7a0a46b92a3fd95/.

Cybersecurity and Infrastructure Security Agency. "Alert (AA20-205A) NSA and CISA Recommend Immediate Actions to Reduce Exposure across Operational Technologies and Control Systems." July 23, 2020. https://us-cert.cisa.gov/ncas/alerts/aa20-205a.

Cybersecurity and Infrastructure Security Agency. "Alert (AA20-258A), Chinese Ministry of State Security-Affiliated Cyber Threat Actor Activity." September 14, 2020. https://us-cert.cisa.gov/ncas/alerts/aa20-258a.

Cybersecurity and Infrastructure Security Agency. "Alert (AA20-301A) North Korean Advanced Persistent Threat Focus: Kimsuky." October 27, 2020. https://us-cert.cisa.gov/ncas/alerts/aa20-301a.

Cybersecurity and Infrastructure Security Agency. "Alert (TA18-074A)." March 15, 2018. https://us-cert.cisa.gov/ncas/alerts/TA18-074A.

Cybersecurity and Infrastructure Security Agency. "Countering Foreign Influence Task Force." Accessed on January 29, 2022. https://www.cisa.gov/cfi-task-force.

Cybersecurity and Infrastructure Security Agency. "Malware Analysis Report (AR20-045A): MAR-10265965-1.v1—North Korean Trojan: BISTROMATH." February 14, 2020. https://us-cert.cisa.gov/ncas/analysis-reports/ar20-045a.

Cybersecurity and Infrastructure Security Agency. "Malware Analysis Report (AR20-303B): MAR-10310246-1.v1—ZEBROCY Backdoor." October 29, 2020. https://us-cert.cisa.gov/ncas/analysis-reports/ar20-303b.

Cybersecurity and Infrastructure Security Agency. Accessed on January 29, 2022. "Traffic Light Protocol Definitions and Usage." https://www.cisa.gov/tlp.

Cybersecurity and Infrastructure Security Agency. *CISA Global.* February 2021. https://www.cisa.gov/sites/default/files/publications/CISA%20Global_2.1.21_508.pdf.

Cybersecurity and Infrastructure Security Agency. April 2017. https://us-cert.cisa.gov/government-users/compliance-and-reporting/incident-definition#:~:text=An%20incident%20is%20the%20act,NIST%20Special%20Publication%20800%2D61.

Cybersecurity and Infrastructure Security Agency. August 5, 2021. https://www.cisa.gov/news/2021/08/05/cisa-launches-new-joint-cyber-defense-collaborative.

Cybersecurity and Infrastructure Security Agency. Accessed on January 29, 2022. https://www.cisa.gov/publication/jcdc-fact-sheet.

Cyberspace Solarium Commission. 2020. https://www.solarium.gov/.

Defense Security Service. "Targeting U.S. Technologies: A Trend Analysis of Reporting from Defense Industry." 2011. https://premium.globalsecurity.org/intell/library/reports/2011/2011-dss-targeting-us-tech.pdf.

Defense Security Service. "Targeting U.S. Technologies: A Trend Analysis of Reporting from Defense Industry." 2012. https://www.dcsa.mil/Portals/69/documents/about/err/2012_Trend_Analysis_Report.pdf.

Defense Security Service. "Targeting U.S. Technologies: A Trend Analysis of Reporting from Defense Industry." 2013. https://premium.globalsecurity.org/intell/library/reports/2011/2011-dss-targeting-us-tech.pdf.

Defense Security Service. "Targeting U.S. Technologies: A Trend Analysis of Reporting from Defense Industry." 2014. https://www.dcsa.mil/Portals/69/documents/about/err/2014_Trend_Analysis_Report.pdf.

Deibert, Ronald J., Rafal Rohizinski, and Masashi Crete-Nishihata. "Cyclone in Cyberspace: Information Shaping and Denial in the 2008 Russia-Georgia War." *Security Dialogue* 43, no. 1 (February 15, 2012): 3–24. https://journals.sagepub.com/doi/full/10.1177/0967010611431079.

Delerue, François, Alix Desforges, and Aude Géry. "A Close Look at France's New Military Cyber Strategy." *War on the Rocks*, April 23, 2019. https://warontherocks.com/2019/04/a-close-look-at-frances-new-military-cyber-strategy/.

Demcheck, Chris C. "Uncivil and Post-Western Cyber Westphalia: Changing Interstate Power Relations of the Cybered Age." *Cyber Defense Review* (Spring 2016): 49–74. https://cyberdefensereview.army.mil/Portals/6/Documents/CDR%20Journal%20Articles/Uncivil%20and%20Post-Western%20Cyber%20Wesphalia_Demchak.pdf?ver=2018-07-31-093713-953.

Diehl, Paul, and Gary Goertz. *War and Peace in International Rivalry.* Ann Arbor: University of Michigan Press, 2000.

"A Digital Geneva Convention to Protect Cyberspace." *Microsoft Policy Papers.* February 14, 2017. https://query.prod.cms.rt.microsoft.com/cms/api/am/binary/RW67QH.

Dollar, David, and Peter Petri. "Why It's Time to End the Tit-for-Tat Tariffs in the U.S.-China Trade War." *Brookings.* October 5, 2018. https://www.brookings.edu/blog/order-from-chaos/2018/10/05/why-its-time-to-end-the-tit-for-tat-tariffs-in-the-u-s-china-trade-war/.

Downs, George W., and David M. Rocke. "Tacit Bargaining and Arms Control." *World Politics: A Quarterly Journal of International Relations* 39, no. 3 (April 1987): 297–325. https://doi.org/10.2307/2010222.

Downs, George W., and David M. Rocke. *Tacit Bargaining, Arms Races, and Arms Control*. Ann Arbor: University of Michigan Press, 1990.

Draft United Nations Convention on Cooperation in Combating Cybercrime, Russia, 2017. Annexed to U.N. General Assembly. "Letter dated 11 October 2017 from the Permanent Representative of the Russian Federation to the United Nations addressed to the Secretary-General, UN Doc. A/C.3/72/12." October 16, 2017. https://undocs.org/A/C.3/72/12.

Dragos. "Allanite, since 2017." https://www.dragos.com/threat/allanite/.

Duyvesteyn, Isabelle. "Between Doomsday and Dismissal: Cyber War, the Parameters of War, and Collective Defense." *Atlantisch Perspectief* 38, no. 7 (2014): 20–24. https://www.jstor.org/stable/e48504447.

Eckstein, Harry. "Case Study and Theory in Political Science." In *Handbook of Political Science, Vol. 7: Strategies of Inquiry*, edited by Fred I. Greenstein and Nelson W. Polsby, 79–137. Reading, MA: Addison-Wesley, 1975.

Egan, Brian J. "International Law and Stability in Cyberspace." *Berkeley Journal of International Law* 35, no. 1 (2017): 169–180. https://www.law.berkeley.edu/wp-content/uploads/2016/12/BJIL-article-International-Law-and-Stability-in-Cyberspace.pdf.

Egloff, Florian J. "Public Attribution of Cyber Intrusions." *Journal of Cybersecurity* 6 (2020). https://doi.org/10.1093/cybsec/tyaa012.

European Union. "Complete Guide to GDPR Compliance." Accessed on January 29, 2022. https://gdpr.eu/.

Executive Office of the President of the United States. *Annual Intellectual Property Report to Congress*. February 2019. https://www.whitehouse.gov/wp-content/uploads/2019/02/IPEC-2018-Annual-Intellectual-Property-Report-to-Congress.pdf.

Executive Order 13694. "Blocking the Property of Certain Persons Engaging in Significant Malicious Cyber-Enabled Activities." Washington, DC: Federal Register, April 2015. https://www.treasury.gov/resourcecenter/sanctions/Programs/Documents/cyber_eo.pdf.

Executive Order 13757. "Taking Additional Steps to Address the National Emergency with Respect to Significant Malicious Cyber-Enabled Activities." Washington, DC: Federal Register, December 2016. https://www.treasury.gov/resourcecenter/sanctions/Programs/Documents/cyber2_eo.pdf.

Executive Order on Strengthening the Cybersecurity of Federal Networks and Critical Infrastructure. May 11, 2017. https://www.whitehouse.gov/presidential-actions/presidential-executive-order-strengthening-cybersecurity-federal-networks-critical-infrastructure/.

Eyerys. "FBI Seizes Domain Responsible for Major Russian Botnet." May 24, 2018. https://www.eyerys.com/articles/timeline/fbi-seizes-domain-responsible-major-russian-botnet.

Fact Sheet on Presidential Policy Directive 20. Accessed on January 29, 2022. https://epic.org/privacy/cybersecurity/Pres-Policy-Dir-20-FactSheet.pdf.

Fang, Frank. "Cybersecurity Firm Details How China Hacked Western Firms to Steal Aviation Tech." *Epoch Times*, October 16, 2019. https://www.theepochtimes.com/cybersecurity-firm-details-how-china-hacked-western-firms-to-steal-aviation-tech_3118899.html.

FBI National Press Office. "The FBI Launches a Combating Foreign Influence Webpage." August 30, 2018. https://www.fbi.gov/news/pressrel/press-releases/the-fbi-launches-a-combating-foreign-influence-webpage.

Fearon, James D. "Rationalist Explanations for War." *International Organization* 49, no. 3 (Summer 1995): 379–414. https://web.stanford.edu/group/fearon-research/cgi-bin/wordpress/wp-content/uploads/2013/10/Rationalist-Explanations-for-War.pdf.

Fearon, James D. "Signaling Foreign Policy Interests: Tying Hands versus Sinking Costs." *Journal of Conflict Resolution* 41, no. 1 (February 1997): 68–90. https://doi.org/10.1177%2F0022002797041001004.

Fearon, James D. "Threats to Use Force: Costly Signals and Bargaining in International Crises." PhD diss., University of California, 1992.

Fell, James. "A Review of Fuzzing Tools and Methods." Originally published in *PenTest*, March 2017. https://dl.packetstormsecurity.net/papers/general/a-review-of-fuzzing-tools-and-methods.pdf.

Ferdinando, Lisa. "DoD Officials: Chinese Actions Threaten U.S. Technological, Industrial Base." *DOD News*, June 21, 2018. https://www.defense.gov/Explore/News/Article/Article/1557188/.

"Financial Systemic Analysis & Resilience Center Appoints Scott DePasquale as President." *Cision PRNewswire*, March 15, 2017. https://www.prnewswire.com/news-releases/financial-systemic-analysis--resilience-center-appoints-scott-depasquale-as-president-300423909.html.

Finnemore, Martha. "Cybersecurity and the Concept of Norms." *Carnegie Endowment for International Peace*. November 30, 2017. https://carnegieendowment.org/2017/11/30/cybersecurity-and-concept-of-norms-pub-74870.

Finnemore, Martha, and Duncan B. Hollis. "Beyond Naming and Shaming: Accusations and International Law in Cybersecurity." *European Journal of International Law* 31, no. 3 (August 2020): 969–1003. https://doi.org/10.1093/ejil/chaa056.

Finnemore, Martha, and Duncan B. Hollis. "Constructing Norms for Global Cybersecurity." *American Journal of International Law* 110, no. 3 (July 2016): 425–479. https://doi.org/10.1017/S0002930000016894.

Finnemore, Martha, and Kathryn Sikkink. "International Norm Dynamics and Political Change." *International Organization* 52, no. 4 (Autumn 1998): 887–917. https://www.jstor.org/stable/2601361.

FireEye. "Advanced Persistent Threat Groups." Accessed on January 29, 2022. https://content.fireeye.com/apt-41/website-apt-groups.

FireEye. "M-Trends 2020, FireEye Mandiant Services Special Report." 2021. https://content.fireeye.com/m-trends/rpt-m-trends-2020.

FireEye. "What Is a Zero-Day Exploit?" Accessed on January 29, 2022. https://www.fireeye.com/current-threats/what-is-a-zero-day-exploit.html.

First. Accessed on January 29, 2022. https://www.first.org/cvss/specification-document.

Fischer, Lucy. "Downing Street Plans New 5G Club of Democracies." *The Times*, May 29, 2020. https://www.thetimes.co.uk/article/downing-street-plans-new-5g-club-of-democracies-bfnd5wj57.

Fischerkeller, Michael P. "Current International Law Is Not an Adequate Regime for Cyberspace." *Lawfare*, April 22, 2021. https://www.lawfareblog.com/current-international-law-not-adequate-regime-cyberspace.

Fischerkeller, Michael P. "The Cyberspace Solarium Commission Report and Persistent Engagement." *Lawfare*, March 23, 2020. https://www.lawfareblog.com/cyberspace-solarium-commission-report-and-persistent-engagement.

Fischerkeller, Michael P. "The Fait Accompli and Persistent Engagement in Cyberspace." *War on the Rocks*, June 24, 2020. https://warontherocks.com/2020/06/the-fait-accompli-and-persistent-engagement-in-cyberspace/.

Fischerkeller, Michael P. "Opportunity Seldom Knocks Twice: Influencing China's Trajectory via Defend Forward/Persistent Engagement in Cyberspace." *Asia Policy Journal* 15, no. 4 (October 2020): 65–89. https://www.nbr.org/publication/opportunity-seldom-knocks-twice-influencing-chinas-trajectory-via-defend-forward-and-persistent-engagement-in-cyberspace/.

Fischerkeller, Michael P. *What Is the Purpose of the Cyber Mission Force?* Alexandria, VA: Institute for Defense Analyses, 2019.

Fischerkeller, Michael P., and Richard J. Harknett. "Cyber Persistence Theory, Intelligence Contests, and Strategic Competition." *Texas National Security Review: Special Issue—Cyber Competition* (September 17, 2020). https://tnsr.org/roundtable/policy-roundtable-cyber-conflict-as-an-intelligence-contest/.

Fischerkeller, Michael P., and Richard J. Harknett. "Initiative Persistence as the Central Approach for US Cyber Strategy." *Kybernau* 1, no. 1 (August 2021). https://www.artsci.uc.edu/content/dam/refresh/artsandsciences-62/departments/political-science/ccsp/pdf_downloadableflyers/Kybernao_PaperSeries_Issue1_Final.pdf.

Fischerkeller, Michael P., and Richard J. Harknett. "Persistent Engagement, Agreed Competition, Cyberspace Interaction Dynamics and Escalation." *Cyber Defense Review—Special Edition* (2019). https://cyberdefensereview.army.mil/Portals/6/CDR-SE_S5-P3-Fischerkeller.pdf.

Fischerkeller, Michael P., and Richard J. Harknett. "Persistent Engagement and Cost Imposition: Distinguishing between Cause and Effect." *Lawfare*, February 6, 2020. https://www.lawfareblog.com/persistent-engagement-and-cost-imposition-distinguishing-between-cause-and-effect.

Fischerkeller, Michael P., and Richard J. Harknett. "A Response on Persistent Engagement and Agreed Competition." *Lawfare*, June 27, 2019. https://www.lawfareblog.com/response-persistent-engagement-and-agreed-competition.

Fischerkeller, Michael P., and Richard J. Harknett. "What Is Agreed Competition in Cyberspace?" *Lawfare*, February 19, 2019. https://www.lawfareblog.com/what-agreed-competition-cyberspace.

Fischerkeller, Michael P., Richard J. Harknett, and Jelena Vićić. "The Limits of Deterrence and the Need for Persistence." In *The Cyber Deterrence Problem*, edited by Aaron Brantly, 21–38. London: Rowman & Littlefield Ltd., 2020.

Ford, Christopher A. "International Security for Cyberspace: New Models for Reducing Risk." *Arms Control and International Security Papers* 1, no. 20 (2020). https://www.state.gov/wp-content/uploads/2020/10/T-paper-series-Cybersecurity-Format-508.pdf.

Foreign, Commonwealth and Development Office. "Application of International Law to States' Conduct in Cyberspace: UK Statement." June 3, 2021. https://www.gov.uk/government/publications/application-of-international-law-to-states-conduct-in-cyberspace-uk-statement/application-of-international-law-to-states-conduct-in-cyberspace-uk-statement.

Freedman, Lawrence. *Deterrence*. Cambridge: Polity Press, 2008.

Freeman, Ambassador Chas W., Jr. "China as a Great Power: Remarks to China Renaissance Capital Investors." *Middle East Policy Council*. https://mepc.org/speeches/china-great-power.

Fruhlinger, Josh. "Ransomware Explained: How It Works and How to Remove It." *CSO*, June 19, 2020. https://www.csoonline.com/article/3236183/what-is-ransomware-how-it-works-and-how-to-remove-it.html.

"Full Transcript: President Obama's Final End-of-Year Press Conference." *Politico*, December 16, 2016. https://www.politico.com/story/2016/12/obama-press-conference-transcript-232763.

G7 Declaration on Responsible States Behavior in Cyberspace. April 11, 2017. https://www.mofa.go.jp/files/000246367.pdf.

G20 Leaders' Communiqué. November 16, 2015. http://www.g20.utoronto.ca/2015/151116-communique.pdf.

Gaddis, John Lewis. *The Long Peace: Inquiries into the History of the Cold War*. Oxford: Oxford University Press, 1989.

"A Game Changer in IT Security." *MIT Technology Review Insights*. September 8, 2021. https://www.technologyreview.com/2021/09/08/1034262/a-game-changer-in-it-security/.

Garfinkel, Ben, and Allan Dafoe. "Artificial Intelligence, Foresight, and the Offense-Defense Balance." *War on the Rocks*, December 19, 2019. https://warontherocks.com/2019/12/artificial-intelligence-foresight-and-the-offense-defense-balance/.

Gartner Glossary. "Bring Your Own Device (BYOD)." Accessed on January 29, 2022. https://www.gartner.com/en/information-technology/glossary/bring-your-own-device-byod.

Gartzke, Erik. "The Myth of Cyberwar: Bringing War in Cyberspace Back Down to Earth." *International Security* 38, no. 2 (Fall 2013): 41–73. https://www.mitpressjournals.org/doi/pdf/10.1162/ISEC_a_00136.

Gartzke, Erik, and John R. Lindsay. "Weaving Tangled Webs: Offense, Defense and Deception in Cyberspace." *Security Studies* 24, no. 2 (2015): 316–348. https://doi.org/10.1080/09636412.2015.1038188.

"GDP Revisions Put China on Target to Double Economy, but Data Doubts Remain." *Reuters*, November 21, 2019. https://www.reuters.com/article/us-china-economy-gdp/gdp-revisions-put-china-on-target-to-double-economy-but-data-doubts-remain-idUSKBN1XW04C.

Geddes, Barbara. "How the Cases You Choose Affect the Answers You Get: Selection Bias in Comparative Politics." In *Political Analysis, vol. 2*, edited by James A. Stimson, 131–150. Ann Arbor: University of Michigan Press, 1990.

Gelernter, David. *Mirror Worlds: Or the Day Software Puts the Universe in a Shoebox . . . How It Will Happen and What It Will Mean*. New York: Oxford University Press, 1991.

Geller, Eric. "Lesson from Log4j: Open-source Software Improvements Need Help from Feds." *Politico*, January 6, 2022. https://www.politico.com/news/2022/01/06/open-source-software-help-526676.

Geller, Eric. "White House Announces Ransomware Task Force—and Hacking Back Is One Option." *Politico*, July 14, 2021. https://www.politico.com/news/2021/07/14/white-house-ransomware-task-force-499723.

Geller, Eric, and Christian Vasquez. "Beware Cyber Command's Twitter Account." *Politico*, November 9, 2018. https://www.politico.com/newsletters/morning-cybersecurity/2018/11/09/beware-cyber-commands-twitter-account-407006.

George, Alexander L. "Strategies for Crisis Management." In *Avoiding War: Problems of Crisis Management*, edited by Alexander L. George, 377–394. Boulder, CO: Westview Press, 1991.

George, Alexander L. "US-Soviet Global Rivalry: Norms of Competition." *Journal of Peace Research* 23, no. 3 (September 1986): 247–262. https://www.jstor.org/stable/423823.

George, Alexander L., David K. Hall, and William R. Simons, *Limits of Coercive Diplomacy*. Boston: Little, Brown & Company, 1971.

"Georgia Hit by Massive Cyber-attack." *BBC News*, October 28, 2019. https://www.bbc.com/news/technology-50207192.

"Germany Issues Arrest Warrant for Russian Suspect in Parliament Hack: Newspaper." *Reuters*, May 5, 2020. https://www.reuters.com/article/us-russia-germany-warrant/germany-issues-arrest-warrant-for-russian-suspect-in-parliament-hack-newspaper-idUSKBN22H0TB.

Gewirtz, Julian Baird. "China's Long March to Technological Supremacy: The Roots of Xi Jinping's Ambition to 'Catch Up and Surpass.'" *Foreign Affairs* 27 (August 27, 2019). https://www.foreignaffairs.com/articles/china/2019-08-27/chinas-long-march-technological-supremacy.

Giddens, Anthony. *Central Problems in Social Theory*. London: Macmillan, 1979.

Gill, Indermit. "Future Development Reads: Xi Jingping, China's People's Party, and the Middle-Income Trap." *Brookings*. October 20, 2017. https://www.brookings.edu/blog/future-development/2017/10/20/future-development-reads-xi-jinping-chinas-peoples-party-and-the-middle-income-trap/.

Gill, Indermit S., and Homi Kharas. "The Middle Income Trap Turns 10." *World Bank Group*. August 2015. http://documents.worldbank.org/curated/en/291521468179640202/pdf/WPS7403.pdf.

Gilli, Andrea, and Mauro Gilli. "Why China Has Not Caught Up Yet: Military-Technological Superiority and the Limits of Imitation, Reverse Engineering, and Cyber Espionage." *International Security* 43, no. 3 (Winter 2018/19): 141–189. https://www.mitpressjournals.org/doi/full/10.1162/isec_a_00337.

Glaser, Charles L. *Rational Theory of International Politics: The Logic of Competition and Cooperation*. Princeton, NJ: Princeton University Press, 2010.

Gleck, James. *Chaos: Making a New Science*. New York: Penguin Books, 1987.

Global Commission on the Stability of Cyberspace. *Advancing Cyber Stability: Final Report—Appendix B.1*. Global Commission on the Stability of Cyberspace. November 2019. https://cyberstability.org/wp-content/uploads/2020/02/GCSC-Advancing-Cyberstability.pdf.

Goldman, Emily O. "The Cyber Paradigm Shift." In *Ten Years In: Implementing Strategic Approaches to Cyberspace*, edited by Jacquelyn G. Schneider, Emily O. Goldman, and Michael Warner, 31–46. Newport, RI: Naval War College Press, 2020. https://digital-commons.usnwc.edu/cgi/viewcontent.cgi?article=1044&context=usnwc-newport-papers.

Goldman, Emily O. "From Reaction to Action: Adopting a Competitive Posture in Cyber Diplomacy." *Texas National Security Review, Special Issue: Cyber Competition* 3, no. 4 (Fall 2020): 84–101. https://tnsr.org/2020/09/from-reaction-to-action-adopting-a-competitive-posture-in-cyber-diplomacy/#_ftn22.

Goldman, Emily O., and Leslie C. Eliason, eds. *The Diffusion of Military Technology and Ideas.* Stanford, CA: Stanford University Press, 2003.

Goldman, Emily, and Richard Harknett. "The Search for Cyber Fundamentals." *Journal of Information Warfare* 15, no. 2 (Spring 2016): 81–88. https://www.jstor.org/stable/26487534.

Goldman, Emily O., and Michael Warner. "Why a Digital Pearl Harbor Makes Sense . . . and Is Possible." *Carnegie Endowment for International Peace.* October 16, 2017. https://carnegieendowment.org/2017/10/16/why-digital-pearl-harbor-makes-sense-.-.-.-and-is-possible-pub-73405.

Goldsmith, Jack. "The Puzzle of the GRU Indictment." *Lawfare*, October 21, 2020. https://www.lawfareblog.com/puzzle-gru-indictment.

Goldsmith, Jack, and Alex Loomis. "Defend Forward and Sovereignty." In *The United States' Defend Forward Cyber Strategy: A Comprehensive Legal Analysis*, edited by Jack Goldsmith, 151–180. Oxford: Oxford University Press, 2022.

Goldsmith, Jack, and Robert D. Williams. "The Failure of the United States' Chinese-Hacking Indictment Strategy." *Lawfare*, December 28, 2018. https://www.lawfareblog.com/failure-united-states-chinese-hacking-indictment-strategy.

Gompert, David C., and Martin Libicki. "Cyber Warfare and Sino-American Crisis Instability." *Survival* 56, no. 4 (2014): 7–22. https://doi.org/10.1080/00396338.2014.941543.

Goodin, Dan. "Massive Denial-of-Service Attack on GitHub Tied to Chinese government." *arsTECHNICA*, March 31, 2015. https://arstechnica.com/information-technology/2015/03/massive-denial-of-service-attack-on-github-tied-to-chinese-government/.

Goodin, Dan. "Microsoft Issues Emergency Patches for 4 Exploited 0-days in Exchange." *arsTECHNICA*, March 2, 2021. https://arstechnica.com/information-technology/2021/03/microsoft-issues-emergency-patches-for-4-exploited-0days-in-exchange/.

Graff, Garrett M. "How a Dorm Room Minecraft Scam Brought Down the Internet." *Wired*, December 13, 2017. https://www.wired.com/story/mirai-botnet-minecraft-scam-brought-down-the-internet/.

Graff, Garrett M. "The Man Who Speaks Softly—and Commands a Big Cyber Army." *Wired*, October 13, 2020. https://www.wired.com/story/general-paul-nakasone-cyber-command-nsa/.

Green, Brendan Rittenhouse. *The Revolution That Failed: Nuclear Competition, Arms Control, and the Cold War.* Cambridge: Cambridge University Press, 2020.

Green, James A. "Fluctuating Evidentiary Standards for Self-Defence in the International Court of Justice." *International & Comparative Law Quarterly* 58, no. 1 (January 2009): 163–180. https://www.jstor.org/stable/20488277.

Greenberg, Andy. "Chinese Hacking Spree Hit an 'Astronomical' Number of Victims." *Wired*, March 5, 2021. https://www.wired.com/story/china-microsoft-exchange-server-hack-victims/.

Greenberg, Andy. "France Ties Russia's Sandworm to a Multiyear Hacking Spree." *Wired*, February 15, 2021. https://www.wired.com/story/sandworm-centreon-russia-hack/.

Greenberg, Andy. "This Map Shows the Global Spread of Zero-Day Hacking Techniques." *Wired*, June 4, 2020. https://www.wired.com/story/zero-day-hacking-map-countries/.

Greenberg, Andy. "The Untold Story of NotPetya, the Most Devastating Cyberattack in History." *Wired*, September 22, 2018. https://www.wired.com/story/notpetya-cyberattack-ukraine-russia-code-crashed-the-world/.

Greico, Joseph. "The Relative-Gains Problem for International Cooperation Comment." *American Political Science Review* 87 (September 1993): 729. https://www.jstor.org/stable/2938747.

Grigsby, Alex. "The United Nations Doubles Its Workload on Cyber Norms, and Not Everyone Is Pleased." *Council of Foreign Relations*. November 15, 2018. https://www.cfr.org/blog/united-nations-doubles-its-workload-cyber-norms-and-not-everyone-pleased.

Groll, Elias. "The U.S. Hoped Indicting 5 Chinese Hackers Would Deter Beijing's Cyberwarriors. It Hasn't Worked." *Foreign Policy*, September 2, 2015. https://foreignpolicy.com/2015/09/02/the-u-s-hoped-indicting-5-chinese-hackers-would-deter-beijings-cyberwarriors-it-hasnt-worked/.

Gross, Judah Ari. "IDF Says It Thwarted a Hamas Cyber Attack during Weekend Battle." *The Times of Israel*, May 5, 2019. https://www.timesofisrael.com/idf-says-it-thwarted-a-hamas-cyber-attack-during-weekend-battle/.

Guerrero-Saade, Juan Andrés, and Costin Raiu. "Walking in Your Enemy's Shadow: When Fourth Party Collection Becomes Attribution Hell." *virusBULLETIN*, October 20, 2017. https://www.virusbulletin.com/blog/2017/10/vb2017-paper-walking-your-enemys-shadow-when-fourth-party-collection-becomes-attribution-hell/.

Ha, Matthew, and David Maxwell. *Kim Jong Un's All-Purpose Sword*." Washington, DC: Foundation for the Defense of Democracies Press. October 2018. https://www.fdd.org/analysis/2018/10/03/kim-jong-uns-all-purpose-sword/.

Hacquebord, Feike. "Two Years of Pawn Storm: Examining an Increasingly Relevant Threat." *TrendMicro*. https://documents.trendmicro.com/assets/wp/wp-two-years-of-pawn-storm.pdf.

Hacking, Ian. "Introductory Essay." In *The Structure of Scientific Revolutions*. 4th ed., x–xxiii. Chicago: University of Chicago Press, 2012.

Hafner, Katie, and John Markoff. *CyberPunk: Outlaws and Hackers on the Computer Frontier*. New York: Simon & Schuster, 1991.

Harknett, Richard J. "Integrated Security: A Strategic Response to Anonymity and the Problem of the Few." *Contemporary Security Policy* 24, no. 1 (April 2003): 13–45. https://doi.org/10.1080/13523260312331271809.

Harknett, Richard J. "The Logic of Conventional Deterrence and the End of the Cold War." *Security Studies* 4, no. 1 (Autumn 1994): 86–114. https://doi.org/10.1080/09636419409347576.

Harknett, Richard J. "The Nuclear Condition and the Soft Shell of Territoriality." In *Globalisation: Theory and Practice*, edited by Eleonore Kaufman and Gillian Youngs, 277–288. 3rd ed. London: Continuum, 2008.

Harknett, Richard J. "SolarWinds: The Need for Persistent Engagement." *Lawfare*, December 23, 2020. https://www.lawfareblog.com/solarwinds-need-persistent-engagement.

Harknett, Richard J. "State Preferences, Systemic Constraints, and the Absolute Weapon." In *The Absolute Weapon Revisited: Nuclear Arms and the Emerging International Order*, edited by T. V. Paul, Richard J. Harknett, and James J. Wirtz, 65–100. Ann Arbor: University of Michigan Press, 1997.

Harknett, Richard J., and Max Smeets. "Cyber Campaigns and Strategic Outcomes: The Other Means." *Journal of Strategic Studies* (Spring 2020): 1–34. https://doi.org/10.1080/01402390.2020.1732354.

Harknett, Richard J., and Hasan Yalcin. "The Struggle for Autonomy: A Realist Structural Theory of International Relations." *International Studies Review* 14, no. 4 (December 2012): 499–521. https://www.jstor.org/stable/41804152.

Harknett, Richard J., John Callaghan, and Rudi Kaufmann. "Leaving Deterrence Behind: Warfighting and National Cybersecurity." *Journal of Homeland Security and Emergency Management* 7, no. 1 (Spring 2010): 1–24. https://doi.org/10.2202/1547-7355.1636.

Haynes, Deborah. "Into the Grey Zone: The 'Offensive Cyber' Used to Confuse Islamic State Militants and Prevent Drone Attacks." SkyNews Podcast, February 8, 2021. https://news.sky.com/story/into-the-grey-zone-the-offensive-cyber-used-to-confuse-islamic-state-militants-and-prevent-drone-attacks-12211740.

Healey, Jason. "Triggering the New Forever War in Cyberspace." *The Cipher Brief*, April 1, 2018. https://www.thecipherbrief.com/triggering-new-forever-war-cyberspace.

Healey, Jason, ed. *A Fierce Domain: Conflict in Cyberspace, 1986 to 2012*. Vienna, VA: Cyber Conflict Studies Association, 2013.

Healey, Jason, and Robert Jervis. "Escalation Inversion and Other Oddities of Situational Cyber Stability." *Texas National Security Review* 3, no. 4 (Fall 2020). https://tnsr.org/2020/09/the-escalation-inversion-and-other-oddities-of-situational-cyber-stability/.

Hearing of the House (Select) Intelligence Committee Subject. "Cybersecurity Threats: The Way Forward." November 20, 2014. https://www.nsa.gov/news-features/speeches-testimonies/Article/1620360/hearing-of-the-house-select-intelligence-committee-subject-cybersecurity-threat/.

Hinck, Garrett, and Tim Maurer. "Persistent Enforcement: Criminal Charges as a Response to Nation-State Malicious Cyber Activity." *Journal of National Security Law and Policy* 10, no. 3 (2020). https://jnslp.com/2020/01/23/persistent-enforcement-criminal-charges-as-a-response-to-nation-state-malicious-cyber-activity/.

Hinsley, F. H. *Sovereignty*. 2nd ed. Cambridge: Cambridge University Press, 1986.

Hoffmann, Wyatt. *AI and the Future of Cyber Competition*. Center for Security and Emerging Technology—Georgetown University, January 2021. https://cset.georgetown.edu/publication/ai-and-the-future-of-cyber-competition/.

Hollis, David M. "Cyberwar Case Study: Georgia 2008." *Small Wars Journal* (January 2011): 1–9. https://smallwarsjournal.com/blog/journal/docs-temp/639-hollis.pdf.

The Honeypot Project. Accessed on January 29, 2022. https://www.honeynet.org/about/.

Horowitz, Michael C. "Artificial Intelligence, International Competition, and the Balance of Power." *Texas National Security Review* 1, no. 3 (May 2018). https://tnsr.org/2018/05/artificial-intelligence-international-competition-and-the-balance-of-power/.

Horowitz, Michael C., Gregory C. Allen, Elsa B. Kania, and Paul Scharre. *Strategic Competition in an Era of Artificial Intelligence*. Washington, DC: Center for a New American Security, 2018. https://www.cnas.org/publications/reports/strategic-competition-in-an-era-of-artificial-intelligence.

Hosenball, Mark, and David Brunnstrom, "To Counter Huawei, U.S. Could Take 'Controlling Stake' in Ericsson, Nokia: Attorney General." *Reuters*, February 6, 2020. https://www.reuters.com/article/us-usa-china-espionage/top-u-s-officials-to-spotlight-chinese-spy-operations-pursuit-of-american-secrets-idUSKBN2001DL.

Hsu, Jeremy. "The Strava Heat Map and the End of Secrecy." *Wired*, January 28, 2018. https://www.wired.com/story/strava-heat-map-military-bases-fitness-trackers-privacy/.

Hsu, Spencer S. "U.S. Cyber Command Helps Prosecutors Seize Stolen Cryptocurrency Traced to Illicit N. Korea Nuclear Weapons Program." *Washington Post*, August 28, 2020. https://www.washingtonpost.com/local/legal-issues/us-cyber-command-helps-prosecutors-seize-stolen-cryptocurrency-traced-to-illicit-n-korea-nuclear-weapons-program/2020/08/28/12e7959c-e886-11ea-970a-64c73a1c2392_story.html.

Hufbauer, Gary, Jeffrey Schott, Kimberly Elliott, and Barbara Oegg. *Economic Sanctions Reconsidered*. 3rd ed. Washington, DC: Peterson Institute for International Economics, 2007.

Hunt, Will. *The Flight to Safety-Critical AI: Lessons in AI Safety from the Aviation Industry*. Berkeley: University of California, August 2020. https://cltc.berkeley.edu/wp-content/uploads/2020/08/Flight-to-Safety-Critical-AI.pdf.

Huth, Paul K. "Deterrence and International Conflict: Empirical Findings and Theoretical Debates." *Annual Review of Political Science* 2, no. 1 (June 1999): 25–48. https://doi.org/10.1146/annurev.polisci.2.1.25.

Implications of Artificial Intelligence for Cybersecurity: Proceedings of a Workshop. Washington, DC: National Academies Press. 2019. http://nap.edu/25488.

Insikt Group (Recorded Future). "How North Korea Revolutionized the Internet as a Tool for Rogue Regimes." February 2020. https://www.recordedfuture.com/north-korea-internet-tool/.

Institute for Security + Technology. Accessed on January 29, 2022. https://securityandtechnology.org/ransomwaretaskforce/.

Institute of Directors. "Cybersecurity: Underpinning the Digital Economy." March 2016. https://www.iod.com/news/news/articles/Cyber-security-underpinning-the-digital-economy.

International Law Commission. "Report of the International Law Commission on the Work of Its Fifty-Third Session, Draft Articles on Responsibility of States for Internationally Wrongful Acts, with Commentaries, art. 22, UN Doc. A/56/10." June 9, 2001. https://legal.un.org/ilc/texts/instruments/english/commentaries/9_6_2001.pdf.

International Law Commission Draft Articles on Responsibility of States for Internationally Wrongful Acts, U.N. Doc. A/56/10. 2001. https://legal.un.org/ilc/texts/instruments/english/commentaries/9_6_2001.pdf.

International Monetary Fund. "Word Economic Outlook Database." Accessed on January 29, 2022. https://www.imf.org/en/Countries/CHN#data.

Internet World Stats. July 3, 2021. https://www.internetworldstats.com/emarketing.htm.

"An Interview with Paul M. Nakasone." *Joint Force Quarterly* 92 (1st Quarter 2019): 4–9. https://ndupress.ndu.edu/JFQ/Joint-Force-Quarterly-92.aspx.

Israel National Cyber Directorate. Accessed on January 29, 2022. https://www.gov.il/en/departments/israel_national_cyber_directorate.

Ittelson, Pavlina, and Vladimir Radunovic. "What's New with Cybersecurity Negotiations? UN Cyber OEWG Final Report Analysis." *Diplo*, March 19, 2021. https://www.diplomacy.edu/blog/whats-new-cybersecurity-negotiations-un-cyber-oewg-final-report-analysis.

Japan. *Cybersecurity Strategy (Provisional Translation).* July 27, 2018. https://nisc.go.jp/eng/pdf/cs-senryaku2018-en.pdf.

Jensen, Eric Talbot. "Tallinn Manual 2.0: Highlights and Insights." *Georgetown Journal of International Law* 48 (2017): 735–778. https://papers.ssrn.com/sol3/papers.cfm?abstract_id=2932110#.

Jensen, Eric Talbot, and Sean Watts. "A Cyber Duty of Due Diligence: Gentle Civilizer or Crude Destabilizer." *Texas Law Review* 95 (2017): 1555–1577. https://texaslawreview.org/wp-content/uploads/2017/11/Jensen.Watts_.pdf.

Jervis, Robert. "Cooperation under the Security Dilemma." *World Politics* 30, no. 2 (1978): 167–214. https://doi.org/10.2307/2009958.

Jervis, Robert. *Perception and Misperception in International Politics.* Princeton, NJ: Princeton University Press, 1976.

Jeutner, Valentin. "The Digital Geneva Convention: A Critical Appraisal of Microsoft's Proposal." *Journal of International Humanitarian Legal Studies* 10, no. 1 (June 2019): 158–170. https://doi.org/10.1163/18781527-01001009.

Junio, Timothy J., "How Probable Is Cyber War? Bringing IR Theory Back In to the Cyber Conflict Debate." *Journal of Strategic Studies* 36, no. 1 (2013): 125–133. https://doi.org/10.1080/01402390.2012.739561.

Kahn, Hasan N., David Hounshell, and Erica Fuchs. "Science and Research Policy at the End of Moore's Law." *Nature Electronics* 1 (January 2018): 14–21. https://doi.org/10.1038/s41928-017-0005-9.

Kahn, Herman, with a new introduction by Thomas C. Schelling. *On Escalation: Metaphors and Scenarios.* London: Routledge, 2017.

Kaljulaid, Kersti (President, Republic of Estonia). "Speech at the Opening of the International Conference on Cyber Conflict (CyCon) 2019." May 29, 2019. https://www.president.ee/en/official-duties/speeches/15241-president-of-the-republic-at-the-opening-of-cycon-2019/index.html.

Kaspersky. "Kaspersky Incident Response Analyst Report—2020." 2021. https://media.kasperskycontenthub.com/wp-content/uploads/sites/43/2020/08/06094905/Kaspersky_Incident-Response-Analyst_2020.pdf.

Keal, Paul. *Unspoken Rules and Superpower Dominance*. London: Macmillan, 1983.

Kello, Lucas. *The Virtual Weapon and International Order*. New Haven, CT: Yale University Press, 2017.

Keohane, Robert, ed., *Neorealism and its Critics*. New York: Columbia University Press, 1986.

Keshet, Lior. "An Aggressive Launch: TrickBot Trojan Rises with Redirection Attacks in the UK." *Security Intelligence*. November 8, 2016. https://securityintelligence.com/an-aggressive-launch-trickbot-trojan-rises-with-redirection-attacks-in-the-uk/.

King, Gary. *Unifying Political Methodology: The Likelihood Theory of Statistical Inference*. Cambridge: Cambridge University Press, 1989.

King, Gary, Robert O. Keohane, and Sidney Verba. *Designing Social Inquiry: Scientific Inference Qualitative Research*. Princeton, NJ: Princeton University Press, 1994.

Kinkade, William. "New Military Capabilities: Propellants and Implications." In *The Uncertain Course: New Weapons, Strategies and Mind-sets*, edited by Carl G. Jacobsen, 69–70. New York: Oxford University Press, 1987.

Klein, James P., Gary Goertz, and Paul F. Diehl. "The New Rivalry Dataset: Procedures and Patterns." *Journal of Peace Research* 43, no. 3 (May 2006): 331–348. https://www.jstor.org/stable/27640320.

Kleine-Ahlbrandt, Stephanie. "North Korea's Illicit Cyber Operations: What Can Be Done?" *38 North*, February 28, 2020. https://www.38north.org/2020/02/skleineahlbrandt022820/.

Koch, Karin. *International Trade Policy and the GATT, 1947–1967*. Stockholm: Almquist and Wiksell, 1969.

Koerner, Brendan I. "Inside the Cyberattack that Shocked the U.S. Government." *Wired*, October 23, 2016. https://www.wired.com/2016/10/inside-cyberattack-shocked-us-government/.

Korzak, Elaine. "UN GGE on Cybersecurity: The End of an Era?" *The Diplomat*, July 31, 2017. https://thediplomat.com/2017/07/un-gge-on-cybersecurity-have-china-and-russia-just-made-cyberspace-less-safe/.

Koskenniemi, Martti. "International Cyber Law: Does It Exist and Do We Need It?" *European Cyber Diplomacy Dialogue, EU Cyber Direct, Opening Lecture*. April 9, 2019. https://eucyberdirect.eu/content_events/professor-martti-koskenniemi-international-cyber-law-does-it-exist-and-do-we-need-it-european-cyber-diplomacy-dialogue-2019-2/.

Kosseff, Jeff. "Collective Countermeasures in Cyberspace." *Notre Dame Journal of Comparative & International Law* 10, no. 1 (January 2020): 18–34. https://scholarship.law.nd.edu/cgi/viewcontent.cgi?article=1115&context=ndjicl.

Kovacs, Eduard. "Dridex Still Active after Takedown Attempt." *SecurityWeek*, October 19, 2015. https://www.securityweek.com/dridex-still-active-after-takedown-attempt.

Kovacs, Eduard. "Over 22,000 Vulnerabilities Disclosed in 2019: Report." *SecurityWeek*, February 18, 2020. https://www.securityweek.com/over-22000-vulnerabilities-disclosed-2019-report.

Krebs, Brian. "Attacks Aimed at Disrupting the Trickbot Botnet." *Krebs on Security*, October 2, 2020. https://krebsonsecurity.com/2020/10/attacks-aimed-at-disrupting-the-trickbot-botnet/.

Krebs, Brian. "Report: U.S. Cyber Command Behind Trickbot Tricks." *Krebs on Security*, October 10, 2020. https://krebsonsecurity.com/2020/10/report-u-s-cyber-command-behind-trickbot-tricks/.

Kuhn, Thomas S. *The Structure of Scientific Revolutions*. 4th ed. Chicago: University of Chicago Press, 2012.

Kulwin, Noah. "Reddit, Elites and the Dream of GameStop to the Moon." *New York Times*, January 29, 2021. https://www.nytimes.com/2021/01/28/opinion/reddit-gamestop-robinhood-hedge-fund.html.

Kyu-sik, Yoon. "North Korea's Cyber Warfare Capability and Threat." *Military Review* [Gunsanondan] 68 (Winter 2011): 64–95.

Largent, William. "New VPNFilter Malware Targets at Least 500K Networking Devices Worldwide." *Cisco Talos,* May 23, 2018. https://blog.talosintelligence.com/2018/05/VPNFilter.html.

Largent, William. "VPNFilter Update—VPNFilter Exploits Endpoints, Targets New Devices." *Cisco Talos,* June 6, 2018. https://blog.talosintelligence.com/2018/06/vpnfilter-update.html.

"Latvia Condemns Cyber-attack against Georgia." February 21, 2020. https://www.mfa.gov.lv/en/news/latest-news/65504-latvia-condemns-cyber-attack-against-georgia.

Laudrain, Arthur P. B. "France's New Offensive Cyber Doctrine." *Lawfare,* February 26, 2019. https://www.lawfareblog.com/frances-new-offensive-cyber-doctrine.

Lawder, David, and Ruby Lian. "U.S. Panel Launches Trade Secret Theft Probe into China Steel." *Business News,* May 26, 2016. https://www.reuters.com/article/us-usa-china-steel-idUSKCN0YH2KX.

Layton, Roslyn. "State Department's 5G Clean Network Club Gains Members Quickly." *Forbes,* September 4, 2020. https://www.forbes.com/sites/roslynlayton/2020/09/04/state-departments-5g-clean-network-club-gains-members-quickly/?sh=49119e7b7536.

Leonard, Ashley. "The Problem with Patching: 7 Top Complaints." *DARKReading,* April 22, 2016. https://www.darkreading.com/endpoint/the-problem-with-patching-7-top-complaints-/a/d-id/1325232.

"Letter of 5 July 2019 from the Minister of Foreign Affairs to the President of the House of Representatives on the International Legal Order in Cyberspace, Appendix: International Law in Cyberspace." 2019. https://www.government.nl/documents/parliamentary-documents/2019/09/26/letter-to-the-parliament-on-the-international-legal-order-in-cyberspace.

Lewis, James A. "Rethinking Cyber Security: Strategy, Mass Effects, and States." *Center for Strategic and International Studies,* January 2018. https://www.csis.org/analysis/rethinking-cybersecurity.

Libicki, Martin C. "Correlations between Cyberspace Attacks and Kinetic Attacks." In *2020 12th International Conference on Cyber Conflict—20/20 Vision: The Next Decade,* edited by T. Jančárková, L. Lindström, M. Signoretti, I. Tolga, and G. Visky, 199–213. Tallinn: NATO CCDCOE, 2020. https://ccdcoe.org/uploads/2020/05/CyCon_2020_11_Libicki.pdf.

Libicki, Martin C. *Crisis and Escalation in Cyberspace.* Santa Monica, CA: Rand Corporation, 2012.

Libicki, Martin C. *Cyberdeterrence and Cyberwar.* Santa Monica, CA: Rand Corporation, 2009.

Libicki, Martin C. "Cyberspace Is Not a Warfighting Domain." *I/S: A Journal of Law and Policy for the Information Society* 8, no. 2 (2012): 325–340. https://moritzlaw.osu.edu/ostlj/?file=2012%2F02%2F4.Libicki.pdf.

Libicki, Martin C. "Norms and Normalization." *Cyber Defense Review* (Spring 2020): 41–52. https://cyberdefensereview.army.mil/Portals/6/CDR%20V5N1%20-%2004_Libicki_WEB.pdf.

Liddell Hart, B. H. *Thoughts on War.* London: Farber, 1944.

Liff, Adam P. "Cyberwar: A New 'Absolute Weapon'? The Proliferation of Cyberwarfare Capabilities and Interstate War." *Journal of Strategic Studies* 35, no. 3 (2012): 401–428. https://doi.org/10.1080/01402390.2012.663252.

Liff, Adam P. "The Proliferation of Cyberwarfare Capabilities and Interstate War, Redux: Liff Responds to Junio." *Journal of Strategic Studies* 36, no. 1 (2013): 134–138. https://doi.org/10.1080/01402390.2012.733312.

Lin, Herbert S. "Escalation Dynamics and Conflict Termination in Cyberspace." *Strategic Studies Quarterly* 6, no. 3 (Fall 2012): 46–70. https://www.jstor.org/stable/26267261.

Lin, Herbert S. "Offensive Cyber Operations and the Use of Force." *Journal of National Security Law and Policy* 4, no. 63 (2010): 63–86. https://jnslp.com/wp-content/uploads/2010/08/06_Lin.pdf.

Lindsay, Jon R. "Tipping the Scales: The Attribution Problem and the Feasibility of Deterrence Against Cyberattack." *Journal of Cybersecurity* 1, no. 1 (2015): 53–67. https://academic.oup.com/cybersecurity/article/1/1/53/2354517.

Lindsay, Jon R., and Erik Gartzke. "Coercion through Cyberspace: The Stability-Instability Paradox Revisited." In *Coercion: The Power to Hurt in International Politics*, edited by Kelly M. Greenhill and Peter Krause, 176–204. New York: Oxford University Press, 2018.

Lindsay, Jon R., and Erik Gartzke, eds. *Cross-Domain Deterrence: Strategy in an Era of Complexity*. Oxford: Oxford University Press, 2019.

Lipson, Charles. "Why Are Some International Agreements Informal?" *International Organization* 45, no. 4 (October 1991): 495–538. https://www.jstor.org/stable/2706946.

Lohn, Andrew J. *Hacking AI: A Primer for Policymakers on Machine Learning Cybersecurity*. Washington, DC: Center for Security and Emerging Technology–Georgetown University, 2020. https://cset.georgetown.edu/wp-content/uploads/CSET-Hacking-AI.pdf.

Lohn, Andrew J. *Poison in the Well: Securing the Shared Resources of Machine Learning*. Washington, DC: Center for Security and Emerging Technology–Georgetown University, June 2021. https://cset.georgetown.edu/publication/poison-in-the-well/.

Longfellow, Henry Wadsworth. *Tales of a Wayside Inn*. 1863.

Lynch, Colum. "U.S. State Department Appoints Envoy to Counter Chinese Influence at the U.N." *Foreign Policy*, January 22, 2020. https://foreignpolicy.com/2020/01/22/us-state-department-appoints-envoy-counter-chinese-influence-un-trump/.

Lyngaas, Sean. "German Intelligence Agencies Warn of Russian Hacking Threats to Critical Infrastructure." *CyberScoop*, May 26, 2020. https://www.cyberscoop.com/german-intelligence-memo-berserk-bear-critical-infrastructure/.

Lyngaas, Sean. "Norwegian Police Implicate Fancy Bear in Parliament Hack, Describe 'Brute Forcing' of Email Accounts." *CyberScoop*, December 8, 2020. https://www.cyberscoop.com/norwegian-police-implicate-fancy-bear-in-parliament-hack-describe-brute-forcing-of-email-accounts/.

Lyngaas, Sean. "NSA Chief Confirms He Set Up Task Force to Counter Russian Hackers." July 23, 2018. https://www.cyberscoop.com/russia-small-group-paul-nakasone-nsa-aspen/.

Lynn, William J., III. "Defending a New Domain: The Pentagon's Cyberstrategy." *Foreign Affairs*, September/October 2010. https://www.foreignaffairs.com/articles/united-states/2010-09-01/defending-new-domain.

Ma, Damien. "Can China Avoid the Middle Income Trap?" *Foreign Policy*, March 12, 2016. https://foreignpolicy.com/2016/03/12/can-china-avoid-the-middle-income-trap-five-year-plan-economy-two-sessions/.

Maayan, Gilad David. "The IoT Rundown for 2020: Stats, Risks, and Solutions." *SecurityToday*, January 13, 2020. https://securitytoday.com/articles/2020/01/13/the-iot-rundown-for-2020.aspx.

Machtiger, Peter. "The Latest GRU Indictment: A Failed Exercise in Deterrence." *JustSecurity*, October 29, 2020. https://www.justsecurity.org/73071/the-latest-gru-indictment-a-failed-exercise-in-deterrence/.

Mahnken, Thomas G. "Cyber War and Cyber Warfare." In *America's Cyber Future: Security and Prosperity in the Information Age, Vol. 2*, edited by Kristin Lord and Travis Sharp, 53–62. Washington DC: Center for New American Security, 2011.

Mandiant. "APT1: Exposing One of China's Cyber Espionage Units." 2013. http://it-report-lb-1-312482071.us-east-1.elb.amazonaws.com/.

Mandiant. "Deep Dive into Cyber Reality: Security Effectiveness Report 2020." 2021. https://content.fireeye.com/security-effectiveness/rpt-security-effectiveness-2020-deep-dive-into-cyber-reality.

Mandiant. "Mandiant Exposes APT1—One of China's Cyber Espionage Units & Releases 3,000 Indicators." February 9, 2013. https://www.fireeye.com/blog/threat-research/2013/02/mandiant-exposes-apt1-chinas-cyber-espionage-units.html.

Maness, Ryan C., Brandon Valeriano, and Benjamin Jensen. *Code Book for the Dyadic Cyber Incident and Dispute Dataset (Version 1.1)*. Accessed on February 1, 2021. http://www.brandonvaleriano.com/uploads/8/1/7/3/81735138/dcid_1.1_codebook.pdf.

Margolin, Josh, and Ivan Pereira. "Outdated Computer System Exploited in Florida Water Treatment Plant Hack." *ABC News*. February 11, 2021. https://abcnews.go.com/amp/US/outdated-computer-system-exploited-florida-water-treatment-plant/story?id=75805550.

Markey, Michael, Jonathon Pearl, and Benjamin Bahney. "How Satellites Can Save Arms Control." *Foreign Affairs*, August 5, 2020. https://www.foreignaffairs.com/articles/asia/2020-08-05/how-satellites-can-save-arms-control.

Marks, Joseph. "McCain Leaves a Rich Cyber Legacy." *Nextgov*, August 27, 2018. https://www.nextgov.com/cybersecurity/2018/08/mccain-leaves-rich-cyber-legacy/150847/.

Mattis, Peter. Written Testimony in "U.S.-China Economic and Security Review Commission, Hearing on Chinese Intelligence Services and Espionage Operations." June 9, 2016. https://www.uscc.gov/sites/default/files/Peter%20Mattis_Written%20Testimony060916.pdf.

Maurer, Tim, and Arthur Nelson. "COVID-19's Other Virus: Targeting the Financial System." *Strategic Europe*, April 21, 2020. https://carnegieeurope.eu/strategiceurope/81599.

Maurer, Tim, and Arthur Nelson. *International Strategy to Better Protect the Financial System against Cyber Threats*. Washington, DC: Carnegie Endowment for International Peace, 2020. https://carnegieendowment.org/2020/11/18/international-strategy-to-better-protect-financial-system-against-cyber-threats-pub-83105.

McConnell, Mike. "Cyberwar Is the New Atomic Age." *New Perspectives Quarterly* 26, no. 3 (Summer 2009): 72–77. https://onlinelibrary.wiley.com/doi/pdf/10.1111/j.1540-5842.2009.01103.x.

McGraw, Gary. "Cyber War Is Inevitable (Unless We Build Security In)." *Journal of Strategic Studies* 36, no. 1 (2013): 109–119. http://doi.org/cp6f.

McKinsey Global Institute. *The China Effect on Global Innovation*. October 2015. https://www.mckinsey.com/~/media/McKinsey/Featured%20Insights/Innovation/Gauging%20the%20strength%20of%20Chinese%20innovation/MGI%20China%20Effect_Full%20report_October_2015.ashx.

McKune, Sarah. "An Analysis of the International Code of Conduct for Information Security." *The Citizen Lab*, September 28, 2015. https://citizenlab.ca/2015/09/international-code-of-conduct/.

McLaughlin, David, Saleha Mohsin, and Jacob Rund. "All About Cfius, Trump's Watchdog on China Dealmaking." *Washington Post*, September 15, 2020. https://www.washingtonpost.com/business/energy/all-about-cfius-trumps-watchdog-on-china-dealmaking/2020/09/15/1fdb46fa-f762-11ea-85f7-5941188a98cd_story.html.

McLelland, Charles. "Optimising the Smart Office: A Marriage of Technology and People." *ZDNet*, March 1, 2018. https://www.zdnet.com/article/optimising-the-smart-office-a-marriage-of-technology-and-people/.

McWhinney, Edward. "General Assembly Resolution 2131 (XX) of 21 December 1965 Declaration on the Inadmissibility of Intervention in the Domestic Affairs of States and the Protection of Their Independence and Sovereignty." https://legal.un.org/avl/pdf/ha/ga_2131-xx/ga_2131-xx_e.pdf.

Mearsheimer, John. *Conventional Deterrence*. Ithaca, NY: Cornell University Press, 1983.

Mearsheimer, John. *The Tragedy of Great Power Politics*. New York: W. W. Norton, 2001.

Mehta, Aaron. "Esper Proclaims Election Security an 'Enduring Mission' for Pentagon." *C4ISRNET*, September 19, 2019. https://www.c4isrnet.com/cyber/2019/09/19/esper-proclaims-election-security-an-enduring-mission-for-pentagon/.

Metrick, Kathleen, Parnian Najafi, and Jared Semrau. "Zero-Day Exploitation Increasingly Demonstrates Access to Money, Rather than Skill—Intelligence for Vulnerability Management, Part One." April 6, 2020. https://www.fireeye.com/blog/threat-research/2020/04/zero-day-exploitation-demonstrates-access-to-money-not-skill.html.

Microsoft. *Microsoft Security Intelligence Report* 24 (January–December 2018). https://info.microsoft.com/SIRv24Report.html.

Microsoft Security. "HAFNIUM Targeting Exchange Servers with 0-day Exploits." March 2, 2021. https://www.microsoft.com/security/blog/2021/03/02/hafnium-targeting-exchange-servers/.

Microsoft Security Response Center. "Corporate IoT—A Path to Intrusion." August 5, 2019. https://msrc-blog.microsoft.com/2019/08/05/corporate-iot-a-path-to-intrusion/.

Miller, James N., and Robert J. Butler. *National Cyber Defense Center: A Key Next Step toward a Whole-of-Nation Approach to Cyber Security.* Laurel, MD: Johns Hopkins Applied Physics Laboratory, 2021. https://www.jhuapl.edu/Content/documents/NationalCyberDefenseCenter.pdf.

Miller, John W. "Steelmaker Alleges Chinese Government Hackers Stole Plans for Developing New Steel Technology." *Wall Street Journal*, April 28, 2016. https://www.wsj.com/amp/articles/u-s-steel-accuses-china-of-hacking-1461859201.

Miller, Maggie. "New Coalition Aims to Combat Growing Wave of Ransomware Attacks." *The Hill*, January 17, 2021. https://thehill.com/policy/cybersecurity/534544-new-coalition-aims-to-combat-growing-wave-of-ransomware-attacks.

Missile Technology Control Regime. Accessed on January 29, 2022. https://mtcr.info/guidelines-for-sensitive-missile-relevant-transfers/.

Mitchell, Billy. "State Department Seeks 'Clean Path' 5G Networks." *FedScoop*, June 9, 2020. https://www.fedscoop.com/state-department-seeks-clean-path-5g-networks/.

Mombauer, Annika. "A Reluctant Military Leader? Helmuth von Moltke and the July Crisis of 1914." *War in History* 6, no. 4 (1999): 417–446. https://journals.sagepub.com/doi/pdf/10.1177/096834459900600403.

Morgan, Clifton, Navin Bapat, and Valentina Krustev. "The Threat and Imposition of Economic Sanctions, 1971–2000." *Conflict Management and Peace Science* 28, no. 1 (2008): 92–110. https://doi.org/10.1177%2F0738894208097668.

Morgan, Patrick. *Deterrence.* Beverly Hills, CA: Sage Publications, 1977.

Morrow, James D. "Capabilities, Uncertainty, and Resolve: A Limited Information Model of Crisis Bargaining." *American Journal of Political Science* 33, no. 4 (November 1989): 941–972. https://www.jstor.org/stable/2111116.

Morrow, James D. "The Strategic Setting of Choices: Signaling, Commitment, and Negotiation in International Politics." In *International Relations: A Strategic Choice Approach*, edited by David Lake and Robert Powell, 86–91. Princeton, NJ: Princeton University Press, 1999.

Motherboard. June 27, 2018. https://www.documentcloud.org/documents/4624362-Cybercom-Operation-Glowing-Symphony-Documents.html.

Moynihan, Harriet. "The Application of International Law to State Cyberattacks." *Chatham House International Law Program.* December 2019. https://www.chathamhouse.org/sites/default/files/publications/research/2019-11-29-Intl-Law-Cyberattacks.pdf.

Mueller, John. "The Escalating Irrelevance of Nuclear Weapons." In *The Absolute Weapon Revisited: Nuclear Arms and the Emerging International Order*, edited by T. V. Paul, Richard J. Harknett, and James J. Wirtz, 73–98. Ann Arbor: University of Michigan Press, 1997.

Müller, Vincent C., and Nick Bostrom. "Future Progress in Artificial Intelligence: A Survey of Expert Opinion." In *Fundamental Issues of Artificial Intelligence*, edited by Vincent C. Müller, 555–572. Berlin: Springer, 2016.

Myers, Adam. "First-Ever Adversary Ranking in 2019 Global Threat Report Highlights the Importance of Speed." *CrowdStrike Blog*, February 19, 2019. https://www.crowdstrike.com/blog/first-ever-adversary-ranking-in-2019-global-threat-report-highlights-the-importance-of-speed/.

Nakashima, Ellen. "Cyber Command Has Sought to Disrupt the World's Largest Botnet, Hoping to Reduce Its Potential Impact on the Election." *Washington Post*, October 9, 2020. https://www.washingtonpost.com/national-security/cyber-command-trickbot-disrupt/2020/10/09/19587aae-0a32-11eb-a166-dc429b380d10_story.html.

Nakashima, Ellen. "New Details Emerge about 2014 Russian Hack of the State Department: It Was 'Hand to Hand Combat.'" *Washington Post*, April 3, 2017. https://www.washingtonpost.com/world/national-security/new-details-emerge-about-2014-russian-hack-of-the-state-department-it-was-hand-to-hand-combat/2017/04/03/d89168e0-124c-11e7-833c-503e1f6394c9_story.html.

Nakashima, Ellen. "NSA and Cyber Command to Coordinate Actions to Counter Russian Election Interference in 2018 amid Absence of White House Guidance." *Washington Post*, July 17, 2018. https://www.washingtonpost.com/world/national-security/nsa-and-cyber-command-to-coordinate-actions-to-counter-russian-election-interference-in-2018-amid-absence-of-white-house-guidance/2018/07/17/baac95b2-8900-11e8-85ae-511bc1146b0b_story.html.

Nakashima, Ellen. "U.S. Military Cyber Operation to Attack ISIS Last Year Sparked Heated Debate over Alerting Allies." *Washington Post*, May 9, 2017. https://www.washingtonpost.com/world/national-security/us-military-cyber-operation-to-attack-isis-last-year-sparked-heated-debate-over-alerting-allies/2017/05/08/93a120a2-30d5-11e7-9dec-764dc781686f_story.html.

Nakashima, Ellen. "U.S. Undertook Cyber Operation against Iran as Part of Effort to Secure the 2020 Election." *Washington Post*, November 3, 2020. https://www.washingtonpost.com/national-security/cybercom-targets-iran-election-interference/2020/11/03/aa0c9790-1e11-11eb-ba21-f2f001f0554b_story.html.

Nakashima, Ellen, and Devlin Barrett. "U.S. Charges Russian Intelligence Officers in Several High-Profile Cyberattacks." *Washington Post*, October 19, 2020. https://www.washingtonpost.com/national-security/russia-cyberattacks-election-interference/2020/10/19/51a84208-1208-11eb-bc10-40b25382f1be_story.html.

Nakasone, Paul M. "A Cyber Force for Persistent Operations." *Joint Force Quarterly* 92 (1st Quarter 2019): 10–14. https://ndupress.ndu.edu/JFQ/Joint-Force-Quarterly-92.aspx.

Nakasone, Paul M., and Michael Sulmeyer. "How to Compete in Cyberspace: Cyber Command's New Approach." *Foreign Affairs*, August 25, 2020. https://www.foreignaffairs.com/articles/united-states/2020-08-25/cybersecurity.

Naraine, Ryan. "Microsoft: Multiple Exchange Server Zero-Days under Attack by Chinese Hacking Group." *SecurityWeek*, March 2, 2021. https://www.securityweek.com/microsoft-4-exchange-server-zero-days-under-attack-chinese-apt-group.

Naraine, Ryan. "Microsoft Office Zero-Day Hit in Targeted Attacks." *SecurityWeek*, September 7, 2021. https://www.securityweek.com/microsoft-office-zero-day-hit-targeted-attacks.

National Academies of Sciences, Engineering, and Medicine. *Implications of Artificial Intelligence for Cybersecurity: Proceedings of a Workshop.* Washington, DC: The National Academies Press, 2019.

National Counterintelligence and Security Center. *Foreign Economic Espionage in Cyberspace.* 2018. https://www.dni.gov/files/NCSC/documents/news/20180724-economic-espionage-pub.pdf.

National Cyber Security Alliance. Accessed on January 29, 2022. https://staysafeonline.org/cybersecurity-awareness-month/.

National Cyber Security Centre. Accessed on January 29, 2022. https://www.ncsc.gov.uk/.

National Cyber Security Centre and National Security Agency. "Advisory: Turla Group Exploits Iranian APT to Expand Coverage of Victims." October 21, 2019. https://www.ncsc.gov.uk/news/turla-group-exploits-iran-apt-to-expand-coverage-of-victims.

National Institute of Standards and Technology. "Glossary of Terms." Accessed on January 29, 2022. https://csrc.nist.gov/glossary/term/vulnerability.

National Intelligence Council. *Global Trends 2040: A More Contested World.* March 2021. https://www.dni.gov/files/ODNI/documents/assessments/GlobalTrends_2040.pdf.

National Research Council. *Computers at Risk: Safe Computing in the Information Age.* Washington, DC: The National Academies Press, 1991. https://doi.org/10.17226/1581.

National Security Institute. "Silverado Debate: Cyber Offense versus Cyber Defense." January 21, 2021. https://nationalsecurity.gmu.edu/nsi-silverado-debate-cyber-offense-vs-cyber-defense/.

National Security Advisor John Bolton on Cyber Strategy, off-camera, on-record press conference. September 20, 2020. https://www.c-span.org/video/?451807-1/national-security-adviser-bolton-briefs-cyber-strategy-audio-only.

National Security Agency. "Russian State-Sponsored Malicious Cyber Actors Exploit Known Vulnerability in Virtual Workspaces." December 7, 2020. https://www.nsa.gov/News-Features/Feature-Stories/Article-View/Article/2434988/russian-state-sponsored-malicious-cyber-actors-exploit-known-vulnerability-in-v/.

National Security Agency | Central Security Service| Central Security Service. "The NSA's Cybersecurity Collaboration Center." Accessed on January 29, 2022. https://www.nsa.gov/what-we-do/cybersecurity/cybersecurity-collaboration-center/.

"NATO Recognises Cyberspace as a 'Domain of Operations' at Warsaw Summit." Accessed on January 29, 2022. https://ccdcoe.org/incyder-articles/nato-recognises-cyberspace-as-a-domain-of-operations-at-warsaw-summit/.

"The Netherlands Considers Russia's GRU Responsible for Cyber Attacks against Georgia." February 20, 2020. https://www.government.nl/ministries/ministry-of-foreign-affairs/documents/diplomatic-statements/2020/02/20/the-netherlands-considers-russia%E2%80%99s-gru-responsible-for-cyber-attacks-against-georgia.

New Zealand. *The Application of International Law to State Activity in Cyberspace.* December 1, 2020. https://www.mfat.govt.nz/assets/Peace-Rights-and-Security/International-security/International-Cyber-statement.pdf.

"New Zealand Condemns Malicious Cyber Activity against Georgia." February 21, 2020. https://www.gcsb.govt.nz/news/gcsb-media-statement/.

Newman, Lily Hay. "100 Million More IoT Devices Are Exposed—and They Won't Be the Last." *Wired,* April 13, 2021. https://www.wired.com/story/namewreck-iot-vulnerabilities-tcpip-millions-devices/.

Newman, Lily Hay. "The NSA Warns That Russia Is Attacking Remote Work Platforms." *Wired,* December 7, 2020. https://www.wired.com/story/nsa-warns-russia-attacking-vmware-remote-work-platforms/.

Newman, Lily Hay. "What We Know About Friday's Massive East Coast Internet Outage." *Wired,* October 21, 2016. https://www.wired.com/2016/10/internet-outage-ddos-dns-dyn/.

"Next China: Tit for Tat." *BloombergNews,* July 16, 2020. https://www.bloomberg.com/news/newsletters/2020-07-17/next-china-tit-for-tat.

Ney, Paul C., Jr. "DOD General Counsel Remarks at U.S. Cyber Command Legal Conference." March 2, 2020. https://www.defense.gov/Newsroom/Speeches/Speech/Article/2099378/dod-general-counsel-remarks-at-us-cyber-command-legal-conference/.

Ng, Alfred. "How the Equifax Hack Happened, and What Still Needs to Be Done." C|*Net,* September 7, 2018. https://www.cnet.com/news/equifaxs-hack-one-year-later-a-look-back-at-how-it-happened-and-whats-changed/.

Nichols, Michelle. "North Korea Took $2 Billion in Cyberattacks to Fund Weapons Program: U.N. Report." *Reuters,* August 5, 2019. https://www.reuters.com/article/us-northkorea-cyber-un/north-korea-took-2-billion-in-cyberattacks-to-fund-weapons-program-u-n-report-idUSKCN1UV1ZX.

Nichols, Michelle, and Raphael Satter. "U.N. Experts Point Finger at North Korea for $281 Million Cyber Theft, KuCoin Likely Victim." *Reuters,* February 9, 2021. https://www.reuters.com/article/us-northkorea-sanctions-cyber-idCAKBN2AA00Q.

Nosco, Timothy, Jared Ziegler, Zechariah Clark, Davy Marrero, Todd Finkler, Andrew Barbarello, and W. Michael Petullo. "The Industrial Age of Hacking." Presented at 29th USENIX Security Symposium, August 12–14, 2020. https://www.usenix.org/conference/usenixsecurity20/presentation/nosco.

Novinson, Michael. "The 11 Biggest Ransomware Attacks of 2020 (So Far)." *CRN*, June 30, 2020. https://www.crn.com/slide-shows/security/the-11-biggest-ransomware-attacks-of-2020-so-far-.

NSA Cybersecurity Information Sheet. "Embracing a Zero Trust Security Model." February 25, 2021. https://www.nsa.gov/News-Features/Feature-Stories/Article-View/Article/2515176/nsa-issues-guidance-on-zero-trust-security-model/.

Nye, Joseph S., Jr. "Deterrence and Persuasion in Cyberspace." *International Security* 41, no. 3 (Winter 2016/17): 44–71. https://www.mitpressjournals.org/doi/pdf/10.1162/ISEC_a_00266.

O'Connor, Sean. "How Chinese Companies Facilitate Technology Transfer from the United States." *U.S.–China Economic and Security Review Commission*. May 6, 2019. https://www.uscc.gov/sites/default/files/Research/How%20Chinese%20Companies%20Facilitate%20Tech%20Transfer%20from%20the%20US.pdf.

Office of the United States Trade Representative. "USTR Issues Tariffs on Chinese Products in Response to Unfair Trade Practices." June 15, 2018. https://ustr.gov/about-us/policy-offices/press-office/press-releases/2018/june/ustr-issues-tariffs-chinese-products.

Office of the United States Trade Representative: Executive Office of President. *Findings of the Investigation into China's Acts, Policies, and Practices Related to Technology Transfer, Intellectual Property, and Innovation Under Section 301 of the Trade Act of 1974*. March 22, 2018. https://ustr.gov/sites/default/files/Section%20301%20FINAL.PDF.

Office of the United States Trade Representative: Executive Office of President. *Section 301 Investigation and Hearing: China's Acts, Policies, and Practices Related to Technology Transfer, Intellectual Property, and Innovation*. October 10, 2017. https://ustr.gov/sites/default/files/enforcement/301Investigations/China%20Technology%20Transfer%20Hearing%20Transcript.pdf.

Ohlin, Jens David. "Did Russian Cyber Interference in the 2016 Election Violate International Law?" *Texas Law Review* 95, no. 7 (2017). https://texaslawreview.org/russian-cyber-interference-2016-election-violate-international-law/.

O'Neill, Patrick Howell. "DOJ Drops Massive Report on Its Efforts to Protect U.S. from Cyberattacks." *CyberScoop*, July 19, 2018. https://www.cyberscoop.com/department-of-justice-internal-cyber-digital-task-force-report/.

Organization of Petroleum Exporting Countries (OPEC). *Declaration of Cooperation*. Accessed on January 29, 2022. https://www.opec.org/opec_web/en/publications/4580.htm.

Osgood, Charles E. *An Alternative to War or Surrender*. Urbana: University of Illinois Press, 1962.

Ottis, Rain. "Analysis of the 2007 Cyber Attacks against Estonia from the Information Warfare Perspective." In *Proceedings of the 7th European Conference on Information Warfare and Security, Plymouth 2008*, 163–168. Reading: Academic Publishing Ltd., 2008. https://ccdcoe.org/library/publications/analysis-of-the-2007-cyber-attacks-against-estonia-from-the-information-warfare-perspective/.

Oxford Institute for Ethics, Law and Armed Conflict (ELAC). "Applying International Law in Cyberspace: Protections and Prevention." May 18–19, 2020. https://www.elac.ox.ac.uk/files/elacvirtualworkshopapplyingiltocyberspacereport18-19may2020pdf.

Pagolini, Pierluigi. "IDF Hit Hamas, It Is the First Time a State Launched an Immediate Physical Attack in Response to a Cyber Attack." *SecurityAffairs*, May 6, 2019. https://securityaffairs.co/wordpress/85022/cyber-warfare-2/idf-hit-hamas.html.

Palandrani, Pedro, and Andrew Little. "A Decade of Change: How Tech Evolved in the 2010s and What's in Store for the 2020s." *Mirae Asset*, February 10, 2020. https://www.globalxetfs.com/a-decade-of-change-how-tech-evolved-in-the-2010s-and-whats-in-store-for-the-2020s/.

Panda, Rajaram. "Sino-US Ties: From Ping Pong Diplomacy to Tit-for-Tat Diplomacy." *ModernDiplomacy*, July 7, 2020. https://moderndiplomacy.eu/2020/07/27/sino-us-ties-from-ping-pong-diplomacy-to-tit-for-tat-diplomacy/.

Panetta, Leon E. "Defending the Nation from Cyber Attack." Speech to Business Executives for National Security. October 11, 2012. https://www.bens.org/document.doc?id=188.

Paul, T. V., Richard J. Harknett, and James J. Wirtz, eds. *The Absolute Weapon Revisited: Nuclear Arms and the Emerging International Order*. Ann Arbor: University of Michigan Press, 1998.

Payne, Kenneth. "Artificial Intelligence: A Revolution in Strategic Affairs?" *Survival* 60, no. 5 (2018). https://doi.org/10.1080/00396338.2018.1518374.

Payne, Kenneth. *Strategy, Evolution, and War: From Apes to Artificial Intelligence*. Washington, DC: Georgetown University Press, 2018.

Pinkston, Daniel A. "North Korean Cyber Threats." In *Confronting an "Axis of Cyber"? China, Iran, North Korea, Russia in Cyberspace*, edited by Fabio Rugge, 89–120. Milan: Ledizioni-LediPublishing, 2018. https://library.oapen.org/bitstream/handle/20.500.12657/23931/1006204.pdf?sequence=1&isAllowed=y.

"Policy Roundtable: Cyber Conflict as an Intelligence Contest." *Texas National Security Review—Special Issue: Cyber Competition*. September 17, 2020. https://tnsr.org/roundtable/policy-roundtable-cyber-conflict-as-an-intelligence-contest/.

Pompeo, Michael R. "Announcing the Expansion of the Clean Network to Safeguard America's Assets." August 5, 2020. https://www.state.gov/announcing-the-expansion-of-the-clean-network-to-safeguard-americas-assets/.

Pompeo, Michael R. "The United States Condemns Russian Cyber Attack against the Country of Georgia." February 20, 2020. https://www.state.gov/the-united-states-condemns-russian-cyber-attack-against-the-country-of-georgia/.

Powell, Robert. "Absolute and Relative Gains in International Relations Theory." *American Political Science Review* 85, no. 4 (December 1991): 1303–1329. https://doi.org/10.2307/1963947.

Powell, Robert. *In the Shadow of Power: States and Strategies in International Politics*. Princeton, NJ: Princeton University Press, 1999.

President's Commission on Critical Infrastructure Protection. *Critical Foundations: Protecting America's Infrastructures*.1997. https://sgp.fas.org/library/pccip.pdf.

Raab, Dominic. "UK Condemns Russia's GRU over Georgia Cyber-attacks." February 20, 2020. https://www.gov.uk/government/news/uk-condemns-russias-gru-over-georgia-cyber-attacks.

Randolph, Kevin. "U.S. Energy, Homeland Security, Defense Departments to Partner on Initiative to Protect Energy Infrastructure from Cyberthreats." *Daily Energy Insider*, February 5, 2020. https://dailyenergyinsider.com/news/24113-u-s-energy-homeland-security-defense-departments-to-partner-on-initiative-to-protect-energy-infrastructure-from-cyberthreats/.

Rapoza, Kenneth. "Westinghouse Electric's Chinese 'Trojan Horse.'" *Forbes*, May 17, 2016. https://www.forbes.com/sites/kenrapoza/2016/05/17/westinghouse-electrics-chinese-trojan-horse/amp/.

Ratnam, Gopal. "Underground Hackers and Spies Helped China Steal Jet Secrets. Crowdstrike Researchers Reveal Beijing's Efforts to Boost Its Own Domestic Aircraft Industry." *Rollcall*, October 15, 2019. https://www.rollcall.com/news/policy/hackers-spies-helped-china-steal-jet-secrets-report-says.

Rattray, Gregory J., and Jason Healey. "Non-State Actors in Cyber Conflict." In *America's Cyber Future: Security and Prosperity in the Information Age, Volume II*, edited by Kristen M. Lord and Travis Sharp. Washington, DC: Center for a New American Security, June 2011. 65–86. https://www.files.ethz.ch/isn/129907/CNAS_Cyber_Volume%20II_2.pdf.

Raymond, David, Gregory Conti, Tom Cross, and Michael Nowatkowski. "Key Terrain in Cyberspace: Seeking the High Ground." In *2014 6th International Conference on Cyber Conflict*, edited by P. Brangetto, M. Maybaum, and J. Stinissen, 287–300. Tallinn: NATO CCDCOE Publications, 2014.

Rendell, Darrel. "Understanding the Evolution of Malware." *Computer Fraud & Security* 1 (2019): 17–19. https://doi.org/10.1016/S1361-3723(19)30010-7.

"Report of the Group of Governmental Experts on Advancing Responsible State Behaviour in Cyberspace in the Context of International Security." May 28, 2021. https://front.un-arm.org/wp-content/uploads/2021/06/final-report-2019-2021-gge-1-advance-copy.pdf.

"Report of the Group of Governmental Experts on Developments in the Field of Information and Telecommunications in the Context of International Security." June 24, 2013. https://www.un.org/ga/search/view_doc.asp?symbol=A/68/98.

"Report of the Group of Governmental Experts on Developments in the Field of Information and Telecommunications in the Context of International Security." July 22, 2015. https://www.un.org/ga/search/view_doc.asp?symbol=A/70/174.

Republic of France, Ministry of the Armies. "International Law Applied to Operations in Cyberspace (*Droit international appliqué aux operations dans le cyberspace*)." 2019. https://www.defense.gouv.fr/content/download/567648/9770527/file/international+law+applied+to+operations+in+cyberspace.pdf.

Rhodes, Edward. *Power and MADness*. New York: Columbia University Press, 1989.

Richard, Admiral Charles A. "Forging 21st-Century Strategic Deterrence." *Proceedings* 147, no. 2 (February 21, 2021). https://www.usni.org/magazines/proceedings/2021/february/forging-21st-century-strategic-deterrence.

Rid, Thomas. "Cyber War Will Not Take Place." *Journal of Strategic Studies* 35, no. 1 (2011): 5–32. https://doi.org/10.1080/01402390.2011.608939.

Rid, Thomas. *Cyber War Will Not Take Place*. Oxford: Oxford University Press, 2013.

Rid, Thomas. "Is Cyberwar Real?" in response to Jarno Limnéll. *Foreign Affairs*. March/April 2014. https://www.foreignaffairs.com/articles/commons/2014-02-12/cyberwar-real.

Rid, Thomas. "Think Again: Cyberwar." *Foreign Policy*. February 27, 2012. https://foreignpolicy.com/2012/02/27/think-again-cyberwar/.

River, Charles, ed. *The Maginot Line: The History of the Fortifications That Failed to Protect France from Nazi Germany During World War II*. N.p.: Create Space Publishing, 2017.

Roberts, Anthea. *Is International Law International?* London: Oxford University Press, 2017.

Rodriguez, Salvador. "The FBI Visits Facebook to Talk about 2020 Election Security, with Google, Microsoft and Twitter Joining." *CNBC*, September 24, 2019. https://www.cnbc.com/amp/2019/09/04/facebook-twitter-google-are-meeting-with-us-officials-to-discuss-2020-election-security.html.

Roff, Heather. *Advancing Human Security through Artificial Intelligence*. London: Chatham House, 2017.

Roff, Heather. "COMPASS: A New AI-Driven Situational Awareness Tool for the Pentagon?" *Bulletin of the Atomic Scientists* 10 (May 10, 2018). https://thebulletin.org/2018/05/compass-a-new-ai-driven-situational-awareness-tool-for-the-pentagon/.

Roguski, Przemysław. "France's Declaration on International Law in Cyberspace: The Law of Peacetime Cyber Operations, Part I." *OpinioJuris*, September 24, 2019. http://opiniojuris.org/2019/09/24/frances-declaration-on-international-law-in-cyberspace-the-law-of-peacetime-cyber-operations-part-i/.

Roguski, Przemysław. "Russian Cyber Attacks against Georgia, Public Attributions and Sovereignty in Cyberspace." *JustSecurity*, March 6, 2020. https://www.justsecurity.org/69019/russian-cyber-attacks-against-georgia-public-attributions-and-sovereignty-in-cyberspace/.

Rose, Scott, Oliver Borchert, Stu Mitchell, and Sean Connelly. *Zero Trust Architecture*. National Institute of Standards and Technology. August 2020. https://nvlpubs.nist.gov/nistpubs/SpecialPublications/NIST.SP.800-207.pdf.

Roth, Erik, Jeongmin Seong, and Jonathan Woetzel. "Gauging the Strength of Chinese Innovation." *McKinsey Quarterly* 4 (2015): 66–73. https://www.mckinsey.com/~/media/McKinsey/McKinsey%20Quarterly/Digital%20Newsstand/2015%20Issues%20McKinsey%20Quarterly/Agility.ashx.

Rotman, David. "We're Not Prepared for the End of Moore's Law." *MIT Technology Review*. February 2020. https://www.technologyreview.com/2020/02/24/905789/were-not-prepared-for-the-end-of-moores-law/.

Rudengren, Jan, Lars Rylander, and Claudia Rives Casanova. "It's Democracy, Stupid: Reappraising the Middle-Income Trap." *Institute for Security and Development Policy*. 2014. https://www.files.ethz.ch/isn/184240/2014-rudengren-rylander-casanova-reappraising-the-middle-income-trap.pdf.

Saavedra, Gary J., Kathryn N. Rodhouse, Daniel M. Dunlavy, and Philip W. Kegelmeyer. "A Review of Machine Learning Applications in Fuzzing." https://deepai.org/publication/a-review-of-machine-learning-applications-in-fuzzing.

Sabbagh, Dan. "UK Unveils National Cyber Force of Hackers to Target Foes Digitally." *The Guardian*, November 19, 2020. https://www.theguardian.com/technology/2020/nov/19/uk-unveils-national-cyber-force-of-hackers-to-target-foes-digitally.

Sacks, David. "China's Huawei Is Winning the 5G Race. Here's What the United States Should Do to Respond." *Council on Foreign Relations*, March 29, 2021. https://www.cfr.org/blog/china-huawei-5g.

Sanger, David E., and Julian E. Barnes. "U.S. Tried a More Aggressive Cyberstrategy, and the Feared Attacks Never Came." *New York Times*, November 9, 2020. https://www.nytimes.com/2020/11/09/us/politics/cyberattacks-2020-election.html.

Sanger, David E., Nicole Perlroth, and Julian E. Barnes. "As Understanding of Russian Hacking Grows, So Does Alarm." *New York Times*, January 2, 2020. https://www.nytimes.com/2021/01/02/us/politics/russian-hacking-government.html?referringSource=articleShare.

Scarfone, Karen, and Peter Mell. *National Institute of Standards and Technology: Guide to Intrusion Detection and Prevention Systems–Special Publication 800-94*. US Department of Commerce, February 2007. https://nvlpubs.nist.gov/nistpubs/Legacy/SP/nistspecialpublication800-94.pdf.

Schelling, Thomas C. *Arms and Influence*. New Haven, CT: Yale University Press, 1966.

Schelling, Thomas C. "Bargaining, Communication, and Limited War." *Conflict Resolution* 1, no. 1 (March 1957): 19–36. https://www.jstor.org/stable/172548.

Schelling, Thomas C. "Reciprocal Measures for Arms Stabilization." *Daedalus* 89, no. 4 (Fall 1960): 892–914. https://doi.org/10.1080/00396336108440237.

Schelling, Thomas C. *The Strategy of Conflict*. Cambridge, MA: Harvard University Press, 1960.

Schelling, Thomas C., and Morton H. Halperin. *Strategy and Arms Control*. McLean, VA: Pergamon Press, 1985.

Schick, Shane. "Dridex Trojan Remains a Risk Even Following Takedown Operation and FBI Arrest." *SecurityIntelligence*, October 19, 2015. https://securityintelligence.com/news/dridex-trojan-remains-a-risk-even-following-takedown-operation-and-fbi-arrest/.

Schmitt, Michael N. "The Defense Department's Measured Take on International Law in Cyberspace." *JustSecurity*, March 11, 2020. https://www.justsecurity.org/69119/the-defense-departments-measured-take-on-international-law-in-cyberspace/.

Schmitt, Michael N. "Grey Zones in the International Law of Cyberspace." *Yale Journal of International Law Online* 42, no. 2 (October 2017): 1–21. https://cpb-us-w2.wpmucdn.com/campuspress.yale.edu/dist/8/1581/files/2017/08/Schmitt_Grey-Areas-in-the-International-Law-of-Cyberspace-1cab8kj.pdf.

Schmitt, Michael N. "Peacetime Cyber Responses and Wartime Cyber Operations under International Law: An Analytical *Vade Mecum*." *Harvard National Security Journal* 8 (2017): 239–282. https://harvardnsj.org/wp-content/uploads/sites/13/2017/02/Schmitt-NSJ-Vol-8.pdf.

Schmitt, Michael N. "The Sixth United Nations GGE and International Law in Cyberspace." *JustSecurity*, June 10, 2021. https://www.justsecurity.org/76864/the-sixth-united-nations-gge-and-international-law-in-cyberspace/.

Schmitt, Michael N. "Taming the Lawless Void: Tracking the Evolution of International Law Rules for Cyberspace." *Texas National Security Review* 3, no. 3 (Autumn 2020). https://tnsr.org/2020/07/taming-the-lawless-void-tracking-the-evolution-of-international-law-rules-for-cyberspace/.

Schmitt, Michael N. "Three International Law Rules for Responding Effectively to Hostile Cyber Operations." *JustSecurity*, July 13, 2021. https://www.justsecurity.org/77402/three-international-law-rules-for-responding-effectively-to-hostile-cyber-operations/.

Schmitt, Michael N., and Liis Vihul. "International Cyber Law Politicized: The UN GGE's Failure to Advance Cyber Norms." *JustSecurity*, June 30, 2017. https://www.justsecurity.org/42768/international-cyber-law-politicized-gges-failure-advance-cyber-norms/.

Schmitt, Michael N. , and Liis Vihul, eds. *Tallinn Manual 2.0 on the International Law Applicable to Cyber Operations*. 2nd ed. Cambridge: Cambridge University Press, 2017.

Schneier, Bruce. "Artificial Intelligence and the Attack/Defense Balance." *Schneier on Security*, March 15, 2018. https://www.schneier.com/blog/archives/2018/03/artificial_inte.html.

Schneier, Bruce. "Machine Learning to Detect Software Vulnerabilities." *Schneier on Security*, January 8, 2019. https://www.schneier.com/blog/archives/2019/01/machine_learnin.html.

Schondorf, Roy. "Israel's Perspective on Key Legal and Practical Issues Concerning the Application of International Law to Cyber Operations." *EJIL:Talk!*, December 9, 2020. https://www.ejiltalk.org/israels-perspective-on-key-legal-and-practical-issues-concerning-the-application-of-international-law-to-cyber-operations/.

Schulze, Matthias. "German Military Cyber Operations Are in a Legal Gray Zone." *Lawfare*, April 8, 2020. https://www.lawfareblog.com/german-military-cyber-operations-are-legal-gray-zone.

Schwartz, Matthew J. "Dridex Malware Campaign Disrupted." *BankInfoSecurity*, October 14, 2015. https://www.bankinfosecurity.com/dridex-malware-campaign-disrupted-a-8590.

Seals, Tara. "Companies Take an Average of 100–120 Days to Patch Vulnerabilities." *InfoSecurity Magazine*, October 1, 2015. https://www.infosecurity-magazine.com/news/companies-average-120-days-patch/.

Seals, Tara. "IoT Attacks Skyrocket, Doubling in 6 Months." *threatpost*, September 6, 2021. https://threatpost.com/iot-attacks-doubling/169224/.

Secretary of Defense Remarks for the US INDOPACOM Change of Command. April 30, 2021. https://www.defense.gov/Newsroom/Speeches/Speech/Article/2592093/secretary-of-defense-remarks-for-the-us-indopacom-change-of-command/.

Segal, Adam. "The Development of Cyber Norms at the United Nations Ends in Deadlock, Now What?" June 19, 2017. https://www.cfr.org/blog/development-cyber-norms-united-nations-ends-deadlock-now-what.

Senate Armed Services Committee Hearing on Encryption & Cyber Matters. September 13, 2016. https://www.armed-services.senate.gov/imo/media/doc/16-68_09-13-16.pdf.

Senate Armed Services Committee Hearing on US Cyber Command. April 5, 2016. https://www.armed-services.senate.gov/imo/media/doc/16-35_4-05-16.pdf.

Shackelford, Scott. "How Far Should Organizations Be Able to Go to Defend against Cyberattacks?" *GCN*, February 19, 2019. https://gcn.com/articles/2019/02/19/hacking-back.aspx.

Shakarian, Paulo. "The 2008 Russian Cyber Campaign against Georgia." *Military Review* 91, no. 6 (November–December 2011): 63–68. https://www.armyupress.army.mil/Portals/7/military-review/Archives/English/MilitaryReview_20111231_art013.pdf.

Shannon, Claude E. "A Mathematical Theory of Communication." *Bell System Technical Journal* 27 (October 1948): 379–423, 623–656. https://ieeexplore.ieee.org/document/6773024.

Shirer, W. L. *The Collapse of the Third Republic: An Inquiry into the fall of France in 1940*. New York: Simon & Schuster, 1969.

Shuu, Hu, Zhu Changzheng, and Yang Zheyu. "Liu He on China's New Transformation Trail." *Caixin Online*, October 28, 2010. http://english.caing.com/2010-11-08/100196829.html.

Simonite, Tom. "Moore's Law Is Dead. Now What?" *MIT Technology Review* 13 (May 2016). https://www.technologyreview.com/2016/05/13/245938/moores-law-is-dead-now-what/.

Skoudis, Edward. "Evolutionary Trends in Cyberspace." In *Cyberpower and National Security*, edited by Franklin D. Kramer, Stuart H. Starr, and Larry K. Wentz, 147–170. Washington,

DC: National Defense University Press, 2009. https://ndupress.ndu.edu/Portals/68/Documents/Books/CTBSP-Exports/Cyberpower/Cyberpower-I-Preface.pdf?ver=2017-06-16-115055-553.

Slantchev, Branislav L. *Military Threats: The Costs of Coercion and the Price of Peace.* Cambridge: Cambridge University Press, 2011.

Slayton, Rebecca. "What Is the Cyber Offense-Defense Balance? Conceptions, Causes, and Assessment." *International Security* 41, no. 3 (Winter 2016/17): 72–109. https://www.belfercenter.org/publication/what-cyber-offense-defense-balance-conceptions-causes-and-assessment.

Smith, Brad. "Growing Consensus on the Need for an International Treaty on Nation State Attacks." *Microsoft.* April 13, 2017. https://blogs.microsoft.com/on-the-issues/2017/04/13/growing-consensus-need-international-treaty-nation-state-attacks/.

Snyder, Glenn H. *Deterrence and Defense.* Princeton, NJ: Princeton University Press, 1961.

Snyder, Glenn H. "Process Variables in Neorealist Theories." *Security Studies* 5, no. 3 (1996): 167–192. https://doi.org/10.1080/09636419608429279.

Snyder, Jack. *The Ideology of the Offensive: Military Decision Making and the Disasters of 1914.* Ithaca, NY: Cornell University Press, 1984.

Society for Worldwide Interbank Financial Telecommunications. Accessed on January 29, 2022. https://www.swift.com/about-us.

Soesanto, Stefan. "Europe's Incertitude in Cyberspace." *Lawfare,* August 3, 2020. https://www.lawfareblog.com/europes-incertitude-cyberspace.

Soesanto, Stefan. "When Does a 'Cyber Attack' Demand Retaliation? NATO Broadens Its View." *DefenseOne,* June 30 2021. https://www.defenseone.com/ideas/2021/06/when-does-cyber-attack-demand-retaliation-nato-broadens-its-view/175028/.

Sophos. "The State of Ransomware 2020." May 2020. https://www.sophos.com/en-us/medialibrary/Gated-Assets/white-papers/sophos-the-state-of-ransomware-2020-wp.pdf.

Starks, Tim. "It's Hard to Keep a Big Botnet Down: TrickBot Sputters Back toward Full Health." *CyberScoop,* November 30, 2020. https://www.cyberscoop.com/trickbot-status-microsoft-cyber-command-takedown/.

"State Council Decision on Accelerating the Development of Strategic Emerging Industries." 2017. https://chinaenergyportal.org/wp-content/uploads/2017/01/Development-Strategic-Emerging-Industries.pdf.

"Statement by Mr. Richard Kadlčák Special Envoy for Cyberspace Director of Cybersecurity Department." February 11, 2020. https://www.nukib.cz/download/publications_en/CZ%20Statement%20-%20OEWG%20-%20International%20Law%2011.02.2020.pdf.

Statement of Admiral Michael S. Rogers, Commander United States Cyber Command, before the Senate Armed Services Committee. April 5, 2016. https://www.armed-services.senate.gov/imo/media/doc/Rogers_04-05-16.pdf.

Statement of General Paul M. Nakasone, Commander, United States Cyber Command, before the House Committee on Armed Services, Subcommittee on Intelligence, Emergent Threats, and Capabilities. March 3, 2019. https://armedservices.house.gov/_cache/files/e/d/ed0549b9-c479-4ae0-943d-66cf8fd933c1/AEDF855100875FF9DBB6F5E7472F6E36.nakasone-cybercom-hasc-posture-statement-final-3-13-19.pdf.

Statement of Mr. Kenneth Rapuano, Assistant Secretary of Defense for Homeland Defense and Global Security and Principal Cyber Advisor, Testimony before the House Armed Services Committee Subcommittee on Intelligence and Emerging Threats and Capabilities. March 4, 2020. https://www.congress.gov/116/meeting/house/110592/witnesses/HHRG-116-AS26-Wstate-RapuanoK-20200304.pdf.

Statista. Accessed on January 29, 2022. https://www.statista.com/topics/1145/internet-usage-worldwide/.

Sterling, Bruce. "Operation Shady RAT." *Wired,* August 3, 2011. https://www.wired.com/2011/08/operation-shady-rat/.

Stone, John. "Cyber War Will Take Place!" *Journal of Strategic Studies* 36, no. 1 (2013): 101–108. https://doi.org/10.1080/01402390.2012.730485.

Stopbadware. Accessed on January 29, 2022. https://www.stopbadware.org/.

Stubbs, Jack, Joseph Menn, and Christopher Bing. "Inside the West's Failed Fight against China's 'Cloud Hopper' Hackers." *Reuters*, June 26, 2019. https://www.reuters.com/investigates/special-report/china-cyber-cloudhopper/.

Sullivan, Laura. "As China Hacked, U.S. Businesses Turned a Blind Eye." *NPR*, April 12, 2019. https://www.npr.org/2019/04/12/711779130/as-china-hacked-u-s-businesses-turned-a-blind-eye.

"SWIFT Attackers' Malware Linked to More Financial Attacks." *Symantec Security Response*, May 26, 2016. https://www.symantec.com/connect/blogs/swift-attackers-malware-linked-more-financial-attacks.

Swinhoe, Dan. "Why Businesses Don't Report Cybercrimes to Law Enforcement." *CSO*, May 30, 2019. https://www.csoonline.com/article/3398700/why-businesses-don-t-report-cybercrimes-to-law-enforcement.html.

Tang, Frank. "China Set to Break Key Economic Barrier despite Trade War, but Can It Avoid the Middle Income Trap?" *South China Morning Post*, January 1, 2020. https://www.scmp.com/economy/china-economy/article/3044124/china-set-break-key-economic-barrier-despite-trade-war-can-it.

"Targeting U.S. Technologies: A Trend Analysis of Reporting from Defense Industry, Defense Security Service, 2013." 2013. https://www.hsdl.org/?abstract&did=757213.

Tcheyan, Lucas, and Sam Bresnick. "Reciprocity Is a Tool, Not a Strategy, against China." *Foreign Policy*, August 20, 2020. https://foreignpolicy.com/2020/08/20/china-reciprocity-trump-tit-for-tat-strategy-trade-war/.

Temple-Raston, Dina. "How the U.S. Hacked Isis." *NPR*, September 26, 2019. https://www.npr.org/2019/09/26/763545811/how-the-u-s-hacked-isis.

ThaiCERT. *Threat Group Cards: A Threat Actor Encyclopedia*. June 19, 2019. https://www.thaicert.or.th/downloads/files/A_Threat_Actor_Encyclopedia.pdf.

Thomas, Andrea. "Germany Points Finger at Russia over Parliament Attack." *Wall Street Journal*, May 13, 2016. https://www.wsj.com/articles/germany-points-finger-at-russia-over-parliament-hacking-attack-1463151250.

Thompson, William. "Identifying Rivals and Rivalries in World Politics." *International Studies Quarterly* 45, no. 4 (December 2001): 557–586. https://www.jstor.org/stable/3096060.

Tiezzi, Shannon. "First Biden-Xi Phone Call Shows Not Much Has Changed in US-China Relations." *The Diplomat*, February 12, 2020. https://thediplomat.com/2021/02/first-biden-xi-phone-call-shows-not-much-has-changed-in-us-china-relations/.

Tikk, Eneken, Kadri Kaska, and Liis Vihul. *International Cyber Incidents: Legal Considerations* Tallinn: NATO CCDCOE, 2010. https://ccdcoe.org/uploads/2018/10/legalconsiderations_0.pdf.

Timberg, Craig, and Ellen Nakashima. "Federal Investigators Find Evidence of Previously Unknown Tactics Used to Penetrate Government Networks," *Washington Post*, December 17, 2020. https://www.washingtonpost.com/business/technology/government-warns-new-hacking-tactics-russia/2020/12/17/bba43fd8-408c-11eb-a402-fba110db3b42_story.html.

"Trading Economics: National Bureau of Statistics of China." Accessed on January 29, 2022. https://tradingeconomics.com/china/gdp-growth-annual.

Unknown Tactics Used to Penetrate Government Networks." *Washington Post*, December 17, 2020. https://www.washingtonpost.com/business/technology/government-warns-new-hacking-tactics-russia/2020/12/17/bba43fd8-408c-11eb-a402-fba110db3b42_story.html.

Title 10 United States Code, Section 394. Accessed on January 29, 2022. https://casetext.com/statute/united-states-code/title-10-armed-forces/subtitle-a-general-military-law/part-i-organization-and-general-military-powers/chapter-19-cyber-matters/section-394-authorities-concerning-military-cyber-operations.

Tor, Uri. "'Cumulative Deterrence' as a New Paradigm for Cyber Deterrence." *Journal of Strategic Studies* 40, no. 1–2 (2017): 92–117. https://www.tandfonline.com/doi/full/10.1080/01402390.2015.1115975.

Townsend, Kevin. "92% of External Web Apps Have Exploitable Security Flaws or Weaknesses: Report." *SecurityWeek*, October 30, 2018. https://www.securityweek.com/92-external-web-apps-have-exploitable-security-flaws-or-weaknesses-report.

Townsend, Kevin. "New Law Will Help Chinese Government Stockpile Zero-Days." *SecurityWeek*, July 14, 2021. https://www.securityweek.com/new-law-will-help-chinese-government-stockpile-zero-days.

Tozzi, Chris. "Avoiding the Pitfalls of the Shared Responsibility Model for Cloud Security." *PaloAlto Networks*, September 24, 2020. https://blog.paloaltonetworks.com/prisma-cloud/pitfalls-shared-responsibility-cloud-security/.

Trap or Wall: Real Challenges and Strategic Choice in China's Economy. 2014. http://en.drc.gov.cn/2014-06/26/content_17617382.htm.

TrendMicro. Accessed on January 29, 2022. https://www.trendmicro.com/vinfo/us/security/definition/exploit.

Trimble, Stephen. "China Completes Assembly of First High-Bypass Turbofan Engine." *Flight Global*, December 29, 2017. https://www.flightglobal.com/systems-and-interiors/china-completes-assembly-of-first-high-bypass-turbofan-engine/126587.article.

Turton, William, and Kartikay Mehrotra. "FireEye Discovered SolarWinds Breach While Probing Own Hack." *Bloomberg*, December 14, 2020 (updated December 15, 2020). https://www.bloomberg.com/news/articles/2020-12-15/fireeye-stumbled-across-solarwinds-breach-while-probing-own-hack.

United Nations. "Report of the Panel of Experts Established Pursuant to Resolution 1874 (2009)—S/2019/691." 2009. https://www.securitycouncilreport.org/atf/cf/%7B65BFCF9B-6D27-4E9C-8CD3-CF6E4FF96FF9%7D/S_2019_691.pdf.

United Nations. "Security Council Strengthens Sanctions on Democratic Republic of Korea, Unanimously Adopting Resolution 2321." 2016. https://www.un.org/press/en/2016/sc12603.doc.htm.

United Nations General Assembly. "Countering the Use of Information and Communications Technologies for Criminal Purposes, Resolution 74/247." December 27, 2019. https://undocs.org/en/A/RES/74/247.

United Nations General Assembly. *Friendly Relations Declaration.* 1970.

United Nations General Assembly. "Letter Dated 9 January 2015 from the Permanent Representatives of China, Kazakhstan, Kyrgyzstan, the Russian Federation, Tajikistan and Uzbekistan to the United Nations addressed to the Secretary-General, U.N. Doc. A/69/273." 2015. http://www.un.org/Docs/journal/asp/ws.asp?m=A/69/723.

United Nations General Assembly. "Open-ended Working Group on Developments in the Field of Information and Telecommunications in the Context of International Security: A/AC.290/2021/CRP.2." March 10, 2021. https://front.un-arm.org/wp-content/uploads/2021/03/Final-report-A-AC.290-2021-CRP.2.pdf.

United Nations General Assembly. "Resolution 2625 (XXV), Annex, Declaration on Principles of International Law Concerning Friendly Relations and Co-operation among States in Accordance with the Charter of the United Nations." October 24, 1970. https://www.un.org/ruleoflaw/files/3dda1f104.pdf.

United Nations General Assembly. "Resolution 56/83 Annex, United Nations Document A/CN.4/L. 778, Responsibility of States for Internationally Wrongful Acts, Chapter II." May 30, 2011. https://documents-dds-ny.un.org/doc/UNDOC/LTD/G11/614/25/PDF/G1161425.pdf?OpenElement.

United Nations Group of Governmental Experts on Developments in the Field of Information and Telecommunications in the Context of International Security. A/70/174, July 22, 2015. https://www.un.org/ga/search/view_doc.asp?symbol=A/70/174.

United Nations Report of the International Law Commission, Sixty-eighth session (May 2–June 10 and July 4–August 12, 2016), A/71/10. https://documents-dds-ny.un.org/doc/UNDOC/GEN/G16/184/25/PDF/G1618425.pdf?OpenElement.

United States District Court for the Central District of California. *United States of America v. PARK JIN HYOK, also known as ("aka") "Jin Hyok Park," aka "Pak Jin Hek," Case No. MJ18-1749*. June 2018. https://www.justice.gov/opa/press-release/file/1092091/download.

United States District Court, Southern District of California. *United States of America v. Zhang Zhang-Gui, Zha Rong, Chai Meng, Liu Chunliang, Gao Hong Kun, Zhuang Xiaowei, Ma Zhiqi, Li Xiao, Gu Gen, Tian Xi, Case No. 13CR3132-H*. October 2018. https://www.justice.gov/opa/press-release/file/1106491/download.

United States District Court, Western District of Pennsylvania. *United States of America v. Wang Dong, Sun Kailiang, Wen Xinyu, Huang Zhenyu, Gu Chunhui, Criminal No. 14-118*. May 2014. https://www.justice.gov/iso/opa/resources/5122014519132358461949.pdf.

United States Government Accountability Office. *DATA PROTECTION: Actions Taken by Equifax and Federal Agencies in Response to the 2017 Breach*. August 2018. https://www.warren.senate.gov/imo/media/doc/2018.09.06%20GAO%20Equifax%20report.pdf.

United States Government Accountability Office. "Economic Sanctions: Agencies Assess Impacts on Targets, and Studies Suggest Several Factors Contribute to Sanctions' Effectiveness." October 2019. https://www.gao.gov/assets/710/701891.pdf.

United States of America v. Internet Research Agency. February 16, 2018. https://www.justice.gov/file/1035477/download.

United States of America v. Li Xiaoyou (a/k/a "OroOlxy") and Dong Jiazhi. July 7, 2020. https://www.justice.gov/opa/press-release/file/1295981/download.

United States of America v. Viktor Borisovich Netyksho, Boris Alekseyevich Antonov, Dmitriy Sergeyevich Badin, Ivan Sergeyevich Yermakov, Aleksey Viktorovich Lukasheb, Sergey Aleksandrovich Morgachev, Nikolay Yuryvich Kozachek, Pavel Vyacheslavovich Yershov, Artem Andreyevich Malyshev, Aleksandr Vladimirovich Osadchuk, Aleksey Aleksandrovich Potemkin, and Anatoliy Sergeyovich Kovalev. July 13, 2018. https://www.justice.gov/file/1080281.

United States Senate. Hearing to Receive Testimony on Counter-ISIL (Islamic State of Iraq and the Levant) Operations and Middle East Strategy. April 28, 2016. https://www.armed-services.senate.gov/imo/media/doc/16-51_04-28-16.pdf.

United States Senate, Committee on Homeland Security and Governmental Affairs. *Federal Cybersecurity: America's Data Still at Risk*. August 2021. https://www.hsgac.senate.gov/imo/media/doc/Federal%20Cybersecurity%20-%20America%27s%20Data%20Still%20at%20Risk%20%28FINAL%29.pdf.

US-China Economic and Security Review Commission. "2019 Report to Congress of the U.S.-China Economic and Security Review Commission." November 2019. https://www.uscc.gov/sites/default/files/2019-11/2019%20Annual%20Report%20to%20Congress.pdf.

US Cyber Command. "Hunt Forward Estonia: Estonia, US Strengthen Partnership in Cyber Domain with Joint Operation." December 3, 2020. https://www.cybercom.mil/Media/News/Article/2433245/hunt-forward-estonia-estonia-us-strengthen-partnership-in-cyber-domain-with-joi/.

US Cyber Command. "New CNMF Initiative Shares Malware Samples with Cybersecurity Industry." November 5, 2018. https://www.cybercom.mil/Media/News/News-Display/Article/1681533/new-cnmf-initiative-shares-malware-samples-with-cybersecurity-industry/.

US Department of Commerce. "Commerce Department Prohibits WeChat and TikTok Transactions to Protect the National Security of the United States." September 18, 2020. https://www.commerce.gov/news/press-releases/2020/09/commerce-department-prohibits-wechat-and-tiktok-transactions-protect.

US Department of Defense. *Department of Defense Cyber Strategy: Summary*. Department of Defense, 2018. https://media.defense.gov/2018/Sep/18/2002041658/-1/-1/1/CYBER_STRATEGY_SUMMARY_FINAL.PDF.

US Department of Defense. *Department of Defense Strategy for Operating in Cyberspace.* July 2011. https://csrc.nist.gov/CSRC/media/Projects/ISPAB/documents/DOD-Strategy-for-Operating-in-Cyberspace.pdf.

US Department of Defense. *DoD Cyber Strategy.* April 2015. https://archive.defense.gov/home/features/2015/0415_cyber-strategy/final_2015_dod_cyber_strategy_for_web.pdf;

US Department of Defense. *Joint Publication 3-0, Operations.* Washington, DC: Department of Defense, 2017. https://www.jcs.mil/Portals/36/Documents/Doctrine/pubs/jp3_0ch1.pdf.

US Department of Defense. *Joint Publication 3-12, Cyberspace Operations.* Washington, DC: Department of Defense, June 8, 2018. https://www.jcs.mil/Portals/36/Documents/Doctrine/pubs/jp3_12.pdf.

US Department of Defense. *Joint Publication 3-13, Information Operations.* Washington, DC: Department of Defense, 2012. https://www.jcs.mil/Portals/36/Documents/Doctrine/pubs/jp3_13.pdf.

US Department of Justice. "Bugat Botnet Administrator Arrested and Malware Disabled." October 13, 2015. https://www.justice.gov/opa/pr/bugat-botnet-administrator-arrested-and-malware-disabled.

US Department of Justice. "China Initiative." September 1, 2020. https://www.justice.gov/usao-edtx/china-initiative.

US Department of Justice. "Information about the Department of Justice's China Initiative and a Compilation of China-Related Prosecutions since 2018." November 12, 2020. https://www.justice.gov/opa/information-about-department-justice-s-china-initiative-and-compilation-china-related.

US Department of Justice. "Justice Department Announces Court-Authorized Effort to Disrupt Exploitation of Microsoft Exchange Server Vulnerabilities." April 13, 2021. https://www.justice.gov/usao-sdtx/pr/justice-department-announces-court-authorized-effort-disrupt-exploitation-microsoft.

US Department of Justice. "North Korean Regime-Backed Programmer Charged with Conspiracy to Conduct Multiple Cyber Attacks and Intrusions." September 6, 2018. https://www.justice.gov/opa/pr/north-korean-regime-backed-programmer-charged-conspiracy-conduct-multiple-cyber-attacks-and.

US Department of Justice. *Report of the Attorney General's Cyber Digital Task Force.* July 2, 2018. https://www.justice.gov/ag/page/file/1076696/download.

US Department of Justice. "United States Files Complaint to Forfeit 280 Cryptocurrency Accounts Tied to Hacks of Two Exchanges by North Korean Actors." August 27, 2020. https://www.justice.gov/opa/pr/united-states-files-complaint-forfeit-280-cryptocurrency-accounts-tied-hacks-two-exchanges.

US Department of Justice. *United States of America v. Zhu Hua and Zhang Shilong.* December 2018. https://www.justice.gov/opa/page/file/1122671/download.

US Department of Justice. "United States Seizes Domain Names Used by Iran's Islamic Revolutionary Guard Corps." October 7, 2020. https://www.justice.gov/usao-ndca/pr/united-states-seizes-domain-names-used-iran-s-islamic-revolutionary-guard-corps.

US Department of Justice. "U.S. Charges Five Chinese Military Hackers for Cyber Espionage Against U.S. Corporations and a Labor Organization for Commercial Advantage: First Time Criminal Charges Are Filed against Known State Actors for Hacking." May 19, 2014. https://www.justice.gov/opa/pr/us-charges-five-chinese-military-hackers-cyber-espionage-against-us-corporations-and-labor.

US Department of State. *Bureau for International Narcotics and Law Enforcement Affairs International Narcotics Control Strategy Report, Volume I.* March 2020. https://www.state.gov/wp-content/uploads/2020/06/Tab-1-INCSR-Vol.-I-Final-for-Printing-1-29-20-508-4.pdf.

US Department of State. "The Clean Network." Accessed on January 29, 2022. https://www.state.gov/the-clean-network/#:~:text=The%20Clean%20Network%20program%20is,as%20the%20Chinese%20Communist%20Party.

US Department of the Treasury. Accessed on January 29, 2022. "The Committee on Foreign Investment in the United States (CFIUS)." https://home.treasury.gov/policy-issues/international/the-committee-on-foreign-investment-in-the-united-states-cfius.

US Department of the Treasury. "Treasury Sanctions North Korean State-Sponsored Malicious Cyber Groups." September 13, 2019. https://home.treasury.gov/news/press-releases/sm774.

US Federal Information Modernization Act of 2014. *Annual Report to Congress* (Executive Office of the President of the United States, 2016). https://www.whitehouse.gov/sites/whitehouse.gov/files/briefing-room/presidential-actions/related-omb-material/fy_2016_fisma_report%20to_congress_official_release_march_10_2017.pdf.

US Federal Information Modernization Act of 2014. *Annual Report to Congress* (Executive Office of the President of the United States, 2017). https://www.whitehouse.gov/wp-content/uploads/2017/11/FY2017FISMAReportCongress.pdf.

US Federal Information Modernization Act of 2014. *Annual Report to Congress* (Executive Office of the President of the United States, 2018). https://www.whitehouse.gov/wp-content/uploads/2019/08/FISMA-2018-Report-FINAL-to-post.pdf.

US Federal Information Modernization Act of 2014. *Annual Report to Congress* (Executive Office of the President of the United States, 2020). https://www.whitehouse.gov/wp-content/uploads/2020/05/2019-FISMARMAs.pdf.

US House Resolution 5576—Cyber Deterrence and Response Act of 2018 (115th Congress 2017–2018). https://www.congress.gov/bill/115th-congress/house-bill/5576/text.

"U.S. Sanctions North Korean Hackers for Swift Hack, Wannacry and Other Cyberattacks That Fund Its Weapons Programs." *Bloomberg*, September 14, 2019. https://www.japantimes.co.jp/news/2019/09/14/asia-pacific/u-s-sanctions-north-korean-hackers-swift-hack-wannacry-cyberattacks-fund-weapons-programs/#.XkAVcjFKjIU.

US Securities and Exchange Commission. "Release Nos. 33-10459, 34-82746, Commission Statement and Guidance on Public Company Cybersecurity Disclosures, Security Exchange Commission." February 26, 2018. https://www.sec.gov/rules/interp/2018/33-10459.pdf.

US Senate Committee on Armed Services. "Hearing to Review Testimony on United States Special Operations Command and United States Cyber Command in Review of the Defense Authorization Request for Fiscal Year 2020 and the Future Years Defense Program." February 14, 2019. https://www.armed-services.senate.gov/imo/media/doc/19-13_02-14-19.pdf.

"US Treasury Sanctions North Korean Hacker, Company for Cyber Attacks." *Reuters*, September 6, 2018. https://www.reuters.com/article/us-cyber-northkorea-sanctions/u-s-treasury-sanctions-north-korean-hacker-company-for-cyber-attacks-idUSKCN1LM2I7.

van Creveld, Martin. *Technology and War*. New York: Free Press, 1989.

Vavra, Shannon. "Cyber Command Deploys Abroad to Fend Off Foreign Hacking ahead of the 2020 Election." *CyberScoop*, August 25, 2020. https://www.cyberscoop.com/2020-presidential-election-cyber-command-nakasone-deployed-protect-interference-hacking/.

Vavra, Shannon. "Cyber Command Has Redeployed Overseas in Effort to Protect 2020 Elections." *CyberScoop*, May 7, 2019. https://www.cyberscoop.com/cyber-command-redeployed-overseas-effort-protect-2020-elections/.

Vavra, Shannon. "Pentagon's Next Cyber Policy Guru Predicts More Collective Responses in Cyberspace." *CyberScoop*, November 21, 2019. https://www.cyberscoop.com/pentagons-next-cyber-policy-guru-predicts-collective-responses-cyberspace/.

Vavra, Shannon. "Trump Administration Expands Economic Restrictions on Huawei." *CyberScoop*, August 17, 2020. https://www.cyberscoop.com/huawei-entity-list-commerce-trump-expanded-38-affiliates/.

Vavra, Shannon. "Why Cyber Command's Latest Warning Is a Win for the Government's Information Sharing Efforts." *CyberScoop*, July 10, 2019. https://www.cyberscoop.com/cyber-command-information-sharing-virustotal-iran-russia/.

VerifiedVoting. Accessed on January 29, 2022. https://verifiedvoting.org/.

Verizon. "Verizon 2019 Data Breach Investigation Report." 2020. https://enterprise.verizon.com/resources/reports/dbir/2019/introduction/.

Verizon. "Verizon Business 2020 Data Breach Investigations Report." 2021. https://enterprise.verizon.com/resources/reports/dbir/.

Vijayan, Jai. "Russian Hackers Using Iranian APT's Infrastructure in Widespread Attacks." *DARKReading*, October 21, 2019. https://www.darkreading.com/attacks-breaches/russian-hackers-using-iranian-apts-infrastructure-in-widespread-attacks/d/d-id/1336134.

von Clausewitz, Carl. *On War, Book 1, Chapter 7*. Edited and translated by Michael Howard and Peter Paret. Princeton, NJ: Princeton University Press, 1976.

Waddell, Kaveh. "The Twitter Bot That Sounds Just Like Me." *The Atlantic*, August 18, 2016. https://www.theatlantic.com/technology/archive/2016/08/the-twitter-bot-that-sounds-just-like-me/496340/.

Waldman, Arielle. "Nation-State Hacker Indictments: Do They Help or Hinder?" *SearchSecurity*, April 2021. https://searchsecurity.techtarget.com/feature/Nation-state-hacker-indictments-Do-they-help-or-hinder.

Waltz, Kenneth. *Theory of International Politics*. Reading, MA: Addison-Wesley, 1979.

Ware, Willis H. *Security Controls for Computer Systems: Report of Defense Science Board Task Force on Computer Security*. Santa Monica, CA: Rand Corporation, 1970. https://www.rand.org/pubs/reports/R609-1.html.

Warner, Michael. "A Brief History of Cyber Conflict." In *Ten Years In: Implementing Strategic Approaches to Cyberspace*, edited by Jacquelyn G. Schneider, Emily O. Goldman, and Michael Warner, 31–46. Newport, RI: Naval War College Press, 2020. https://digital-commons.usnwc.edu/cgi/viewcontent.cgi?article=1044&context=usnwc-newport-papers.

Warner, Michael. "Invisible Battlegrounds: On Force and Revolutions, Military and Otherwise." In *The Palgrave Handbook of Security, Risk and Intelligence*, edited by Robert Dover, Huw Dylan, and Michael S. Goodman, 247–264. London: Palgrave Macmillan, 2017.

Warner, Michael. *The Rise and Fall of Intelligence: An International Security History*. Washington, DC: Georgetown University Press, 2014.

Warner, Michael. "US Cyber Command's First Decade." *Hoover Institution / Aegis Series Paper No. 2008*. November 2020. https://www.hoover.org/sites/default/files/research/docs/warner_webready.pdf.

Warner, Michael. "US Cyber Command's Road to Full Operational Capability." In *Stand Up and Fight! The Creation of US Security Organizations, 1942–2005*, edited by Ty Seidule and Jacqueline E. Whitt. Carlisle, PA: US Army War College, 2015: chapter 7. https://press.armywarcollege.edu/monographs/11/.

Watson, Ben. "Is the U.S. Ready to Escalate in Cyberspace?" *Defense One*, November 21, 2018. https://www.defenseone.com/feature/is-the-us-ready-to-escalate-in-cyberspace.

Weedon, Jan. Written testimony in "Hearing before the U.S.-China Economic and Security Review Commission, Commercial Cyber Espionage and Barriers to Digital Trade in China." June 15, 2015. https://www.uscc.gov/sites/default/files/Weedon%20Testimony.pdf.

Wertime, David. "Unpacking Xi Jinping's Pet Phrase for U.S.-China Ties." *Foreign Policy*, September 23, 2015. https://foreignpolicy.com/2015/09/23/unpacking-xi-jinpings-pet-phrase-new-model-of-great-power-relations-us-china-explainer/.

Wheeler, Tom, and David Simpson. "Why 5G Requires New Approaches to Cybersecurity." *Brookings*. September 3, 2019. https://www.brookings.edu/research/why-5g-requires-new-approaches-to-cybersecurity/.

The White House. 2017. https://www.whitehouse.gov/briefings-statements/remarks-homeland-security-advisor-thomas-p-bossert-cyber-week-2017/.

The White House, *Cybersecurity Policy Review*. 2009. https://obamawhitehouse.archives.gov/cyberreview/documents/.

The White House. *Interim National Security Strategic Guidance*. March 2021. https://www.whitehouse.gov/wp-content/uploads/2021/03/NSC-1v2.pdf.

The White House. *International Strategy for Cyberspace: Prosperity, Security, and Openness in a Networked World*. May 2011. https://obamawhitehouse.archives.gov/sites/default/files/rss_viewer/international_strategy_for_cyberspace.pdf.

The White House. "Message to the Congress on Securing the Information and Communications Technology and Services Supply Chain." May 15, 2019. https://trumpwhitehouse.archives.gov/briefings-statements/message-congress-securing-information-communications-technology-services-supply-chain/.

The White House. *National Cyber Strategy of the United States of America*. September 2018. https://www.whitehouse.gov/wp-content/uploads/2018/09/National-Cyber-Strategy.pdf.

The White House. *National Security Decision Directive Number 145*. September 17, 1984. https://fas.org/irp/offdocs/nsdd145.htm.

The White House. Presidential Policy Directive 41—United States Cyber Incident Coordination. July 26, 2016. https://obamawhitehouse.archives.gov/the-press-office/2016/07/26/presidential-policy-directive-united-states-cyber-incident.

The White House. "White House Report to Congress on Cyber Deterrence Policy." December 29, 2015. https://insidecybersecurity.com/sites/insidecybersecurity.com/files/documents/dec2015/cs2015_0133.pdf.

Whitley, Jarred. "Washington Bureaucrats Are Handing China Keys to 5G Kingdom." *The Hill*, February 22, 2019. https://thehill.com/opinion/technology/430801-washington-bureaucrats-are-handing-china-keys-to-5g-kingdom.

Whyte, Christopher. "Problems of Poison: New Paradigms and 'Agreed' Competition in the Era of AI-Enabled Cyber Operations." In *2020 12th International Conference on Cyber Conflict—20/20 Vision: The Next Decade*, edited by T. Jančárková, L. Lindström, M. Signoretti, I. Tolga, and G. Visky, 215–232. Tallinn: NATO CCDCOE, 2020.

Williams, Lauren C. "NSA Chief Explains New Cyber Directorate." October 10, 2019. https://fcw.com/articles/2019/10/10/nsa-nagasone-cyber-group-williams.aspx.

Winder, Davey. "Google Chrome Update Gets Serious: Homeland Security (CISA) Confirms Attacks Underway." *Forbes*, November 15, 2020. https://www.forbes.com/sites/daveywinder/2020/11/15/google-chrome-update-gets-serious-homeland-security-cisa-confirms-attacks-underway/amp/.

Wolff, Josephine. "The SolarWinds Hack Is Unlike Anything We Have Ever Seen Before." *Slate*, December 18, 2020. https://slate.com/technology/2020/12/solarwinds-hack-malware-active-breach.html.

Wong, Chun Han. "China Restricts U.S. Diplomats' Movements in Latest Tit-for-Tat." *Wall Street Journal*, September 11, 2020. https://www.wsj.com/amp/articles/china-restricts-u-s-diplomats-movements-in-latest-tit-for-tat-11599834078.

Wong, Edward. "China Aims for 6.5% Economic Growth over Next 5 Years, Xi Says." *New York Times*, November 3, 2015. https://www.nytimes.com/2015/11/04/world/asia/china-economic-growth-xi.html.

Work, J. D. "Evaluating Commercial Cyber Intelligence Activity." *International Journal of Intelligence and CounterIntelligence* 33, no. 2 (2020): 278–308. https://doi.org/10.1080/08850607.2019.1690877.

Work, J. D., and Richard J. Harknett. "Troubled Vision: Understanding Recent Israel-Iranian Offensive Cyber Exchanges." *Atlantic Council of the United States Issue Brief*. July 22, 2020. https://www.atlanticcouncil.org/in-depth-research-reports/issue-brief/troubled-vision-understanding-israeli-iranian-offensive-cyber-exchanges/.

The World Bank. https://data.worldbank.org/country/china.

The World Bank and Development Research Center of the State Council, the People's Republic of China. *China 2030: Building a Modern, Harmonious and Creative Society*. 2013. https://www.worldbank.org/en/news/feature/2012/02/27/china-2030-executive-summary.

World Economic Forum. "This Is How Many Websites Exist Globally." Accessed on January 29, 2022. https://www.weforum.org/agenda/2019/09/chart-of-the-day-how-many-websites-are-there/.

Wright, Jeremy (Attorney General of the United Kingdom). "Cyber and International Law in the 21st Century." *Chatham House*. May 23, 2018. https://www.chathamhouse.org/event/cyber-and-international-law-21st-century.

Yew, Richard. "How Fast Can You Patch? How to Buy Time during the Next Zero-Day Vulnerability." *Medium*, October 1, 2018. https://medium.com/@VZMediaPlatform/how-fast-can-you-patch-how-to-buy-time-during-the-next-zero-day-vulnerability-6c5772ee3ba8.

Yould, Rachel. "Beyond the American Fortress: Understanding Homeland Security in the Information Age." In *Bombs and Bandwidth: The Emerging Relationship between Information Technology and Security*, edited by Robert Latham, 74–97. New York: The New Press, 2003.

Young, Kong Ji, Kim Kyoung Gon, and Lim Jong In. "The All-Purpose Sword: North Korea's Operations and Strategies." *2019 11th International Conference on Cyber Conflict*. https://ccdcoe.org/uploads/2019/06/Art_08_The-All-Purpose-Sword.pdf.

Zetter, Kim. "Legal Experts: Stuxnet Attack on Iran Was Illegal 'Act of Force.'" *Wired*, March 25, 2013. https://www.wired.com/2013/03/stuxnet-act-of-force/.

Zetter, Kim. "That Insane, $81M Bangladesh Bank Heist? Here's What We Know." *Wired*, May 17, 2016. https://www.wired.com/2016/05/insane-81m-bangladesh-bank-heist-heres-know/.

Zimpanu, Catalin. "DarkHotel Hackers Use VPN Zero-Day to Breach Chinese Government Agencies." *ZDNet*, April 6, 2020. https://www.zdnet.com/article/darkhotel-hackers-use-vpn-zero-day-to-compromise-chinese-government-agencies/.

INDEX

For the benefit of digital users, indexed terms that span two pages (e.g., 52–53) may, on occasion, appear on only one of those pages.